Medical Dosimetry Certification
Study Guide

Second Edition

Medical Dosimetry Certification Study Guide

Second Edition

K.N. Govinda Rajan, Ph.D.

Medical Physics Publishing
Madison, Wisconsin

Library of Congress Control Number: 2015947292
ISBN soft cover: 978-1-930524-80-4
ISBN eBook: 978-1-930524-81-1

Published by:
Medical Physics Publishing
4555 Helgesen Drive
Madison, WI 53718
(608) 224-4508 or 1-800-442-5778
mpp@medicalphysics.org
www.medicalphysics.org

Printed in the United States of America

Contents

I **Radiation Physics** 1
  A. Radioactivity 1
  B. Production of X-rays 6
  C. Interaction of Radiation with Matter—Photons and Electrons 8
  D. Treatment Machine Characteristics 12
  E. Radiation Quantities, Units, and Measurements 23
  Answers 35

II. **Localization** **65**
  A. Treatment Planning Concepts 65
  B. Acquisition of Patient Data 68
  C. Patient Positioning and Immobilization 89
  D. Ancillary Treatment Devices 94
  E. Treatment Simulation and Verification 95
  Answers 107

III. **Treatment Planning, Techniques, and Delivery** **127**
  A. Treatment Planning Concepts 127
  B. Photon Beam Isodose Curve Parameters 129
  C. Photon Beam Isodose Distributions 132
  D. Electron Beam Dose Distributions 134
  E. Treatment Planning Systems 142
  F. Radiobiology and Clinical Oncology 146
  G. Treatment Techniques and Treatment Delivery 158
  Answers 171

IV. **Dose Calculation Methods** **191**
  A. Applied Mathematics 191
  B. Beam Calibrations (Photon and Electron Beams) 197
  C. Basic External Beam Calculations (Photon Beams) 208
  D. Basic External Beam Calculations (Electron Beams) 256
  E. Effects of Beam-modifying Devices 268
  F. Irregular Field Calculations 278
  G. Special Calculations 281
  H. Corrections for Tissue Inhomogeneities and Surface Obliquities 288
  Answers 295

V. **Brachytherapy** **333**
  A. Basic Concepts 333
  B. Brachytherapy Source Characteristics 337
  C. Dose Distributions 346

    D.   Dose Calculations...352

    E.   Source Localization 353

    F.   Interstitial Dosimetry...359

    G.   Intracavitary Therapy Dosimetry 371

    Answers... **375**

VI. Radiation Protection 391

    A.   General Radiation Protection 392

    B.   Maximum Permissible Dose Equivalent (Dose and Dose Limits)............................395

    C.   Time, Distance, and Shielding (and ALARA Concepts)...397

    D.   Brachytherapy Source Handling and Storage..399

    E.   Structural Shielding..404

    F.   Radiation Monitoring 410

    Answers... **417**

VII. Quality Assurance 427

    A.   Treatment Delivery Equipment (External Beam Therapy and Brachytherapy).....................427

    B.   Imaging Systems for Treatment Planning..441

    C.   Treatment Planning System (TPS) 447

    D.   Record and Verify Systems 451

    E.   QA in Advanced Treatment Techniques 452

    F.   Display Monitor QA..459

    G.   Measurement Equipment 461

    Answers... **463**

Preface

The updated second edition of the *Medical Dosimetry Certification Study Guide* comes after a gap of nearly 10 years, really a long time for a rapidly changing field like radiation oncology, where the fast-changing newer technologies try to push the cure rate to higher digits all the time. My teaching assignments kept me away from updating the guide for a long time, and it was by chance that I happened to look at the recent syllabus on medical radiation dosimetry and was surprised by the significant changes in its contents in the last 10 years. So I decided to revise the contents of the study guide as per the new syllabus. Questions have been accommodated into various sections as per the new syllabus, and I have added more questions and new topics in the second edition.

In this second edition I have tried to add more problems in the chapter dealing with dose calculations, which will help students do dose calculations for the various treatment situations and calibration conditions. I do not know how long these dose calculations are going to be taught in radiation oncology. With the advent of IMRT and stereotactic radiotherapy, even larger conventional fields are now summations of much smaller dynamic fields, and dosimetry of these fields is now being performed by the treatment planning systems. Even treatment verifications have become more complex. However, the problems discussed in this guide will help one understand the dosimetric concepts in a much better way, and they will also help in following the calculation methods adopted in the TPS systems in conventional radiotherapy. Time did not permit me to concentrate on the field of small field dosimetry in a significant way, but I have made a reference to this topic in the guide. I have not included any material on ^{60}Co teletherapy units and ^{137}Cs LDR remote afterloading equipment since not many of these units are in operation today.

In some sections I have given an introduction to the topic with some illustrative pictures. In the answer sections, I have tried to work out the smaller problems and indicate the method of solving the larger problems. I have tried to avoid any mistakes, but if readers happen to find some errors in any part of the guide, I request they e-mail me at kngrajan@gmail.com with the correct answers. We can then make corrections.

I have made an effort to do most of the figures in this book myself. In the case of few figures that I have adopted, I have obtained permission from the authors. Most of the figures are traceable to the Internet or conference presentations. In case I have not acknowledged the source of any figure by oversight or by not being able to find its source, the author or the reader may kindly e-mail me the source and we will readily acknowledge the same by inserting a correction sheet in the book.

I have benefited a great deal by reading the contributions of many excellent medical physicists and teachers of other related subjects whose works have appeared in various conference proceedings, university websites, and other Internet sites. All of their excellent explanations and illustrations have very much influenced my thoughts and writings. I have also prepared lots of notes from these contributors, and many of these notes do not have detailed references since they were kept for my own understanding. As a result, I might have included a few questions or problems created by other authors whose references I could not locate on the Internet. If any readers or authors bring to my notice such instances, I will include the references in a correction sheet in the book or will incorporate them into the book during the next reprint.

I look forward to feedback from readers regarding the coverage of various topics and where they would expect to see more questions or explanations or improvements in the guide to make it more useful to them. I really enjoyed writing this second edition of the guide. I sincerely hope you, too, get the same enjoyment reading it. And I hope it helps dosimetrists in the profession who are hoping to clear the board certification exam.

I am grateful for all the help from the staff at Medical Physics Publishing. They've done a great job changing my raw text into a wonderful book.

Finally, I am dedicating this edition to my grandchildren, Arjun Rao and Rohan Rao, who did not complain at all when I told them I would not be able to play with them in the evenings because of my writing responsibility. They only kept asking, "Grandpa, when will you be done with this work?"

K.N. Govinda Rajan
Visiting Faculty
PSG Hospitals
Coimbatore, India

I. Radiation Physics

A. Radioactivity

Circle the right answer (Yes or No):

1. (Yes / No) An alpha particle is identical to a helium nucleus with a mass number of 4 and an electrostatic charge of +2.

2. (Yes / No) Alpha particles are usually emitted by low-Z radioactive elements during radioactive decay.

3. (Yes / No) Alpha emission changes the identity of the radionuclide.

4. (Yes / No) ^{226}Ra is an alpha emitter.

5. (Yes / No) ^{60}Co is an alpha emitter.

6. (Yes / No) A beta particle is an electron emitted by the atomic nucleus during radioactive decay.

7. (Yes / No) ^{60}Co is a beta emitter.

8. (Yes / No) ^{192}Ir is a beta emitter.

9. (Yes / No) Emission of beta changes the identity of the radionuclide.

10. (Yes / No) A positron is a positively charged particle but identical to an electron in all other respects.

11. (Yes / No) Beta decay is usually associated with proton rich radionuclides.

12. (Yes / No) Positron decay is usually associated with neutron rich radionuclides.

13. (Yes / No) Emission of gamma radiation does not change the identity of the radionuclide.

14. (Yes / No) Gamma usually follows beta particle emission in radioactive decay.

15. (Yes / No) There are no pure beta emitters.

16. (Yes / No) There are no pure gamma emitters.

17. (Yes / No) Electron capture usually occurs in high-Z radioactive elements.

18. (Yes / No) Electron capture and beta emission are competing modes of decay.

19. (Yes / No) Beta particles emitted in radioactive decay are monoenergetic.

1

20. (Yes / No)   A neutrino is a particle of negligible mass and zero charge postulated to account for the nonconservation of energy during beta decay.

21. (Yes / No)   A neutrino is easy to detect.

22. (Yes / No)   Mass is conserved in radioactive decay.

Match the following:

23. Match the radionuclide to the nature of its emitter.
    a. ^{222}Rn   _____   i.   beta emitter
    b. ^{32}P   _____   ii. alpha emitter
    c. ^{59}Ni   _____   iii. pure beta emitter
    d. ^{90}Sr   _____   iv. electron capture radionuclide
    e. ^{125}I   _____   v.   positron emitter

24. Match the gamma emissions to radionuclides.
    a. 1.17 and 1.33 MeV gammas   _____   i. ^{222}Rn
    b. 0.662 MeV gamma   _____   ii. ^{60}Co
    c. Several gammas of mean energy
        around 400 keV   _____   iii. ^{137}Cs
    d. Several gammas of mean energy
        around 0.8 MeV   _____   iv. ^{125}I
    e. Mean energy 28 keV   _____   v. ^{192}Ir

25. Match the half life to its radionuclide.
    a. 5.26 years   _____   i. ^{226}Ra
    b. 30 years   _____   ii. ^{60}Co
    c. 74 days   _____   iii. ^{137}Cs
    d. 1626 years   _____   iv. ^{125}I
    e. 59.6 days   _____   v. ^{192}Ir

Choose the right answer:

26. The mean life of ^{192}Ir source is given by _____ days.
    a. 106.5
    b. 90
    c. 120
    d. 350

27. After two half lives, the initial activity of a given radioisotope would have reduced to _____.
    a. one half
    b. no reduction
    c. one third
    d. one fourth

28. The mean life of a radioactive source is given by _____.
    a. $1.5\,T_{1/2}$
    b. $2.4\,T_{1/2}$
    c. $1.44\,T_{1/2}$
    d. $2\,T_{1/2}$

29. The SI unit of activity is the _____.
    a. becquerel
    b. curie
    c. hertz
    d. roentgen

30. One becquerel (Bq) corresponds to _____ nuclear transformations/second.
    a. 100
    b. 1000
    c. 1
    d. 3.7×1010

31. One curie corresponds to _____.
    a. 1 Bq
    b. 100 Bq
    c. 1010 Bq
    d. 37 GBq

32. Isomeric transition is characterized by _____.
    a. an increase in the atomic number
    b. an increase in the mass number
    c. a decrease in the mass number
    d. no change in the atomic or mass number

33. The SI unit of radioactivity is the _____.
    a. curie
    b. gray
    c. becquerel
    d. rutherford

34. Effective decay constant is _____.

ogical decay constants

diological and biological decay constants

regarding the becquerel?

r second.

el.

of 1 µg of radium.

    f. All of the above are true.

37. Radionuclides are produced by irradiating the element in _____.
     a. a nuclear reactor
     b. a linac
     c. a cyclotron
     d. a furnace kept at very high temperature
     e. all the above ways

38. The ^{137}Cs radionuclide is _____.
     a. produced by irradiating ^{136}Cs in a reactor
     b. produced by bombarding ^{138}Cs in a proton beam in a particle accelerator
     c. naturally occurring
     d. a byproduct of fission in a reactor and hence is extracted from the spent fuel elements
     e. not useful in medical applications

39. Match the following terms to their definitions.
     a. isotopes   _____ i.   nuclides having the same number of neutrons
     b. isotones   _____ ii. nuclides having the same number of protons
     c. isomers   _____ iii. nuclides having the same mass numbers
     d. isobars   _____ iv. nuclides having same Z,A but existing in
                                             different energy states

40. The radioactive isotope produced in a nuclear reactor _____.
     a. is neutron rich
     b. decays emitting β^- particles
     c. decrease the atomic number Z by 1 unit
     d. is usually very short lived
     e. is not preferred in medical applications

41. The $^{192}_{77}$Ir isotope _____.
     a. is one of the stable isotopes or Ir
     b. has 77 protons
     c. has 115 neutrons
     d. has more electrons than protons
     e. all of the properties above are true

42. A nuclide (Z,A) may transform into a nuclide (Z–1,A) by the emission of _____.
     a. an alpha
     b. a β^-
     c. a β^+
     d. EC (electron capture)
     e. gamma emission

43. Positron emission occurs in radionuclides that have an excess of _____.
     a. neutrons
     b. protons
     c. positrons
     d. electrons
     e. π mesons

44. Of the following radionuclides, _____ are produced in accelerators.
    a. ^{60}Co
    b. ^{192}Ir
    c. ^{125}I
    d. ^{18}F
    e. ^{14}C

45. Match the change in atomic number to the decay mode.
    Change in Z   *Decay Mode*
    a. Z Z+2   _____ i.   α
    b. Z Z+1   _____ ii. β$^+$
    c. Z Z   _____ iii. β$^-$
    d. Z Z–1   _____ iv. isomerism
    e. Z Z–2   _____ v.   electron capture

46. Match each radionuclide with its method of production.
    Radionuclide   *Method of Production*
    a. ^{131}I   _____ i.   produced in a reactor by irradiating stable element
    b. ^{192}Ir   _____ ii. produced in a cyclotron
    c. ^{57}Co   _____ iii. naturally occurring
    d. ^{226}Ra   _____ iv. eluted from a generator
    e. ^{99m}Tc   _____ v.   separated from spent fuel element

47. Which of the following statements are true regarding positron emission?
    a. Each positron emission is accompanied by a neutrino.
    b. A minimum of 1.02 MeV equivalent mass difference must exist between the parent and daughter atoms.
    c. It is followed by two annihilation photons of energy 0.511 MeV each.
    d. Positrons are monoenergetic.
    e. All the above are true.

48. Which of the following statements are true regarding electron capture (EC)?
    a. It is a competing mode for positron decay.
    b. When positron decay is energetically not possible, only EC can occur.
    c. It results in characteristic x-ray emission.
    d. It results in Auger electrons emission.
    e. It results in the emission of a neutrino.
    f. All of the above are true.

49. Characteristic x-rays are emitted following _____.
    a. positron decay
    b. alpha decay
    c. internal conversion
    d. electron capture
    e. all of the above occur

50. A radionuclide decaying by internal conversion emits _____.
    a. betas
    b. gammas
    c. characteristic x-rays
    d. Auger electrons
    e. all of the above

51.   What happens during internal conversion?
    a.   A beta is emitted.
    b.   There is no change in Z or A.
    c.   L and M shell electrons may also be emitted.
    d.   The shell electrons are ejected by the gamma ray emitted in the isomeric transition.
    e.   All the above are true.

B. Production of X-rays

Circle the right answer (Yes or No):

1.   (Yes / No)   X-rays are produced by decelerating high speed electrons in a target.

2.   (Yes / No)   The mechanism of x-ray production is the same in a kV x-ray unit and in an accelerator unit.

3.   (Yes / No)   The major fraction of the electron energy is converted to x-rays.

4.   (Yes / No)   The x-ray spectrum produced is influenced by the thickness of the target.

5.   (Yes / No)   The x-rays produced would heat up the target.

6.   (Yes / No)   The electrons impinge on a large area of the target in a clinical accelerator.

7.   (Yes / No)   Therapy x-ray beams used in the kV region are medium filtered.

8.   (Yes / No)   The beam quality of kilovoltage therapy x-ray beams can be adequately represented by the HVL.

9.   (Yes / No)   For the accelerator photon beams, AAPM TG-51 recommends $TPR_{20/10}$ as the beam quality specifier.

10.   (Yes / No)   The (Bremsstrahlung) x-ray spectrum produced in the target has a continuous spectrum.

11.   (Yes / No)   The maximum energy of the photons in the x-ray spectrum exceeds the energy of the electrons incident on the target.

12.   (Yes / No)   Characteristic x-rays are also produced in the target along with the continuous x-ray spectrum.

13.   (Yes / No)   The intensity of x-ray spectrum produced by a linear accelerator is maximum along the central axis of the beam and decreases considerably with increasing angle with respect to the central axis.

Choose the right answer(s) (more than one may be correct):

14.   The kV x-ray unit used in orthovoltage therapy must have a target _____.
    a.   of high Z
    b.   of high melting point
    c.   of low efficiency for Bremsstrahlung production
    d.   at 90° with respect to the direction of the incident electrons

15. The output of kilovoltage therapy unit is roughly proportional to _____.
    a. (tube current)^2
    b. (tube voltage kV)^2
    c. Z of target

16. What distinguishes an x-ray photon from a gamma ray photon?
    a. velocity
    b. energy
    c. field
    d. origin
    e. all of the above

17. When a high-energy electron (in the MeV range) collides with a target, any one of the following types of interactions can occur:
    a. inelastic collision with electrons
    b. elastic collision with nuclei
    c. inelastic collision with nuclei
    d. elastic collision with neutron
    e. inelastic collision with proton

18. The Bremsstrahlung production probability for the target material varies as _____.
    a. Z
    b. Z^2
    c. Z^4
    d. 1/Z
    e. $1/Z^2$

19. The efficiency of Bremsstrahlung production for the target material is proportional to _____.
    a. Z^3
    b. Z E
    c. $Z^2 E^2$
    d. Z / E
    e. Z^2 / E

20. The maximum photon energy of the Bremsstrahlung spectrum is determined by the _____.
    a. Z of the target
    b. Z of the added filter
    c. tube construction details
    d. kV across the tube
    e. maximum mA
    f. all of the above

21. The Bremsstrahlung x-ray spectrum produced in a given x-ray target depends on _____.
    a. kinetic energy of the incident electron
    b. binding energy of the electrons of the target atoms
    c. target Z
    d. target thickness
    e. target design
    f. all the above

22. The following figure shows the x-ray spectrum produced by the electron beams of various energies.

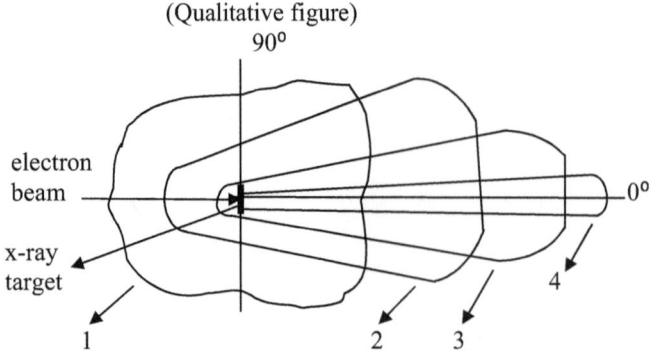

Angular distribution of x-ray emission from the target

Match the spectrum number to the spectrum.

Number		*Spectrum*		
a.	1	_____	i.	4 MV
b.	2	_____	ii.	20 MV
c.	3	_____	iii.	100 kV
d.	4	_____	iv.	400 kV

C. Interaction of Radiation with Matter—Photons and Electrons

Circle the right answer (Yes or No):

1. (Yes / No) Photons lose energy in small increments, thus gradually slowing down in the medium.

2. (Yes / No) Primary photons are the ones transmitted by and have not had any interaction with the medium traversed.

3. (Yes / No) Megavoltage photon interactions can release electrons of significant energy.

4. (Yes / No) Electrons are exponentially attenuated while traversing matter.

5. (Yes / No) Electrons do not produce Bremsstrahlung photons while interacting with matter.

6. (Yes / No) Electrons in their interactions can release secondary electrons of finite range.

7. (Yes / No) Compton interactions are like billiard ball type collisions.

8. (Yes / No) All inelastic collisions of electrons with atomic electrons lead to ionization of atoms.

9. (Yes / No) A pencil beam of electrons incident on a foil spreads into a beam of larger cross-section due to multiple Coulombic interactions with nuclei of atoms.

10. (Yes / No) A narrow beam of photons incident on a foil spreads into a beam of larger cross section due to multiple interactions with nuclei of atoms

11. (Yes / No)   High-energy electrons scatter more than low-energy electrons.

12. (Yes / No)   High-energy electrons scatter more in high-Z materials.

Choose the right answer(s) (more than one may be correct):

13. To produce an electron–positron pair in the vicinity of a nucleus, the interacting photon must have a minimum energy of _____.
    a. 0.5 MeV
    b. 1 MeV
    c. 1.02 MeV
    d. any energy

14. To interact by photoelectric effect, the interacting photon's energy must be _____.
    a. less than the binding energy of the electron in question
    b. equal to or just greater than the binding energy of the electron
    c. any energy
    d. a minimum energy of 1 MeV

15. To undergo a Compton effect, the energy of the interacting photon must be _____.
    a. less than the binding energy of the electron in question
    b. equal to the electron binding energy
    c. very much larger than the electron binding energy
    d. any energy

16. In a photoelectric interaction, the photon loses _____.
    a. all the energy
    b. part of the energy
    c. no energy

17. In a Compton interaction, the photon loses _____.
    a. all its energy
    b. part of its energy
    c. no energy

18. The probability of Compton interaction (i.e., the electronic Compton coefficient) _____.
    a. increases with increasing photon energy
    b. decreases with an increase in photon energy
    c. is independent of photon energy.
    d. increases and then decreases with photon energy.

19. As the photon energy increases, the Compton electron gets ejected _____.
    a. more and more isotropically
    b. more and more in the forward direction
    c. more and more in the backward direction
    d. more at 90° compared to the forward direction.

20. In the megavoltage energy range, the predominant interaction in a patient is _____.
    a. pair production
    b. Compton
    c. photoelectric effect
    d. Thompson scattering

21. The probability of pair production (atomic cross section) in the interacting medium varies as _____.
    a. Z
    b. Z^2
    c. Z^3
    d. independent of Z

22. The probability of photoelectric effect (atomic cross section) in a medium roughly varies as _____.
    a. Z
    b. Z^2
    c. Z^3
    d. independent of Z

23. As the photon energy increases, the probability of photoelectric effect roughly varies as _____.
    a. $1/E$.
    b. $1/E^2$
    c. $1/E^3$
    d. independent of E

24. In pair production, after expending energy for the creation of the pair, the excess photon energy _____.
    a. appears as a scattered photon
    b. is shared by the electron and positron
    c. goes only to electron
    d. goes only to positron

25. At the end of its range in the medium in pair production, the positron _____.
    a. comes to rest
    b. is annihilated, resulting in two annihilation photons
    c. is converted to a photon
    d. is converted to another particle

26. Photoelectric effect involves _____.
    a. a bound electron
    b. an unbound electron
    c. a nuclear interaction

27. A Compton interaction involves a photon interacting with _____.
    a. a free electron
    b. a tightly bound electron
    c. more than one electron
    d. all the electrons of the atom

28. Energy losses by electrons (in their collision interactions) are _____.
    a. independent of Z
    b. dependent on Z

29. Energy losses in radiative collisions, when electrons are stopping in medium, vary as _____.
    a. Z
    b. Z^2
    c. Z^3
    d. independent of Z

30. The probability for photoelectric absorption is maximum when the gamma energy is _____.
    a. equal to the electron binding energy
    b. much greater than the electron binding energy
    c. slightly greater than the electron binding energy
    d. just less than the electron binding energy

31. The threshold energy (in MeV) for pair production in the Coulombic field of the nucleus is _____.
    a. 0.5
    b. 1.02
    c. 10
    d. 25

32. The photoelectric cross section _____.
    a. is independent of the photon energy
    b. is independent of the atomic number of the medium
    c. depends only on the atomic number of the medium
    d. depends on both the photon energy and the atomic number of the medium

33. The Compton mass attenuation coefficient for a given gamma energy depends on _____.
    a. the gamma energy only
    b. the atomic number of the medium only
    c. the electron density of the medium
    d. the square of the atomic number

34. Elastic scattering is characterized by _____.
    a. conservation of kinetic energy
    b. nonconservation of kinetic energy
    c. nonconservation of angular momentum
    d. nonconservation of momentum

35. In Compton scattering, the energy of the incident photon _____.
    a. should be slightly greater than the electron binding energy
    b. is equal to the energy of the scattered photon and the recoil electron (BE is negligible)
    c. increases with scattering angle
    d. gets divided between a pair of charged particles

36. The threshold energy (in MeV) for pair production in the vicinity of an electron is _____.
    a. 0.55
    b. 1.02
    c. 2.04
    d. 10

37. The Compton (electronic) scattering cross section depends on _____.
    a. the gamma energy only
    b. the atomic number of the medium
    c. the binding energy of the medium
    d. none of these parameters

38. The mass attenuation coefficient due to Compton interactions _____.
    a. depends on Z of the medium
    b. increases with energy
    c. decreases with energy
    d. depends on electron density of medium

39. For photoelectric absorption to take place, the gamma energy should be _____.
    a. exactly equal to or slightly larger than the electron binding energy
    b. much greater than the electron binding energy
    c. much less than the electron binding energy
    d. just less than the electron binding energy

40. When a photon traverses a medium, which of the following events can occur?
    a. no interaction
    b. complete absorption
    c. scatter
    d. none of the above

41. When an electron traverses a medium, which of the following events can occur?
    a. no interaction
    b. complete absorption in an inelastic collision
    c. complete conversion into a Bremsstrahlung photon in a radiative collision
    d. none of the above

42. In Compton scatter, the energy of the back-scattered photon (in MeV) is _____.
    a. 0.511
    b. 0.255
    c. 0
    d. none of the above

43. The energy loss of electrons in water or tissue is roughly given (in MeV/cm) by _____.
    a. 2
    b. 1
    c. 0.5
    d. none of the above

44. Multiple scattering of a pencil beam of electrons in a scattering foil results in _____.
    a. its angular spread
    b. its energy degradation
    c. Bremsstrahlung contamination
    d. none of these effects

D. Treatment Machine Characteristics

Circle the right answer (Yes or No):

1. (Yes / No) In a kV x-ray unit, the electrons are accelerated by a constant or pulsating positive DC potential.

2. (Yes / No) In an accelerator, the electrons are accelerated by an alternating AC voltage.

3. (Yes / No) In an accelerator the source of microwave energy is a klystron or a magnetron.

4. (Yes / No) In a microtron the electrons are accelerated in a straight path.

5. (Yes / No) Much of the electron energy deposited in the x-ray target reappears as heat.

6. (Yes / No) An efficient target cooling system is necessary in therapy machines.

7.   (Yes / No)   A flattening filter is used in a ^{60}Co beam to produce a uniform (flat) beam profile.

8.   (Yes / No)   In a ^{60}Co unit the treatment time is controlled using a transmission monitor.

9.   (Yes / No)   There is no radiation from a ^{60}Co machine when the machine is switched off.

10. (Yes / No)   There is no radiation from an accelerator when the machine is switched off.

11. (Yes / No)   A ^{60}Co source is produced by irradiating a ^{59}Co cylinder of source dimensions in a nuclear reactor.

12. (Yes / No)   The output of a ^{60}Co unit has to be measured only during source loading. The output on any treatment day can be determined using the exponential decay law.

13. (Yes / No)   The kV x-ray beam used in radiation therapy is monoenergetic.

14. (Yes / No)   The filter inserted in the beam path of a kV x-ray unit to preferentially absorb the soft components of the x-ray spectrum is known as the "inherent filtration."

15. (Yes / No)   Two fields have the same area. Will they have the same equivalent square field?

Choose the right answer(s) (more than one may be correct):

16. A flattening filter is used in an accelerator to _____.
     a.   attenuate the clinical photon beams and to reduce the beam output
     b.   change the spectra of the photon beam
     c.   filter the low energy photons and to harden the beam
     d.   get a flat beam profile at the clinical depth

17. The flattening filter used in a linear accelerator _____.
     a.   makes the beam quality uniform across the field width
     b.   hardens the beam more in the central region compared to peripheral regions
     c.   produces a horn in the cross-beam profile at shallower depths
     d.   is cone shaped to attenuate more in the central region compared to the peripheral region

18. The flattening filter position in the beam path is _____.
     a.   immaterial
     b.   critical
     c.   not very critical

19. Any error in the reproducible positioning of the flattening filter will affect _____.
     a.   beam flatness
     b.   beam symmetry
     c.   beam output
     d.   none of the above

20. An added filter used with a therapy kV unit _____.
     a.   increases the output
     b.   "softens" or reduces the effective energy of the beam
     c.   achieves desired filtration for therapeutic purposes
     d.   does none of the above

21. The effective energy of an x-ray beam is the energy of that monoenergetic beam that would _____.
    a. give the same exposure rate at a reference distance as the x-ray beam in question
    b. give the same HVL as the x-ray beam in question
    c. have the same inherent and added filtration in the beam path
    d. none of the above

22. The target of a therapy x-ray tube must have _____.
    a. a high Z
    b. a high melting point
    c. low thermal conductivity
    d. low specific heat

23. The efficiency of x-ray production in the kV x-ray therapy tubes is about _____.
    a. 0.1%
    b. 1%
    c. 5%
    d. 10%

24. The focal spot of a kV x-ray therapy tube is about _____.
    a. 1 mm
    b. 3 mm
    c. 10 mm
    d. none of the above

25. The focal spot of a clinical linac is about _____.
    a. 1 mm
    b. 3 mm
    c. 10 mm
    d. none of the above

26. For a therapy x-ray unit operating at 150 kVp, the maximum energy (in keV) of the photon produced in the target is _____.
    a. 150
    b. <150
    c. >150
    d. none of the above

27. The source of electrons in an electron accelerator is the _____.
    a. klystron or magnetron
    b. accelerating waveguide
    c. electron gun
    d. none of the above

28. Some important features of a linac compared to a ^{60}Co unit are _____.
    a. higher beam output
    b. sharper beam
    c. less maintenance cost
    d. less complex in design

29. Modern accelerators can produce _____.
    a. asymmetric fields
    b. circular fields
    c. custom fields
    d. none of the above

30. A _____ target is used in a high-energy linear accelerator.
    a. transmission type
    b. reflection type
    c. refraction type
    d. none of the above

31. The skin sparing effect will be highest for _____.
    a. a diagnostic x-ray beam
    b. an orthovoltage x-ray beam
    c. an electron beam
    d. a megavoltage x-ray beam

32. A _____ power source is used to generate electromagnetic waves for the accelerator guide in a linear accelerator.
    a. magnetron
    b. betatron
    c. cyclotron
    d. none of the above

33. The clinically acceptable limit of flatness over the central 80% of the x-ray beam used in radiotherapy is _____.
    a. ±1%
    b. ±2%
    c. ±3%
    d. ±5%

34. Of the following energies, beam penetration will be highest for _____.
    a. ^{60}Co beam
    b. deep therapy x-rays
    c. 10 MV x-rays
    d. 10 MeV electron beam

35. The best absorbing material for neutrons is _____.
    a. tungsten
    b. polyethylene
    c. lead
    d. copper

36. _____ are used in accelerators as a monitor chamber.
    a. Proportional counters
    b. Ionization chambers
    c. Diodes
    d. Photographic films

37. Match the accelerating mechanism to the machines.

Mechanism of Acceleration	*Machine*
a. direct DC voltage	i. microtron
b. magnetic induction	ii. kV units
c. microwaves	iii. betatron
d. US waves	iv. linac

38. The limitations of kV units in radiotherapy are _____.
    a. increased skin dose
    b. less penetrating giving shallow treatment depths
    c. increased dose to bone
    d. increased patient dose outside the field due to scatter
    e. increasing cost
    f. all of the above

39. _____ can produce photon beams in the mega voltage region.
    a. Linacs
    b. Betatrons
    c. Microtron
    d. Teletherapy machines (^{60}Co)
    e. All of the above

40. The magnetron _____.
    a. is a device generally used in low-energy linacs (about 6 MV or less)
    b. produces microwaves
    c. is usually made of copper
    d. has peak power of about 10 MW
    e. properties given above are all correct

41. A drawing of a klystron is shown in the following figure:

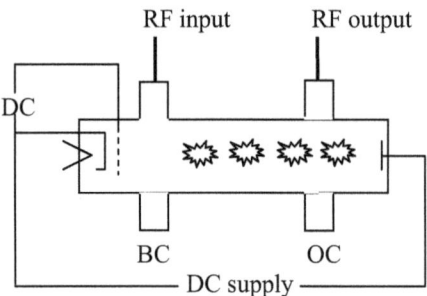

BC: bunching cavity   OC: output cavity

    Match the parts of klystron to their function.

Parts	Function
a. electron gun (e-gun)	_____ i. modulates the velocity of electrons
b. bunching cavity	_____ ii. produces the streaming electrons
c. output cavity	_____ iii. electrons are bunched by velocity modulation
d. collector	_____ iv. excites microwaves

42. Which of the following statements are true regarding a klystron?
    a. To increase the amount of velocity modulation and the power output, intermediate cavities can be added between the input and output cavities.
    b. A large negative pulse applied to the cathode accelerates the electron beam of a three-cavity klystron toward the drift tube.
    c. The output cavity of a three-cavity klystron causes most of the velocity modulation.
    d. In a multi-cavity klystron, tuning all the cavities to the same frequency reduces the bandwidth of the tube.
    e. All of the above are true.

43. In any kind of electron accelerator, having _____ are common conditions?
    a. a source of electrons
    b. a mechanism to inject the electrons into the accelerator structure
    c. an electric field to accelerate the electrons
    d. a magnetic field to slow down the electrons
    e. a vacuum in the accelerator structure
    f. all the above conditions are true

44. The _____ are the four most important components of a linac?
    a. electron gun
    b. accelerator waveguide
    c. source of microwaves
    d. target
    e. beam guidance system
    f. collimators

45. The electron gun _____.
    a. is the source of electrons in the linac
    b. can be a diode or triode
    c. is usually a triode gun assembly that has two filaments to increase the electron emission
    d. cathode is made of thermionic material (thorium) which easily gives off electrons on heating due to lower work function
    e. grid is connected to pulsed voltage source to control the injection of electrons into the accelerator by keeping the grid more negative than the cathode
    f. properties above are all true

46. Which of the following statements are true regarding the linac x-ray beams?
    a. In the MV region, most of the x-rays are emitted in the forward direction.
    b. The x-ray spectrum is a superposition of the Bremsstrahlung and characteristic x-ray spectra.
    c. The linac spectrum does not show characteristic peaks since they are produced only in kilovoltage x-ray units.
    d. The maximum size of the beam is controlled by a primary collimator in the linac head.
    e. All of the above are true.

47. In clinical dosimetry, the beam quality of the x-ray spectrum is generally specified by _____.
    a. HVL
    b. effective energy (of the beam)
    c. nominal accelerating potential (NAP)
    d. $TPR_{20/10}$
    e. PDD(10)
    f. all of the above parameters

48. The half value layer (HVL) _____.
    a. concept is generally used for kV x-ray beams
    b. must be measured with a field just large enough to cover the chamber to avoid influence of scatter from the attenuator
    c. can be measured with any small ionization chamber
    d. measurement requires placing the chamber not too near to the attenuator
    e. is less useful in clinical dosimetry compared to the actual x-ray spectrum

49. For MV x-ray beams, _____.
    a. NAP is a good choice for beam quality specification
    b. $TPR_{20/10}$ (10×10, SAD) is used as the beam quality specifier in European countries and in IAEA and European protocols
    c. PDD (10, 10×10, SSD_{100}) for pure photon beam is used as beam qualifier in the United States and in AAPM protocols
    d. HVL also can used as a beam qualifier

50. Kilovoltage x-ray machines _____.
    a. are used for treating superficial tumors and tumors at shallow depths
    b. generally define treatment fields with cones or applicators
    c. can deliver IMRT treatments at shallow depths
    d. deliver treatment dose using a timer control which needs no correction
    e. use lower current, larger exposure times, larger focal spot, and fixed anode compared to diagnostic kV x-ray machines.

51. A cyclotron can accelerate _____.
    a. protons
    b. neutrons
    c. electrons
    d. gamma rays
    e. all of the above

52. Match the accelerator characteristics to the nature of accelerating waveguides used.

    Accelerator Characteristics   *Waveguides Used*
    a. The microwaves enter on the gun side and propagate   i.   traveling wave
       toward the high-energy end.   ii. standing wave
    b. The microwaves reflected from both ends of waveguide
       form a standing wave pattern.
    c. Only one in four cavities is, at any given moment,
       suitable for acceleration.
    d. Every second cavity carries no electric field and thus
       produces no energy gain for the electron.

53. The typical microwave frequency used in medical linacs is _____.
    a. 0.1 GHz
    b. 1 GHz
    c. 3 GHz
    d. 10 GHz
    e. 20 GHz

54. The beam bending angles used in the beam transport systems of medical linacs are _____.
    a. 30°
    b. 90°
    c. 112.5°
    d. 270°
    e. 330°

55. Focusing coils _____.
    a. are wrapped around the linac accelerating structure
    b. stop the divergence of the beam due to mutual repulsion between electrons
    c. confine electrons to a pencil beam
    d. constantly confine the beam to the central axis of the accelerator structure
    e. properties given above are all true

56. Match the numbers in the graphic to the components in the linac head.

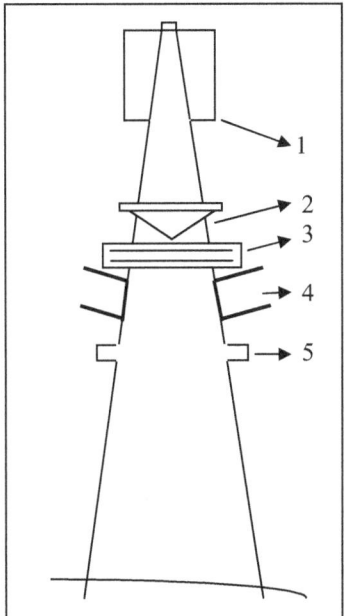

Number		Component
a. 1	_____ i.	ionization chamber
b. 2	_____ ii.	tertiary collimator
c. 3	_____ iii.	secondary collimator
d. 4	_____ iv.	flattening filter
e. 5	_____ v.	primary collimator

57. The flattening filter used in the x-ray beam _____.
    a. preferentially attenuates the photons in the forward direction to produce a uniform beam profile
    b. reduces the beam output considerably
    c. gives rise to "horns" or hotspots in the cross beam profile at shallow depths due to overflattening
    d. can be removed by the user to increase the beam output
    e. can also be used with electron beams

58. In a linac, the photon beam collimation is achieved using _____.
    a. a primary collimator
    b. a secondary collimator (or jaws)
    c. an MLC
    d. cones or applicators
    e. all of the above

59. The transmission ion chambers used in the linac head serve the purposes of _____.
    a. monitoring beam flatness
    b. monitoring beam symmetry
    c. controlling treatment delivery
    d. controlling delivered dose through MUs
    e. changing the treatment modality
    f. all the above

60. MLC designs available on various linac models include _____.
    a. upper jaw as MLC
    b. lower jaw as MLC
    c. tertiary MLC
    d. both upper and lower jaws as MLC
    e. primary collimator as MLC

61. Match the MLC design to the linac make.
    MLC Design   *Linac Make*
    a. upper jaw (UJ) as MLC   _____ i.   Siemens
    b. lower jaw (LJ) as MLC   _____ ii. Varian
    c. tertiary MLCS   _____ iii. Philips
    d. both UJ and LJ as MLCs   _____ iv. Elekta

62. Which of the following statements are true regarding MLCs?
    a. Typical MLCs have 40 pairs of leaves in clinical linacs.
    b. The leaf width at isocenter is 1 cm.
    c. The leaf thickness (in the ray direction) is 6 to 7.5 cm.
    d. The leaf transmission is ≈2%.
    e. The interleaf transmission is ≈3%.
    f. All of the above are true.

63. In the case of MLCs, _____.
    a. there is no limit for the MLC leakage
    b. both UJ and LJ MLCs must have the same leakage as standard collimators
    c. the tertiary jaw must have the same leakage as standard collimators
    d. the interleaf leakage is typically about 2% or less
    e. all the above are true

64. The MLC leaf ends do not have identical designs in all the linacs in the market. Which of the following statements are true regarding the leaf end design?
    a. The leaf ends follow "focused leaf design," i.e., they follow the beam divergence.
    b. The leaf ends are rounded and do not follow the beam divergence, i.e., they are non-focused.
    c. Rounded end penumbra are marginally greater compared to focused leaf end.
    d. There is greater leaf transmission for abutting rounded end leaves.
    e. The light field underestimates the radiation field for rounded end leaves.
    f. All of the above are true.

65. Which of the following statements regarding IGRT are true?
    a. The IGRT systems can be based on planar imaging or CT.
    b. The imaging system is always gantry mounted.
    c. The planar imaging can be kV imaging or MV imaging, but they are always gantry-mounted units.
    d. Pretreatment images are compared with reference images from the treatment planning system for realignment of patient if necessary.
    e. All of the above are true.

66. The Tomotherapy unit uses _____.
    a. a 6 MV beam
    b. a 10 MV beam
    c. a 6 and a 10 MV beam
    d. 6 MeV electron beam
    e. a 10 MeV electron beam

67. Which of the following statements regarding serial tomotherapy are true?
    a. Tomotherapy treatment cannot be delivered using a linac as it requires a special tomotherapy unit.
    b. Treatment is delivered serially slice by slice by moving the couch one to two slices at a time.
    c. A potential problem with serial tomotherapy is the possibility of mismatch between adjacent slice pairs, causing hot or cold spots.
    d. Non-coplanar treatments are possible with a tomotherapy unit.
    e. All of the above are true.

68. Which of the following statements regarding the TomoTherapy Hi-Art® system are true?
    a. It is a helical tomotherapy unit.
    b. A xenon ion chamber CT detector array is used for acquiring images with the patient in treatment position.
    c. The unit acquires 6 MV CT pretreatment images.
    d. Pretreatment images are automatically registered to the radiotherapy planning kV CT using either bony anatomy or soft tissue and bony anatomy.
    e. Tomotherapy has many scan acquisition slice thickness options depending on the resolution required.
    f. All of the above are true.

69. With respect to the differences between tomotherapy and linac treatment delivery systems, the tomotherapy system _____.
    a. has higher beam output
    b. uses no flattening filter
    c. has less scatter contamination
    d. has scatter mainly from the beam and very little from the head
    e. has all the above differences

70. The important features of the binary MLC used in tomotherapy units are _____.
    a. leaf width at isocenter is 6.25 cm
    b. leaf thickness (in the ray direction) is 10 cm
    c. leaf positions is a large number
    d. maximum field width (FOV) is 40 cm
    e. all the above features are true

71. The advantages of CyberKnife® over other linac systems (Novalis, Varian Trilogy, and TomoTherapy) are that it _____.
    a. continuously adapts to lesion motion due to organ motion
    b. continuously adapts to lesion motion due to breathing more accurately compared to linac systems
    c. can treat any part of the body with radiosurgical precision in a less complex way
    d. needs no brain or body frame for accurate treatment
    e. all the above are true

72. The CyberKnife system _____.
    a. is used only for the treatment of intracranial lesions
    b. does not require any immobilization devices
    c. uses different methods for patient position tracking
    d. produces any field shape for treatment delivery
    e. features given above are all true

73. The CyberKnife unit uses a _____.
    a. 6 MV beam
    b. 10 MV beam
    c. 6 MV and a 10 MV beam
    d. 6 MeV electron beam
    e. 10 MeV electron beam

74. Which of the following statements regarding CyberKnife characteristics are true?
    a. This linac, like all other linacs, is an S-band linac.
    b. It operates at normal SSD of 80 cm, though this is variable over a 65 to 100 cm range.
    c. It uses 10 to 20 beams per fraction.
    d. During both pretreatment and treatment, kV imaging helps to monitor the alignment of the target wit respect to the treatment fields.
    e. All the above statements are true.

75. The different models of Gamma Knife® units produced since the introduction of the first model in 1968 are the _____.
    a. model S and model U
    b. model B and model C
    c. model D and model E
    d. model 4C and Perfexion
    e. model Universal

76. Which of the following statements regarding the Gamma Knife 4 model series are true?
    a. It is intended only for the treatment of brain lesions.
    b. 201 ^{60}Co sources focus at the lesion through the helmet worn by the patient.
    c. The lesion is destroyed with very little dose to adjacent normal brain tissues.
    d. Beams in certain directions can be blocked to save critical structures from excessive dose.
    e. All of the above are true.

77. The purpose of the stereotactic head frame is to _____.
    a. increase patient comfort
    b. prevent head motion during the procedure
    c. localize the lesion for targeting during delivery
    d. move the patient's head to the focal point of irradiation
    e. achieve all the above

78. How many secondary collimators are there in the Gamma Knife model 4C and earlier models?
    a. 201
    b. 150
    c. 131
    d. 120
    e. 100

79. The collimation provided by the four focusing helmets of a Gamma Knife unit are of the following diameters (in mm):
    a. 2
    b. 4
    c. 6
    d. 8
    e. 14
    f. 18

80. The improved features of the Perfexion model compared to earlier Gamma Knife models include _____.
    a. replacement of ^{60}Co sources by longer-half-life ^{138}Cs sources
    b. a single collimating system replacing four different collimator setups
    c. providing 4, 8, and 16 mm treatment beams
    d. the beams can be mixed and matched to get various distribution shapes
    e. patient movement by couch motion
    f. all of the above

E. Radiation Quantities, Units, and Measurements

Circle the right answer (Yes or No):

1. (Yes / No) Radiation that has enough energy to ionize the atoms and thus dissipate energy in matter is referred to as ionizing radiation (IR).

2. (Yes / No) Indirectly ionizing radiations (IDIRs) cannot directly produce ionization of atoms in the medium.

3. (Yes / No) Kerma, K, at a particular point is defined as the kinetic energy released (e.g., in the form of charged particle energies) per unit mass, in an infinitesimal volume, around that point by IDIR.

4. (Yes / No) Kerma can be defined for electrons.

5. (Yes / No) All of kerma ends up as energy deposition in the medium.

6. (Yes / No) Absorbed dose is defined for both photons and electrons.

7. (Yes / No) That part of kerma (i.e., electron kinetic energy) that leads to subsequent collision interactions and, hence, energy deposition is known as collision kerma, K_C.

8. (Yes / No) Part of the energy of electrons interacting with a medium appears as Bremsstrahlung.

9. (Yes / No) Collision part of kerma is equal to absorbed dose.

10. (Yes / No) Exposure is the ionization equivalent of collision kerma in air.

11. (Yes / No) Any radiation dosimeter can be used for radiation therapy beam calibration.

12. (Yes / No) Therapy chambers can be calibrated in terms of exposure, air kerma, or "absorbed dose to water."

13. (Yes / No) Primary standards of exposure, air kerma, or water absorbed dose have been established by some primary standards dosimetry laboratories (PSDLs) for a ^{60}Co beam.

14. (Yes / No) Free air ionization chamber is the primary standard of exposure/air kerma for ^{60}Co energy.

15. (Yes / No) Graphite cavity ionization chamber is the primary standard of exposure/air kerma at ^{60}Co energy.

16. (Yes / No) Graphite calorimeter or water calorimeter serves as the primary standard of absorbed dose.

17. (Yes / No) Calorimetric standard of absorbed dose is not available for accelerator beam qualities.

18. (Yes / No) Hospital chambers must be calibrated in terms of exposure or air kerma or absorbed dose to water in an accredited dosimetry calibration laboratory, which is traceable to the primary standards laboratory.

19. (Yes / No) Relative dosimetry can be carried out with other types of radiation dosimeters if its energy dependence is known.

20. (Yes / No) Farmer-type chambers can be used for the dosimetry of accelerator-produced photon and electron beams.

21. (Yes / No)   Some plane-parallel chambers exhibit significant perturbation at low electron energies.

Choose the right answer(s) (more than one may be correct):

22. The range of accelerator photon beam qualities commonly used in radiation therapy is _____.
    a. 6–21 MV
    b. 2–6 MV
    c. 25–50 MV
    d. none of the above

23. The range of accelerator electron beam qualities commonly used in radiation therapy is _____.
    a. 3–15 MeV
    b. <3 MeV
    c. 20–50 MeV
    d. none of the above

24. Beam calibration at the hospital must be carried out using _____.
    a. a Fricke dosimeter
    b. film
    c. a semiconductor dosimeter
    d. an ionization chamber

25. The wall thickness of the chamber used for beam output measurements in the kV region is _____.
    a. 60 to 90 mg/cm^2
    b. a few mg/cm^2
    c. about 500 mg/cm^2
    d. any thickness

26. The equilibrium wall thickness required for the chamber used for beam output measurements at ^{60}Co energy is about _____.
    a. 60 to 90 mg/cm^2
    b. a few mg/cm^2
    c. 500 mg/cm^2
    d. any thickness

27. Farmer chambers are usually calibrated with a buildup cap at ^{60}Co energy to _____.
    a. increase attenuation in the wall
    b. cut off the scatter electrons
    c. reduce the chamber response
    d. provide equilibrium wall thickness

28. A Farmer-type chamber used for exposure or air kerma measurements in a hospital must have _____.
    a. an air equivalent wall
    b. a water equivalent wall
    c. a tissue equivalent wall
    d. a high-Z wall

29. The effective Z of air in the photoelectric region is about _____.
    a. 4.
    b. 7.6
    c. 10.4
    d. 13

30. The SI unit of kerma is _____.
    a. J/kg
    b. kJ/kg
    c. J/g
    d. none of the above

31. J/kg is given the special name of _____.
    a. rad
    b. CGy
    c. Gy
    d. none of the above

32. One Gy is equal to _____.
    a. 10 cGy
    b. 100 cGy
    c. 100 rads
    d. none of the above

33. An exposure of 1 roentgen corresponds to a charge release of _____.
    a. 1 esu/cm^3 at NTP
    b. 1 C/m^3 at NTP
    c. 2.58×10^{-4} C/kg
    d. none of the above

34. Exposure is _____.
    a. defined only for x-rays and gamma rays
    b. defined only for air
    c. defined only for up to 3 MeV
    d. measured according to the definition in a free air ionization chamber
    e. being replaced by air kerma
    f. features given above are all true

35. Air kerma is _____.
    a. defined only for MV x-rays
    b. defined for an infinitesimal mass or volume element
    c. defined for all media
    d. measured by measuring the kinetic energy of all the electrons contributing to kerma
    e. a more fundamental quantity than exposure

36. Exposure is measured in (SI units) as _____.
    a. R/C
    b. C/mg
    c. R/cGy
    d. C/kg
    e. mGy

37. Absorbed dose (in SI) units is measured in _____.
    a. roentgens
    b. Coulombs
    c. grays
    d. sieverts
    e. kilograms

38. The exposure rate at a point in the beam is 10 R/min. The exposure rate in SI units = _____ (C/kg) / min.
    a. 25.8×10^{-4}
    b. 10
    c. 25.8
    d. 52.6
    e. 96

39. The exposure rate at a point is 100 R/min. The air kerma rate (AKR) at this point (in cGy/min) is given by
    _____.
    a. 100
    b. 95.2
    c. 87.6
    d. 50
    e. 43.4

40. The primary standard of absorbed dose is _____.
    a. free air ionization chamber
    b. graphite ionization chamber
    c. graphite calorimeter
    d. water calorimeter
    e. all of the above

41. Match the name of the units to the quantities.

Unit Name		Quantity
a. becquerel	_____ i.	absorbed dose
b. gray	_____ ii.	dose equivalent
c. sievert	_____ iii.	activity
	_____ iv.	air kerma

42. Absorbed dose is measured in _____.
    a. roentgens
    b. becquerels
    c. grays
    d. sieverts
    e. J/kg

43. A dosimeter generally measures _____.
    a. exposure / air kerma
    b. absorbed dose
    c. dose equivalent
    d. effective dose
    e. all of the above

44. Which of the following statements regarding dosimetric measurements are true?
    a. Accuracy and precision mean the same thing.
    b. A measurement can be very precise and still inaccurate.
    c. A measurement can be accurate and at the same highly imprecise.
    d. Any system that exhibits a reproducible and measurable response to radiation can be used as a dosimeter.
    e. All of the above are true.

45. Any measured quantity in clinical dosimetry has uncertainties associated with it. According to the ICRU, these uncertainties are classified as _____.
    a. type A
    b. type B
    c. type C
    d. random
    e. systematic

46. Some of the important properties of a dosimeter include _____.
    a. linearity
    b. energy dependence
    c. dose rate dependence
    d. directional dependence
    e. all of the above

47. Which of the statements regarding ionization dosimeters are true?
    a. An ionization chamber designed for exposure measurements must be preferably made of high-Z wall material.
    b. The dosimeter must be as small as possible to accurately measure the dosimetric quantity.
    c. The dosimeter can be of any shape for dose measurements.
    d. The dosimeter requires proper calibration to measure the dosimetric quantity.
    e. All of the above are true.

48. A 0.6 cc Farmer-type chamber reads 384.5 pA at a point where the AKR is 120 cGy/min. It exhibits linearity of response in the range 100 to 350 cGy/min. You would expect a current of _____ from the chamber if the AKR at the measurement point is 240 cGy/Min.
    a. same current
    b. 2×384.5 pA
    c. much less than 384.5 pA due to ion recombination at higher AKR
    d. zero due to over saturation
    e. it depends on the volume of the chamber

49. A 0.125 cc chamber measures 80 pA at a point where the AKR is 130 cGy/min. A 0.66 cc chamber kept at the same point would measure _____.
    a. zero due to over saturation
    b. the same current since the AKR at the point is the same
    c. $(0.125/0.66) \times 80$ pA
    d. $(0.66/0.125) \times 80$ pA
    e. depends on the chamber wall

50. The _____ is utilized by many dosimeters to detect radiation.
    a. energy of the radiation
    b. temperature rise in the dosimeter
    c. ionization caused
    d. chemical reactions in the dosimeter
    e. all the above properties simultaneously cause a response

51. The dosimeter of highest precision and accuracy in clinical dosimetry is _____.
    a. the ionization chamber
    b. the TLD
    c. film
    d. the diode
    e. the chemical dosimeter

52. For clinical dosimetry of accelerator photon beams, only _____ must be made use of.
    a. calorimeters
    b. Farmer-type ionization chambers
    c. extrapolation chambers
    d. well type ionization chambers
    e. diodes

53. For clinical dosimetry of accelerator electron beams, only _____ must be made use of.
    a. calorimeters
    b. Farmer-type ionization chambers
    c. plane parallel (PP) ionization chambers
    d. well type ionization chambers
    e. diodes

54. Film dosimetry, compared to ionization dosimetry, gives _____.
    a. higher resolution
    b. higher accuracy
    c. dose in 2D in a single exposure
    d. higher sensitivity
    e. dose and dose rate
    f. all the above advantages

55. The most suitable and commonly used dosimeters for *in vivo* dosimetry are _____.
    a. ionization chambers
    b. chemical dosimeters
    c. TLDs
    d. Si diodes
    e. MOSFET dosimeters

56. A single dosimeter that can give the dose distribution in 3D is the _____.
    a. ionization chamber
    b. TLD
    c. semiconductor dosimeter
    d. film dosimeter
    e. gel dosimeter

57. The reader system used with an ionization chamber is _____.
    a. a densitometer
    b. a luminescence reader
    c. an electrometer
    d. an ammeter
    e. a voltmeter

58. The volume (in cm^3) of the Farmer-type chamber used for beam output calibration in radiation oncology is about _____.
    a. 0.05
    b. 0.6
    c. 1.0
    d. 6.6
    e. 20

59. The chamber of high sensitivity required for the measurement of AKS of HDR ^{192}Ir brachytherapy source is generally a _____.
    a. plane parallel chamber
    b. Farmer-type chamber
    c. large volume cylindrical chamber
    d. well type chamber
    e. survey meter

60. The electrometer used with an ionization chamber in clinical dosimetry must have good _____.
    a. accuracy
    b. stability
    c. linearity
    d. leakage characteristics (very low leakage)
    e. all the above

61. Radiochromic film is _____.
    a. sensitive to visible light
    b. self-developing
    c. less sensitive compared to other conventional films
    d. nearly tissue equivalent
    e. very much suited for measuring 2D dose distributions in the clinical photon beams

62. The advantages of radiochromic film in linac beam dosimetry are _____.
    a. no processing is required and hence no chemicals
    b. no darkroom facility is needed
    c. low energy dependence
    d. fractionation and dose rate independent
    e. all of the above

63. Which of the following properties regarding radiochromic film are true?
    a. It has very high resolution.
    b. It is useful for radiotherapy QA.
    c. It is useful for stereotactic fields.
    d. It can be used for dose mapping around brachytherapy sources.
    e. It is highly dose rate dependent.
    f. All of the above are true.

64. Dosimeters based on luminescence property are _____.
    a. TLD dosimeters
    b. optically stimulated luminescence system (OSL)
    c. plastic scintillators
    d. diamond dosimeters
    e. all of the above

65. Which of the following statements regarding thermoluminescence (TL) dosimetry are true?
    a. By proper doping, suitable traps can be created in the forbidden gap between the valency and conduction bands.
    b. Irradiation causes ionization in the material causing free electrons and holes.
    c. The free carriers recombine, but a small fraction get trapped.
    d. When the trapped carriers are excited to the conduction band, they de-excite through luminescent centers, emitting luminescence.
    e. All of the above are true.

66. A TLD dosimeter _____.
    a. is an integrating dosimeter
    b. is very useful in clinical dosimetry since it is available in various forms like powder, chip, rods, etc.
    c. exhibits a linear response over a very wide range
    d. can be chosen to be tissue equivalent for clinical dosimetry work
    e. exhibits all the above properties

67. _____ dosimeters are most suitable for clinical dosimetry.
    a. $LiF : Mg,Ti$
    b. $Li_2B_4O_7 : Mn$
    c. $Li_2B_4O_7 : Mg,Cu,P$
    d. $CaSO_4 : Dy$
    e. $CaF_2 : Mn$

68. One can measure the light emitted by a luminescence dosimeter by using _____.
    a. a photomultiplier and an electrometer
    b. a voltmeter
    c. an ammeter
    d. a galvanometer
    e. a light meter

69. Which of the following statements are true regarding TLD annealing?
    a. Annealing refers to keeping the TLDs at high temperature and then letting them cool in a controlled way.
    b. Annealing is performed to remove the residual signal in the TLD after a readout.
    c. The annealing temperature must be much higher than the readout temperature.
    d. Annealing increases the life of the TLD.
    e. All of the above are true.

70. State which of the following statements regarding Si diodes are true?
    a. N-type Si are produced by doping pure Si with phosphorous, which releases an extra electron as free.
    b. N-type Si contain electrons as free carriers.
    c. P-type Si contain "holes" as free carriers.
    d. By counter-doping the surface of an n- or p-type Si, a p-n junction diode dosimeter is produced.
    e. In a p-n junction diode, the electrons and holes combine to produce neutral Si crystal.
    f. All of the above are true.

71. A Si diode _____.
    a. cannot conduct if the diode circuit is reverse biased
    b. can conduct if a forward voltage is applied to neutralize the diode intrinsic voltage
    c. forward current increases with increasing forward bias
    d. can conduct in the reverse direction if a reverse voltage is applied
    e. is useful as a radiation dosimeter under option (b)

72. State which of the following statements regarding Si diode dosimeters are true?
    a. Si diode dosimeters are simple n-p junction diodes.
    b. They have much higher sensitivity compared to ionization chamber-based dosimeters.
    c. Voltage across the depletion region is very high.
    d. They can measure the ionization produced by the ionizing radiation in the depletion region.
    e. All of the above are true.

73. Which of the points given below are true regarding the use of diode dosimeters in clinical dosimetry?
    a. Diode sensitivity increases with accumulated dose.
    b. Pre-irradiation reduces the change of response.
    c. The buildup encapsulation of the diode must be appropriate for the beam quality in use.
    d. Periodic calibration is necessary for high accuracy.
    e. The diodes must have adequate buildup material to avoid the influence of scatter photons and contamination electrons.
    f. All of the points above are true.

74. The entrance dose measured with a diode depends on geometric factors like SSD, field size, etc. The reasons for this dependence are _____.
    a. dose nonlinearity of the diode
    b. inadequate response at very large SSDs or field sizes
    c. diode leakage
    d. the diode's temperature dependence on dose rate
    e. all of the above

75. A MOSFET dosimeter _____.
    a. is a diode dosimeter made of metal oxides
    b. is a metal oxide semiconductor field effect transistor
    c. is usable in almost all advanced external beam therapy techniques
    d. is particularly useful in high dose gradient fields and in *in vivo* dosimetry
    e. has also been used in beam calibrations in some centers

76. MOSFET dosimeters _____.
    a. can operate with or without gate bias
    b. that can operate wirelessly are available commercially
    c. that operate without bias can be used as implantable dosimeters for daily dose verification
    d. designed as implantable dosimeters can also serve the dual purpose of fiducial markers
    e. all of the above are true

77. The different types of MOSFET dosimeters available for *in vivo* dosimetry work are _____.
    a. single bias, single MOSFET
    b. dual bias, dual MOSFET
    c. unbiased, single MOSFET
    d. multiple bias, multiple MOSFET
    e. all of the above are true

78. A MOSFET dosimeter is shown in the following figure. Which of the following statements are true regarding the dosimeter shown below?

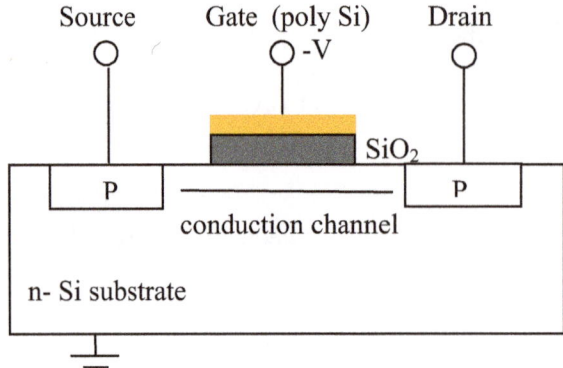

Design of MOSFET dosimeter

    a. This is a p-channel MOSFET dosimeter.
    b. Holes constitute channel current in this dosimeter.
    c. It is a three-terminal PNP transistor, the three terminals being the "source," the "drain," and the "gate."
    d. The region below the gate oxide layer is SiO_2, an insulating layer.
    e. The p-type region below the oxide layer is called the channel.
    f. All of the above are true.

79. In MOSFET dosimeters, the response to radiation is measured as _____.
    a. current produced in the system
    b. the total charge accumulated in the system
    c. the difference in threshold voltages between preexposure and after exposure
    d. cGy
    e. roentgens

80. The advantages of MOSFET dosimeters are _____.
    a. they function as point detectors
    b. their instantaneous readout make them useful for on-line dosimetry
    c. they need no energy dependence correction factors for MV radiation beams
    d. their axial anisotropy is small
    e. all of the above

81. Which of the following properties of MOSFET dosimeters are true?
    a. They have extremely small volumes and very high spatial resolution.
    b. They are ideally suited for measuring in very small radiation fields.
    c. Some MOSFET dosimeters are isotropic and temperature independent.
    d. They are ideal for beam calibrations.
    e. All of the above are true.

82. Dosimeters that can be used for *in vivo* dosimetry are _____.
    a. radiochromic film
    b. TLDs
    c. diode dosimeters
    d. MOSFET dosimeters
    e. all of the above

83. Which of the following statements regarding scintillation dosimeters are true?
    a. Scintillators can be organic or inorganic.
    b. Scintillation is an intrinsic property of all scintillation dosimeters.
    c. All scintillator detectors are tissue equivalent.
    d. Scintillation detectors are widely used in nuclear medicine.
    e. All of the above are true.

84. The advantages of plastic scintillators are _____.
    a. They are water (or tissue) equivalent above 100 keV, so they are very useful for MV dosimetry and brachytherapy dosimetry.
    b. No temperature/pressure corrections are necessary.
    c. They have high spatial resolution.
    d. They can be easily connected to thin copper wires for readout purposes.
    e. All of the above are true.

85. Plastic scintillation dosimetric systems _____.
    a. are linear in dose response in the dose range of therapeutic interest
    b. have significant directional characteristics
    c. exhibit significant energy dependence
    d. show good reproducibility
    e. have all of the above properties

86. Gel dosimeters _____.
    a. are true 3D dosimeters with very high spatial resolution
    b. are not tissue or water equivalent
    c. can record dose over an entire IMRT delivery
    d. make use of Fricke gels or polymer gels
    e. have readout that is the NMR relaxation rate, which is a measure of the dose

I. Radiation Physics ANSWERS

A. Radioactivity

1.   Yes

2.   No   Alpha emission is found only in high-Z, highly unstable large nuclei that exist in natural radioactive series.

3.   Yes   Because Z change results in a different radionuclide.

4.   Yes

5.   No   It is a beta emitter. The emission results in an excited state of ^{60}Ni which goes to ground state by gamma emissions. The gamma energies 1.17 and 1.33 MeV usually attributed to the ^{60}Co source thus actually are emitted from ^{60}Ni.

6.   Yes

7.   Yes

8.   Yes

9.   Yes

10.   Yes   They are practically identical.

11.   No   In a beta decay, a neutron becomes a proton, so there is a tendency to increase proton number. Nuclides decaying by beta decay are neutron rich (having too many neutrons compared to the number of protons).

12.   No   Positron decay transforms a proton into a neutron, so they are proton-rich (neutron-deficient) radionuclides.

13.   Yes   Gamma emissions take the nuclides only from excited states to ground state.

14.   Yes   Because beta emissions often leave daughter nuclides in excited states.

15.   No   There are many pure beta emitters (e.g., ^{32}P and ^{90}Sr nuclides). In these decays, beta emissions simply leave the daughter nuclides in their ground state.

16.   No   In some sense, there are pure gamma emitters. When the excited state of the daughter nuclide is having a long half life, the nuclides behave as pure gamma emitters (e.g., ^{99m}Tc).

17.   Yes   This is because the K-shell is very close to the nucleus as a result of the strong electrostatic attraction of the nucleus.

18.   No   Electron capture and positron emission are competing modes of decay (and they lead to the same daughter nuclide by converting a proton into a neutron).

19. No   Beta particles emitted have a spectrum. This is because beta decay is accompanied by the emission of an antineutrino which can have different energies.

20. Yes   Precisely.

21. No   Detecting a neutrino is an Herculean task, because particles are usually detected by their electrostatic, electromagnetic, or nuclear interactions and neutrinos are incapable of these interactions.

22. Yes   Of course.

23. a. ii.   b. iii.   c. iv.   d. i.   e. iv.

24. a. ii.   b. iii.   c. v.   d. i.   e. iv.

25. a. ii.   b. iii.   c. v.   d. i.   e. iv.

26. a

27. d

28. c

29. a

30. c

31. d

32. d   Isomers are the same atoms differing only in their nuclear energy states, e.g., atoms in metastable state and ground state, respectively.

33. c

34. d

35. c

36. a, b, c, d

37. a, b, c

38. d

39. a. ii.   b. i.   c. iv.   d. iii.
    The third letter from the right denotes what the terms refer to. We can write those letters in bold as shown here: Isoto**p**es; Iso**t**o**n**es; Isom**e**rs; Isob**a**rs **p** stands for (same no. of) protons, **n** for neutrons, **e** for excited state, and **a** for same mass number A.
    Examples:
    ^{1}H and ^{3}H (same **p** means same Z or same element)
    ^{37}Cl and ^{39}K (same **n** means different Z, must be usually adjacent elements)
    ^{99}Tc and ^{99m}Tc (must be same element, one being the metastable state)
    ^{58}Fe and ^{58}Ni (Isobars are different elements but must be adjacent or nearby elements since A is the same)

40. a, b

Much of the phenomenon of radioactivity can be understood from the chart of the nuclides, which is a plot of nuclei as a function of proton number, Z, and neutron number, N (shown in the figure below). All stable nuclei and known radioactive nuclei—both naturally occurring and man-made—are shown on this chart, along with their decay properties. The fact that stable isotopes are found only in the region of stability (thick red line) shows any isotope that is outside this region must be unstable. It is easy to see that to leave the island of stability one must disturb the N/P ratio of the stable elements. This can be done by irradiating the stable elements with neutrons (in a reactor) or protons (in an accelerator or cyclotron) which will produce neutron-rich and proton-rich nuclides. These nuclides will then decay toward the island of stability by changing neutrons into protons or protons into neutrons, with the emission of β^- and β^+ respectively. See the equations below.

$$^A_Z X \rightarrow\ ^A_{Z+1} Y + e^- + \bar{v}_e\,(\beta^- decay)$$

$$^A_Z X \rightarrow\ ^A_{Z-1} Y + e^+ + v_e\,(\beta^+ decay)$$

$$e^- +\ ^A_Z X \rightarrow\ ^A_{Z-1} Y + v_e\,(electron\,capture)$$

β^- and β^+ decays involve the following:

$$n \rightarrow p + e^- + \bar{v}_e$$

$$p \rightarrow n + e^+ + v_e$$

Electron capture and positron decay are competing modes of decay since both reduce the atomic number by one unit. When the mass change is insufficient to cause spontaneous positron emission, a neutron can form by this alternate process where an outside electron is captured by the nucleus to combine with a proton to become a neutron, emitting only a neutrino. (This happens in heavier elements where the mass change is small and the K shell orbitals are closer to nucleus.)

$$^1_1 p +\ ^0_{-1} e \rightarrow\ ^1_0 n + v$$

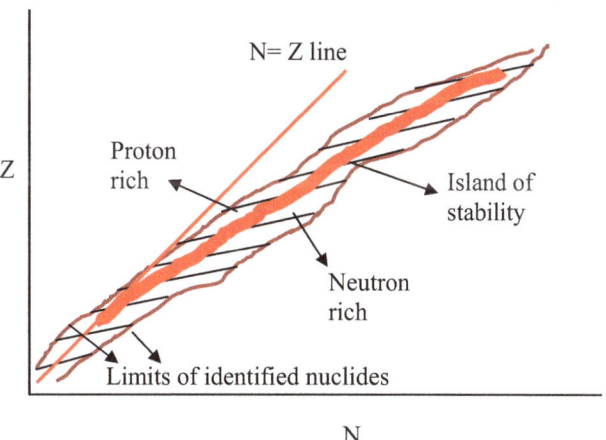

Chart of the nuclides

41. b, c   It is a radioactive isotope of Ir. The number of electrons is equal to the number of protons in the nucleus.

42. c

43. b

44. d

45. a. -   b. iii.   c. iv.   d. ii. and v.   e. i.

46. a. v.   b. i.   c. ii.   d. iii.   e. iv.

47. **a, b, c**   A brief explanation will familiarize you with many concepts in this area. The rest energy of positron is 0.511 MeV. When a positron is emitted the daughter atom must also emit an orbital electron since the atomic number of the daughter atom is now (Z–1). So we must account for the energy of two electrons or a minimum of 1.02 MeV must be available, i.e., $(M_p - M_d) c^2 \geq 1.02$ MeV. Energy in excess of 1.02 MeV is shared between the positron and the neutrino. No such condition exists for beta decay or even electron capture, a competing mode of decay for positron emission. In beta decay, the daughter atom captures an electron from the environment—daughter atomic number is (Z+1)—for becoming neutral which compensates for the beta emitted by the parent atom. In the case of electron capture, the daughter Z goes down by one unit, so it does not have to compensate for the captured electron. The three decays are shown below for comparison:

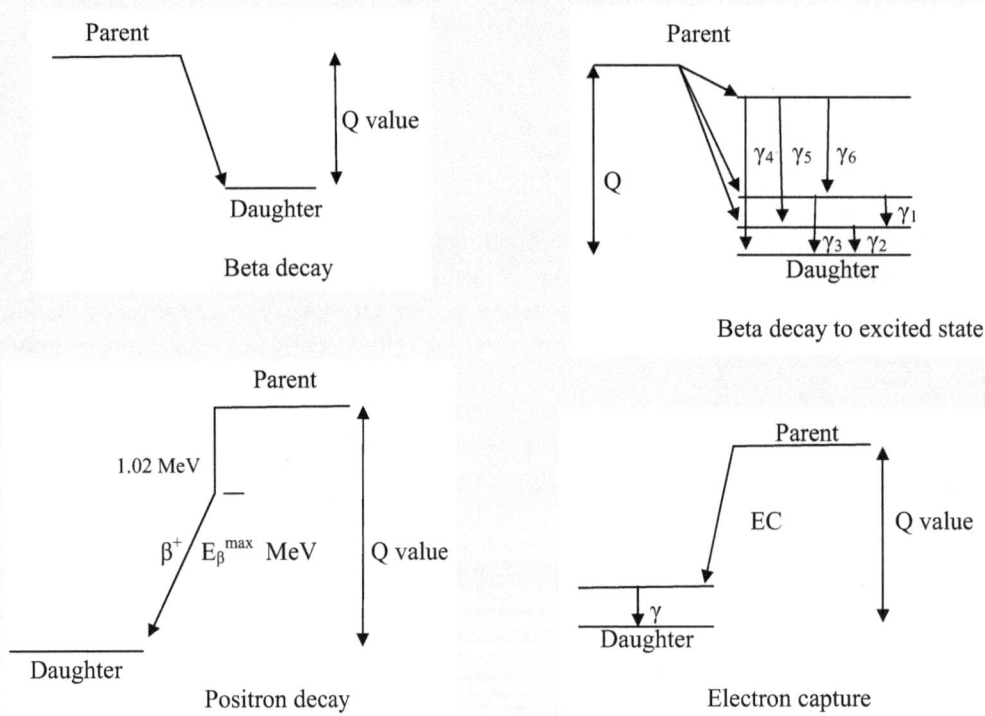

Positrons are not monoenergetic, but exhibit a spectrum since the energies are shared between the recoil atom, the positron, and the neutrino. The positron, being an anti particle, at the end of its range meets with an electron, resulting in annihilation. The total rest mass energy of 1.02 MeV is shared by the annihilation photons. Since they annihilate at rest (at the end of the range) the resultant momentum is zero, so the two photons are emitted in opposite directions (see the figure below).

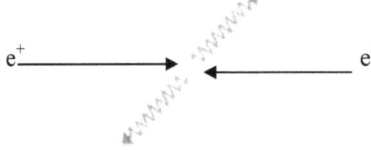

Positron-electron annihilation

48. f   A brief explanation on characteristic x-ray and Auger electron emission may be useful. When EC creates a K shell vacancy, this vacancy is filled by an electron falling from higher levels since the electrons tend to be closest to the nucleus for maximum binding. Since the binding energy for the K-shell is higher, this transition leaves the atom with an excess energy = $E_K - E_U$, with u referring to the upper level involved in the transition. This energy is generally released as a photon, the frequency depending on the Z of the atom. It is characteristic x-ray emission since the photon energy is determined by $(E_K - E_U)$, which is characteristic of the atom.

Sometimes this excess energy does not appear as a photon since the atom can transfer this energy to an electron in the higher level (with much less binding energy) and knock it out of the atom. This electron is known as an Auger electron. The residual excess energy = $E_K - E_U$ appears as the kinetic energy of the Auger electron. What is important to understand is that this is not an internal photoelectric effect, i.e, a photon being produced and knocking out an upper level electron. There is no photon at all in this instance. The same mechanism involved in causing the transition is responsible for Auger effect. The creation of characteristic x-ray and Auger electron are independent of the nature of the mechanism involved, i.e., it has more to do with the consequence of filling a shell vacancy than what caused the vacancy. (There are other mechanisms like photoelectric effect, internal conversion, etc. that remove the shell electron, creating a vacancy.) The figure below illustrates the phenomenon in a nice way.

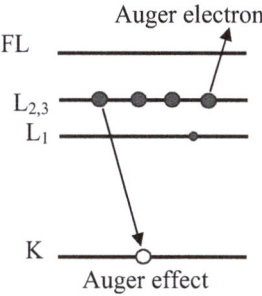

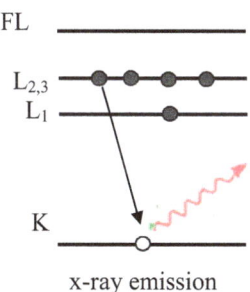

Auger electron and x-ray emission following K capture

49. c, d   The phenomena that create shell vacancy are the ones that cause characteristic x-ray emissions.

50. b, c, d

51. b, c   A brief explanation of internal conversion is given here for the benefit of the readers. Internal conversion occurs in isomeric radionuclides that decay from a metastable state by gamma emission. Generally these nuclides decay by gamma de-excitation. But in a very small fraction of the radionuclides, no gammas are emitted, so how do they de-excite? They have to come to the ground state as per the decay probability. It is found that in these radionuclides the de-excitation energy is directly transferred to the orbital electrons ejecting them out of the atom. Many mistakenly think that in these cases, too, gammas are emitted, but they are absorbed by the electrons, which gain energy and get ejected. This is not true. The electron kinetic energy = (decay energy – BE) where decay energy = $(E_i – E_f)$, the difference in the energies of the radionuclide in its initial and final states, respectively. BE is the electron binding energy. Since BE_K, BE_L, and BE_M are well defined, the kinetic energies of electrons are also well defined, and for each shell, they are monoenergetic. Since electron ejections create vacancies, characteristic x-rays and Auger electrons are also emitted, as in electron capture decay. But note one difference. In electron capture, Z changes, and so the characteristic x-rays are characteristic of the daughter atoms. In isometric transitions, no Z change is involved since both parent and daughter are the same radionuclide in different states. Here the characteristic x-rays are representative of the same atom. So internal conversion and gamma decay are competing modes in the decay of isomeric radionuclides.

B. Production of X-rays

1. Yes   As in kV x-ray units.

2. No   In a kilovoltage machine, the potential difference V between cathode and anode accelerates the filament electrons. In an accelerator, however, the electrons, accelerated to near velocity of light, are bunched together and injected into the microwave carrying cavities to be accelerated by the waves in their phase stable position.

3. No   Only a small fraction is converted to x-rays (fraction of a percent, in kV x-ray units).

4. Yes

5. No   The electron absorption in the target causes the heating up of the target.

6. No   They impinge on a small circular area of about 3 mm diameter, in the target, in a clinical accelerator.

7. Yes   They are not very highly filtered, but medium filtered just to remove very low-energy photons (preferentially) that would only increase the skin dose.

8. No   A heavily filtered kV x-ray beam and a medium filtered (and somewhat higher kV) x-ray beam can have the same HVL. So kVp, HVL would specify beam quality with much less ambiguity.

9. No   AAPM TG-51 recommends PDD $(10, 10 \times 10, SSD)_X$ as beam quality specifier. (Refer to TG-51 protocol for more details.)

10. Yes

11. No   An electron interaction cannot produce a photon of energy exceeding the energy of the electron itself.

12. Yes

13. Yes

14. a, b

15. b, c   See Johns and Cunningham (1983).

16. d   The only difference is the origin. Otherwise they are the same.

17. a, b, c   The definition of elastic collision is one where there is no loss of energy, and all the energy appears as kinetic energies of the products. Ionization/excitation interactions are inelastic since the binding energies delivered to cause ionization or excitation do not appear as kinetic energies of the products. The collisions with nuclei that cause deflection of electrons are elastic collisions since there is no loss of energy to the system. "Elastic" does not mean there is no loss of electron energy because a small energy will be lost as recoil energy, but it still appears as kinetic energy of the product. This energy, however, is negligible considering the mass difference between the electron and the nucleus. Occasional

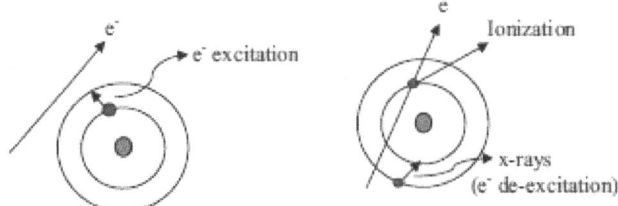

Inelastic collisions causing excitation / ionization in atoms

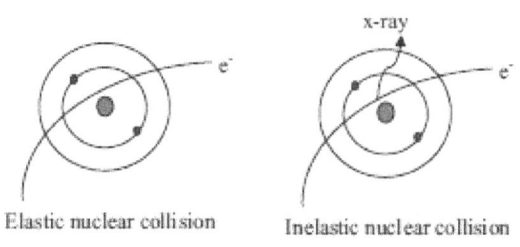

Elastic / inelastic nuclear collisions

nuclear collisions cause the emission of Bremsstrahlung photons. These are inelastic since the energy is lost as Bremsstrahlung photon and will not appear among the kinetic energies of the products. The following figures illustrate the concepts.

18. b

19. b

20. d

21. a, c, d

22. a. iii.   b. iv.   c. i.   d. ii.

C. Interaction of Radiation with Matter—Photons and Electrons

1.   No   Photons either interact or they don't. They do not lose energy continuously.

2.   Yes

3.   Yes   In fact. the contamination electrons accompanying the primary beam incident on the patient are produced in these (Compton) interactions.

4.   No   Strictly electron beams exhibit finite range in a medium.

5.   No   They do, and that is how x-rays are produced in an x-ray machine.

6.   Yes   These are called delta rays, and they are partly responsible for the buildup exhibited by electron beam depth dose curves.

7. Yes   Because the binding energies are small compared to photon energies, the photon sees the electron as a "free" electron in a Compton interaction.

8. No   A significant amount of energy goes into excitation, which reappears as heat in the medium. This results in the heating up of the x-ray target.

9. Yes   This is the principle behind the production of large field sizes for clinical treatment with electrons.

10. No   A photon beam cannot spread in this way since photons do not lose energy like electrons, and they are easily transmitted by foils.

11. No   High-energy electrons are faster than low-energy electrons and hence interact for less time. Therefore they scatter less than low-energy electrons.

12. Yes

13. c

14. b

15. c

16. a

17. b

18. b

19. b

20. b

21. b

22. c

23. c

24. b

25. b

26. a

27. a

28. b

29. b   See Johns and Cunningham, 1983.

30. a

31. b

32. d

33. c

34. a

35. b

36. c

37. a

38. c, d

39. a

40. a, b, c

41. c

42. b

43. a

44. a, b, c

D. Treatment Machine Characteristics

1.   Yes

2.   No   In an accelerator, electrons are accelerated by the electric field vectors of the microwaves (in a phase stable position).

3.   Yes

4.   No

5.   Yes

6.   Yes

7.   No   The emission is not forward peaked but more isotropic in the case of a ^{60}Co beam.

8.   No   The dose rate is constant with no likelihood of intensity fluctuations, so a timer can control the output.

9.   No   The source is still there in the parked position giving some leakage radiation.

10. Yes

11. No   To avoid self shielding, which will lead to nonuniform activity in the source, small pellets or thin discs are activated and tightly packed into the source capsule to produce a ^{60}Co source.

12. No   The output needs to be checked periodically. If the source does not come to the same position every time, the output can vary due to partial shielding.

13. No   The beam has a spectral distribution.

14. No   It is known as "added filter."

15. No   For the same area, the more elongated the field is, the less will be the side of the equivalent square. For instance, an 18 x 6 cm^2 field and a square field of 10.4 cm have the same area, but the side of the equivalent field of the elongated field is 9 cm, not 10.4 cm.

16. d

17. b, c, d

18. b

19. a, b, c

20. c

21. b

22. a, b

23. b

24. b

25. b

26. a

27. c

28. a, b

29. a   Asymmetric fields are produced with a pair of jaws that can move asymmetrically or using a pair of independent jaws. Irregular fields can be produced if the accelerator has an MLC or a micro MLC provision. Custom fields can be produced only by designing custom blocks.

30. a

31. d   This is because of the long range of electrons produced by photon beams.

32. a

33. b   Refer AAPM TG-40 (AAPM Report 46, 1994).

34. c

35. b

36. b

37. a. ii.   b. iii.   c. i. and iv.   d. – (not an accelerating mechanism)

38. a, b, c, d

39. e

40. a, b, c   The magnetron peak power is about 2 MV.

41. a. ii.   b. i.   c. iv.   d. iii.
        The electrons are accelerated toward the anode by the cathode pulse. They are velocity modulated in
        the buncher cavity. You may wonder how? Note the RF signal input. In the microwave the electric field
        is modulating. So the streaming electrons find themselves in different phases of the wave cycle, and
        they get accelerated to different degrees, so the faster ones at the back catch up with slower ones in
        front and eventually become bunches of electrons while drifting between the cavities. As a result,
        bunches of electrons arrive at the output cavity at the proper instant during each cycle of the RF field
        and deliver energy to the output cavity, amplifying the RF output. The electrons continue to travel to
        the collector or the beam stopper to get collected and deliver their residual energy. The energy given up
        by the electrons is the kinetic energy that was originally absorbed from the cathode pulse during initial
        acceleration. To improve the bunching process and to increase amplifier gain, more intermediate
        cavities are added to the klystron.

42. a, b, d   The middle cavity causes most of the velocity modulation.

43. a, b, c, e
        The four conditions are common in all accelerators: a source, an injection mechanism, an electric field,
        and a vacuum to ensure that the beam is not deflected or scattered by the gas molecules. A magnetic
        field cannot accelerate or decelerate charged particles. It can only deflect them into a certain path, and
        the magnetic field is used for that purpose in many accelerators. Once the electron beam acquires
        sufficient energy, there must be a mechanism to extract the beam out of the accelerator or direct them
        to a target for x-ray production. Though the electric field is the accelerating agent in any accelerator,
        the mechanism itself can be vastly different. In a kV machine, the field is produced by a DC voltage. In
        a linac, the field vector of microwaves accelerate the electrons (see the figure below). In a betatron, the
        electric field created by a changing magnetic field (electromagnetic induction) accelerates the
        electrons.

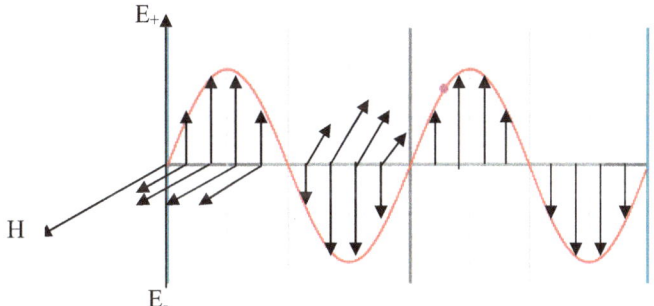

Changing E and H vectors in a microwave

44. a, b, c, d

45. a, b, d, e
        A triode gun assembly has a grid which can be connected to a pulsed voltage source to control the
        injection of electrons into the accelerator. The voltage applied to the grid must be synchronized with
        the pulses applied to the microwave generator so that the electron bunches entering the accelerator
        waveguide and the microwaves are in proper phase.

46. a, b, d   In the MV region, the contribution of characteristic x-rays to the total clinical spectrum is negligible.
        and most of the low-energy photons are attenuated and removed from the linac spectrum.

47. f

48. a, b, d   The HVL must be measured using an air-equivalent chamber that has flat energy response.

49. a, b, c

50. a, b, e   The kV machines can deliver only conventional treatments. In kV therapy treatments, the treatment time must be corrected for shutter timer error. This corrects the timer reading for the finite time required for the shutter to open and close.

51. a   Cyclotrons can accelerate only heavy charged particles. This is because the particle can reach the gap between the Dees at the correct phase for acceleration only when the velocity changes (and hence the kinetic energy) and not the mass. The time period T to reach the gap can be shown to be equal to ($\pi m/Bq$) which depends on mass. When the particle energy reaches relativistic energies, the mass tends to increase and the velocity tends to saturate. This happens when the energy is near the rest mass energy of the particle. For electron it is about 500 keV. So it is not possible to accelerate electrons in a cyclotron.

52. a. i.   b. ii.   c. i. and ii.   d. ii.
For a given energy, the standing wave linac has shorter length compared to the traveling wave linac. This is because the cavities—which have zero electric field and do not contribute to electron acceleration, but only for microwave coupling—can be side coupled.

53. c   This gives a λ of 10 cm.

54. b, c, d   Three systems of bending have been developed for linacs as shown below:

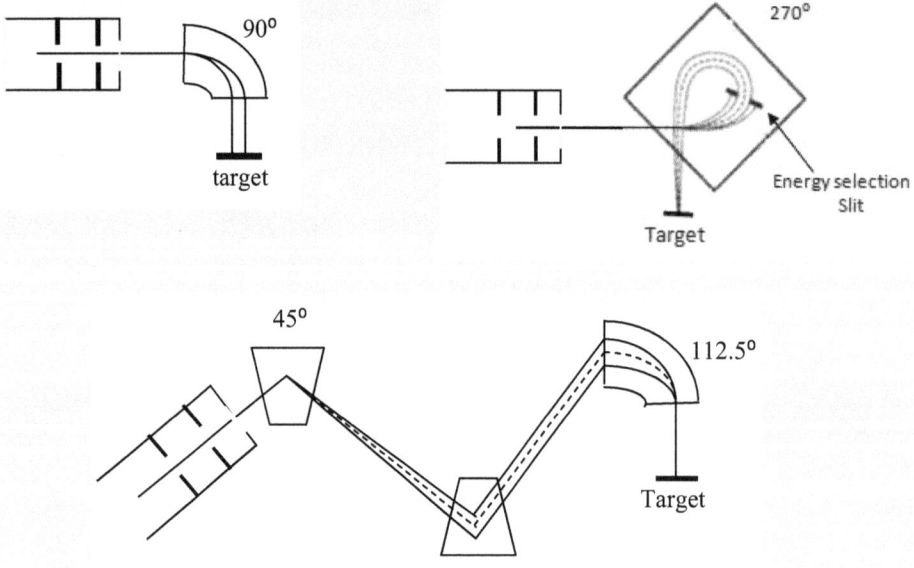

Beam bending designs adopted by linac manufacturers

55. a, b, c   The electron beam defocuses due to mutual repulsion between electrons and small radial field component and needs to be refocused. To confine the beam along the central axis, one needs additional steering coils, which cannot be done by the focusing coils. As the energy increases, the beam becomes more rigid and less easily gets defocused. The beam deviates from the central axis alignment due to structural imperfections, the influence of earth's magnetic field, and the rotational motion of the linac. The steering coil steers the beam to keep the central axis alignment.

56. a. v.   b. iv.   c. i.   d. iii.   e. ii.

57. a, b, c The flattening filter is an integral part of the linac and nor user accessible. The filter is made of copper, brass, or aluminum. They are too thick to modify any electron beam in any way. In the electron beam mode, the filter moves away from the beam path along with the target and a scattering foil comes in the beam path. The effect of beam flattening on the beam profile is shown in the figure below:

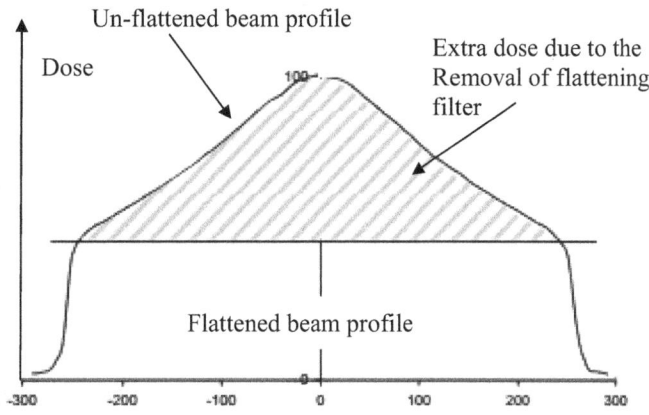

Unflattened and flattened beams of a linac
(Based on Jason Cashmore's thesis, "Operation, characterisation & physical modeling of unflattened medical linear accelerator beams and their application to radiotherapy treatment planning," University of Birmingham, UK, 2013.)

58. a, b, c Conventional treatment techniques make use of primary and secondary collimators. An MLC is employed in advanced treatment techniques for beam shaping or for IMRT treatments by subdividing the secondary collimator defined photon beams into beamlets or segments. Cones and applicators are used for electron collimation.

59. a, b, c, d
The transmission chamber is sealed and so needs no temperature or pressure correction. It can be noticed that there is a pair of transmission chambers (the primary and a secondary) for treatment delivery. If for any reason the primary monitor chamber fails to turn off the beam at the end of treatment, the secondary monitor chamber would turn the beam off after a small delay. There is a third transmission chamber divided into four quadrants for controlling the beam's flatness and symmetry.

60. a, b, c   The figure below shows a tertiary MLC. The other models have either an upper jaw or a lower jaw as MLCs.

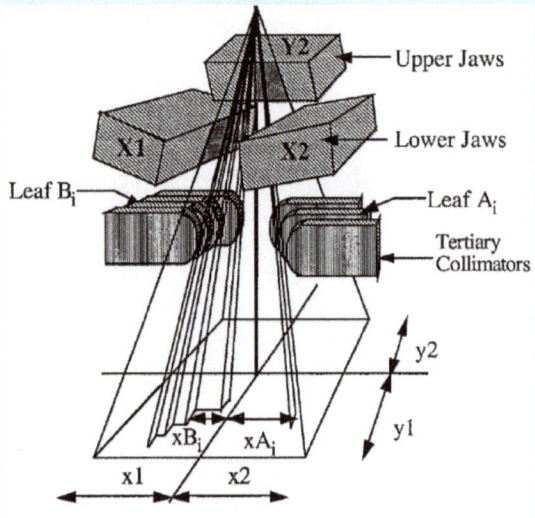

Tertiary collimator as MLC
(From AAPM Report No. 72)

Varian Millennium MLC
(Courtesy of the Varian image gallery)

Tertiary collimator (MLC) attachment for a linac

61. a. iv.   b. i.   c. ii.   d. -

62. f

63. b, d   Standard collimator leakage must be less than 2% of the primary. This value is typically about 1% for the clinical linac. For a tertiary MLC, leakage requirements are the same as for custom blocks (<5%). However, the leaves must be thick enough to provide adequate intraleaf attenuation to compensate for interleaf transmission, to meet this criterion. The interleaf transmission must comply with the manufacturer's recommendations. The measured values must be used in the TPS.

64. f   The focused end design follows the divergence of the beam and hence exhibits slightly smaller penumbra. Siemens uses this design. In the nonfocused design, the leaf moves in a single plane. Both Varian and Elekta use this design. Since rounded ends show more partial transmission, there is potential for more transmission for abutting leaves (see the figure below, adapted from a presentation by Dr. T. Solberg). This shape also leads to an underestimate of the radiation field by 0.5 to about 1 mm.

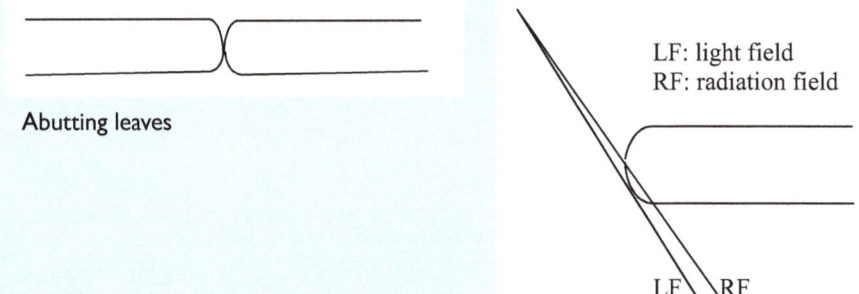

Abutting leaves

LF: light field
RF: radiation field

LF   RF

Rounded leaf design for a multi-leaf collimator

65. a, d   The MV portal imagers and the kV imagers can either be gantry mounted (i.e., EPIDs) or positioned independently of the gantry (i.e., kV x-ray tubes mounted elsewhere in the treatment room), as in the Novalis ExacTrac system. The CyberKnife system also has its imaging systems externally mounted. There is also a Siemens system with CT as a separate system, a CT on rails, with the linac sharing the same couch for patient positioning. This avoids the patient positioning errors that may occur if the patient had to be shifted from one system to the other. The greatest advantage of image guidance is that inconvenient and traumatic stereotactic frames become unnecessary for IGRT if the same accuracy can be achieved in this method.

66. a

67. b, c   Tomotherapy treatment can be delivered using a linac if a binary collimator can be mounted onto the linac.

68. a, b, d, e

          To improve image quality, the Hi-Art TomoTherapy system images the patient at a lower beam energy (3.5 MeV). The same linac source used for treatment (6 MV) can also produce a low-intensity 3.5 MV (nominal) energy x-ray beam for imaging.

69. e   The Hi-Art TomoTherapy system dispenses with the flattening filter and gives a treatment delivery output of about 850 cGy/min (at d_{max}) at the isocenter distance. In-beam scatter and head scatter are less because it is a fan beam. There is, however, a beam hardener, which hardens the beam and removes the low-energy photons preferentially from the beam.

70. a, b, d   Being a binary MLC the leaf has only two positions, open or closed.

71. e   Though Varian Trilogy offers respiratory-gated radiotherapy, it does not track with radiosurgical precision.

72. c   The CyberKnife system can be used for treating any part of the body, including intracranial lesions. Intracranial treatments require simple devices like aquaplast mask, but no stereotactic head frame is required. For extracranial treatments Alpha Cradle (vacuum bag) is made use of. Patient positioning is traced by implanted fiducial markers (metal), spine tracking (Xsight spine module), lung shadow tracking (motion tracking), or soft tissue lung tracking (Xsight lung module), depending upon the nature of the tumor. The system makes use of 12 interchangeable circular collimators covering a diameter range of 6 to 60 mm. The newer model comes with a variable aperture that can replicate the sizes of the fixed external collimators.

73. a

74. b, d   This is the only linac in radiotherapy that operates in the X band (9.3 GHz) as compared to other linacs that operate in the S band (3 GHz). This reduces the size and weight of the linac, making it easy for the 6-DOF robotic arm to carry it and position it at any point in the robot workspace with great accuracy (better than 1 mm).

          While other linac systems use a relatively smaller number of beams or segments per fraction, CyberKnife uses a large number of non-coplanar beams, which improves conformality in treatment and reduces dose to normal tissues. The CyberKnife system defines a spherical (for intracranial targets) or ellipsoidal (for extra cranial targets) workspace around the treatment site with a large number of grids and nodes. A typical treatment would use more than 100 nodes, optimally chosen to realize the treatment objective of maximum dose to target with maximum sparing of the normal critical structures. The system can deliver dose from the same node at different orientations or angles.

75. a, b, d   Model S was introduced in 1968. The Model U, which reduced body dose and improved treatment planning, was introduced in 1986. The Model B was introduced in the following year and had improved collimator design. The next model, Model 4, was introduced in 1999 and had improved patient positioning and improved dose conformity. The next model, released in 2004, came with better image fusion software. The most recent model, the Perfexion, came in the year 2006 and features a larger cavity, very low body dose, faster treatment, and better features.

76. e   This question also illustrates the basic principles behind the use of the Gamma Knife unit to treat intracranial lesions. The name implies that it is a non-invasive "gamma" surgery (Gamma Knife) to eradicate the lesion.

77. b, c   The head frame adds to the patient discomfort and is an invasive procedure to fix it to the skull with screws. However, it's primary role is for the localization of the lesion and to prevent the involuntary movement of the head which will interfere with localization and treatment delivery. (To identify the lesion position with respect to the frame a localization box is attached to the head frame). The head cannot be moved inside the helmet but the helmet itself can be moved by Δx, Δy, Δz to move the focal point (which is fixed for the machine and cannot be moved) to other regions of the lesion when the lesion to be treated is larger than the size of the beam spot.

78. a   There are 201 sources and, hence, there are 201 collimators when all the beams are used to irradiate the lesion. However, any of the one or more secondary collimators can be blocked (plugged) to protect sensitive structures.

79. b, d, e, f

80. b, c, d, e   The new type of collimation used in the Perfexion Gamma Knife system is illustrated in the following figures. There are 5 rings, 8 sectors per ring, and 6, 4, 5, 4, 5 sources in each ring of a sector (as can be easily counted). This enables us to calculate the number of sources per ring (8×6, 8×4, 8×5, etc.). One can mix and match the sectors—all sectors' 16 mm beam gives a spherical distribution, while alternate sectors with 8 mm beams gives a diamond-shaped distribution.

E. Radiation Quantities, Units, and Measurements

1. Yes

2. No   They can, but they cause ionization in single interactions producing charged particles (e.g., electrons or positrons), which are the agents for causing further ionizations and excitations in the medium. For instance, if we assume that in a Compton interaction a photon produces an electron of energy 1 MeV in air, this amounts to one ionization in air by the photon. However, the 1 MeV electron goes on to produce (1,000,000/33.97) ionizations in air. So photons or neutrons are indirectly ionizing radiation.

3. Yes

4. No   See the definition in the earlier question. When, for example, photons interact with matter at a point, kerma tells you how much of the photon energy has been converted into electron energy (E_{el} in the figure below).

5. No   Part of kerma (electron kinetic energy produced) is converted into Bremsstrahlung which escapes from the region of interest. So only that part of kerma (called the collision part of kerma), $E_{el} - E_{br}$ in the figure below, is energy that is dissipated in the medium by ionization and excitation events. (This energy dissipation is not at a point but over the range of the electrons).

6. No   Actually, absorbed dose refers to the energy deposited per unit mass in an infinitesimal mass in the medium. Note, first of all, it is a point quantity and refers always to an infinitesimal mass. Secondly, it refers to energy deposition and so characterizes charged particle interactions. In the figure below, $D(P') = E_d /(\delta m)$.

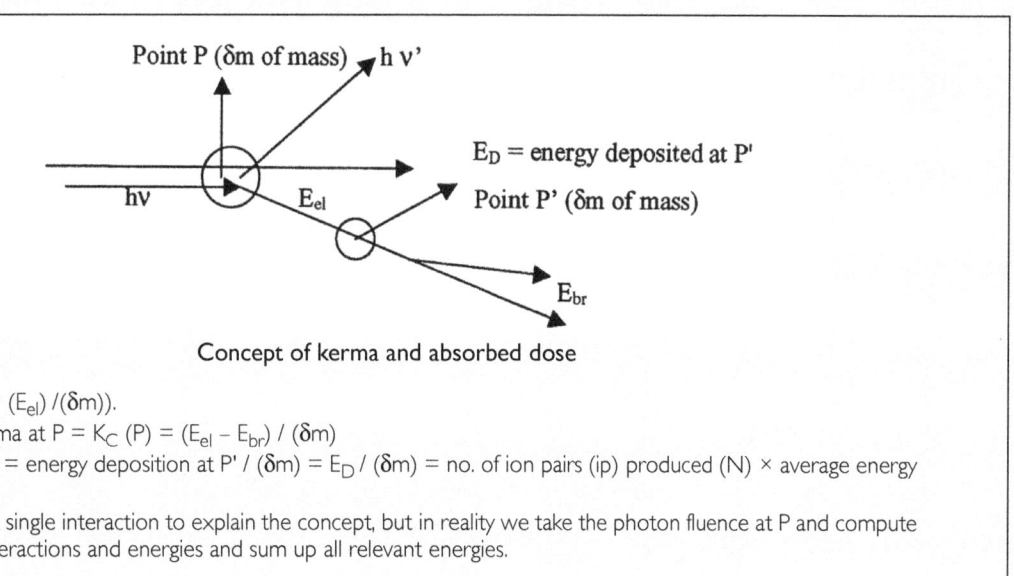

Concept of kerma and absorbed dose

Kerma at P = $K(P) = (E_{el}) /(\delta m))$.
Collision part of Kerma at P = $K_C (P) = (E_{el} – E_{br}) / (\delta m)$
Absorbed dose at P' = energy deposition at P' / (δm) = $E_D / (\delta m)$ = no. of ion pairs (ip) produced (N) × average energy required / ip (W)
We have considered single interaction to explain the concept, but in reality we take the photon fluence at P and compute the number of interactions and energies and sum up all relevant energies.
$K(P) = \Sigma (E_{el})/(\delta m)$
$K_C(P) = \Sigma (E_{el} – E_{br}) / (\delta m)$ and
$D(P') = \Sigma (E_D)/(\delta m) = N \times W/(\delta m)$
All the definitions refer to a particular medium. Air kerma = K_a, etc.

7. Yes   See question 6 above.

8. Yes   This is the mechanism of x-ray production.

9. No   Not always. See question 6 above. But when charged particle equilibrium exists, for example, at P in the above figure, electron energy escaping from P will be compensated by electron energy entering (δm) at P, in which case $K_C(P)$ can be equated to $D(P)$, with a small correction that accounts for very small attenuation of photons over electron ranges. Since any dosimetric system, whether it is a primary standard or a therapy dosimeter, responds to energy absorbed in the sensitive volume, kerma (or exposure) of indirectly ionizing particles can be measured only under electronic equilibrium conditions.

10. Yes   In the figure above, exposure at P = $X(P) = Kc(P)/W/e)(J/kg)/(J/C)$. To obtain exposure in roentgens we must use the conversion factor 2.58×10^{-4} (C/kg)/R.

11. No   Farmer-type ionization chambers (volume about 0.6 cm^3) must be made use of.

12. Yes   Accredited calibration laboratories offer such calibrations.

13. Yes   Hospital chambers used for beam calibration must be traceable to these standards, either directly or indirectly.

14. No   Free air ionization chamber is the primary standard of exposure or air kerma for kV x-ray beams.

15. Yes

16. Yes

17. No   They are available, but not at all PSDLs.

18. Yes

19. Yes

20. Yes   Except for incident electron energies, Eo < 10 MeV.

21. Yes   For example, Markus chamber, but the Roos chamber designed by PTB, Germany, is a modification over the Markus chamber and does not exhibit fluence perturbation at low electron energies. Its stability, however, depends on the reliability of aquadac coating.

22. a

23. a

24. d

25. a

26. c

27. d

28. a

29. b

30. a

31. c

32. b, c

33. a, c

34. f   Since the concept of exposure (or air kerma) is a bit difficult to understand for a beginner, a brief explanation is given here. In the case of interactions involving indirectly ionizing radiations (x, gamma, neutrons) the first step involves the production of charged particles. This is denoted by kerma. The second step involves the interaction of charged particles giving rise to absorbed dose. Exposure is a special case of measuring the ionization produced in air before they recombine. So, questions like, "Why is exposure not defined for charged particles?" become meaningless. Secondly, it is not possible to collect ionization produced by these electrons because of their large ranges in air, especially beyond few hundred keV. So, up to a few hundred kV, the free air ionization chamber measures exposure as per the definition to some extent. (See the figure below which shows equilibrium thickness on axial direction to compensate outflow by inflow of electrons and on lateral direction uncompensated outflow. (The x-ray field is confined only to the shaded region, in the radial direction, requiring all the laterally out-flowing electron fluence to be collected by the electrode.

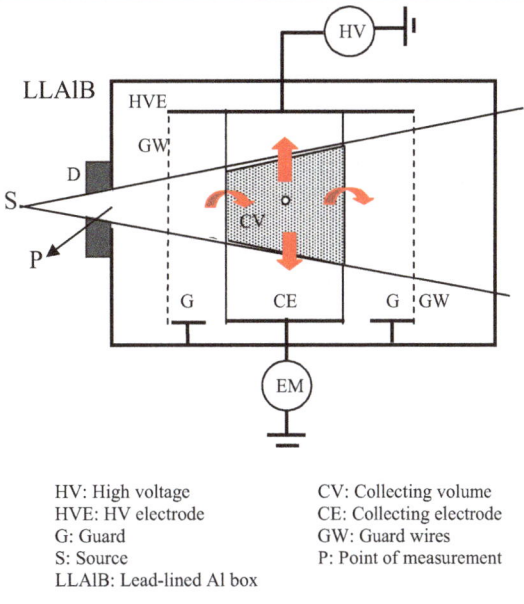

HV: High voltage   CV: Collecting volume
HVE: HV electrode   CE: Collecting electrode
G: Guard   GW: Guard wires
S: Source   P: Point of measurement
LLAlB: Lead-lined Al box

Measurement concept for exposure

Primary standard of exposure in the kV region
(Courtesy of NPL, UK)

Beyond about 300 kV energy, it is more practical to measure in a defined volume by collecting ionization **under charged particle equilibrium in all conditions**, so that the electrons leaving the volume will be compensated by the electrons entering the volume and depositing their ionizations. If the wall is air equivalent, then density reduces by 10^3 and a thin wall would suffice.

  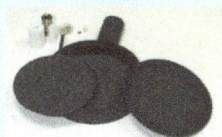 

Primary standard of exposure (also absorbed dose standard)
(Courtesy Standards lab of Taiwan for plane parallel chambers
and PTB, Germany, for cylindrical chamber)

When the wall is not exactly air equivalent, then a non air equivalency correction must be applied to the measurements. The electronic equilibrium is exact to the extent the attenuation of photons in the equilibrium thickness is negligible, or a correction must be applied for it. This difficult-to-evaluate correction increases with energy, and so ICRU limited the definition of exposure arbitrarily to 3 MeV. Practically, exposure standards were required only for kV energies and ^{226}Ra, ^{137}Cs and ^{60}Co sources.

35.   b, e   A brief explanation on air kerma and the relationship with other dosimetric quantities will be very useful to the beginners. So I include a brief explanation.

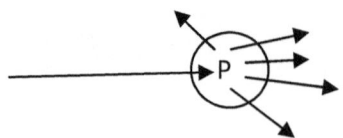

Air kerma at P

A simple figure shown above explains the concepts. X-rays interact in δv of volume at P. The tracks of electrons produced by the x-rays are shown in the figure above. The tracks are ionization tracks. (The electrons produced by x-rays interact with atoms and lose their energies gradually, until they have no more energy to ionize.)

Concept of Air Kerma:
Let Ei be the kinetic energy of ith electron.
Total kinetic energy released = Σ Ei ; Air Kerma at P = (Σ Ei / ρ δv)

Concept of Exposure:
If W is the average energy required to produce an ion pair,
Total ion pairs produced by the electrons = (Σ Ei / W)
Total charge produced by the electrons = (Σ Ei / W) $\times$ 1.6×10^{-19} Coulomb
Exposure at P in SI units = X = [{(Σ Ei / W) $\times$ 1.6×10^{-19}} / (ρ δv)] C/kg

Relating Air Kerma and Exposure:
From the above equation, you can easily see Ka = X (W/e). This requires a small correction which can be neglected in clinical dosimetry, but must be included when one talks of primary standards. Not all electron energy goes as ionization energy. A few electron interactions will produce Bremsstrahlung, which will escape from the point of production and must be taken out of Ka. So, if Kc refers to the collision part of air kerma that causes collision (ionization) interactions, and if g is the Bremsstrahlung fraction, Kc = Ka (1−g). Only this relates to exposure. So the correct equation is

Ka (1−g) = X (W/e) or Ka = X (W/e) / (1−g)

"g" is zero in kV energies and for ^{60}CO it is about 0.003. An important point: If you can measure X with a primary standard, it also measures K_a since one has to only multiply the measured X by [(W/e) / (1−g)]. So the exposure standard and the air kerma standard are one and the same. You don't do anything different to set up air kerma standards. But exposure is an archaic quantity and air kerma is an SI quantity, so exposure has been replaced by air kerma in many European countries.

But there are still some complications. One cannot chase all electrons and collect the ionization. Their ranges will be in meters. (And where will you apply the collecting potential?) You also cannot isolate δv in the path of the beam, as shown in the above figure. What about electrons produced in the proximal volumes that enter (ρ δv)? The only practical method of measuring is to compensate for the outflow of electron fluence from δv by the inflow of electron fluence into δv by providing an envelope of equilibrium thickness (approximately d_{max}), as explained earlier.

Concept of Absorbed Dose:
Each electron loses a small amount of energy ε_i in δv and rest of the energy outside δv.
Total kinetic energy lost in $\delta v = \Sigma \, \varepsilon_i$; absorbed dose at P = $(\Sigma \, \varepsilon_i / \rho \, \delta v)$.
Under CPE conditions, $K_c(P) = D(P)$.

36. d

37. c

38. a   The conversion factor is 2.58×10^{-4} (C/kg)/R.

39. c   AKR to exposure conversion factor = 0.876 cGy/R. This can be derived from the equation $K_a = X \, k_U$ (W/e) / (1−g) in SI units. The main conversion factor is (W/e) = 33.97 J/C. Since X is in R, we need another unit conversion factor R to SI, i.e., 2.58×10^{-4} C/kg)/R. (This factor is obtained from the definition of roentgen, i.e., 1 esu/cc at NTP and converting this into C/kg.) Substituting both factors you get the conversion factor air kerma to exposure (in conventional units). Bremsstrahlung correction $g \approx 0.997$ for ^{60}Co and 0 for kV radiations.

40. b, c, d   A few things about absorbed dose standard are given below. Since a graphite ionization chamber, as a primary standard, measures equilibrium dose, to determine graphite kerma, K_{gr}, it can also be used as an absorbed dose standard.

     The most fundamental property of absorbed dose is the temperature rise in a simple material where the energy will not be dissipated in any other manner (e.g., bond breakage, etc.). We say heat defect is zero

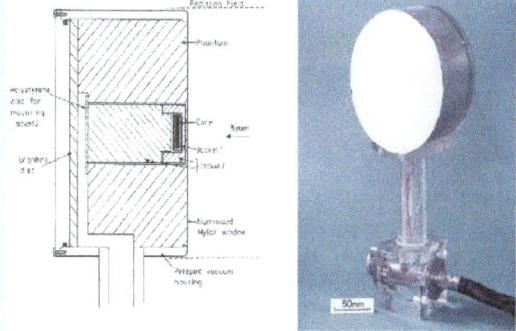

Graphite calorimeter

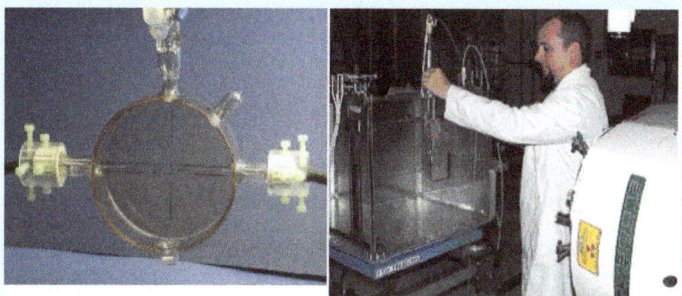

Water calorimeter and water absorbed dose measurement

Absorbed dose calorimeter – the basic components

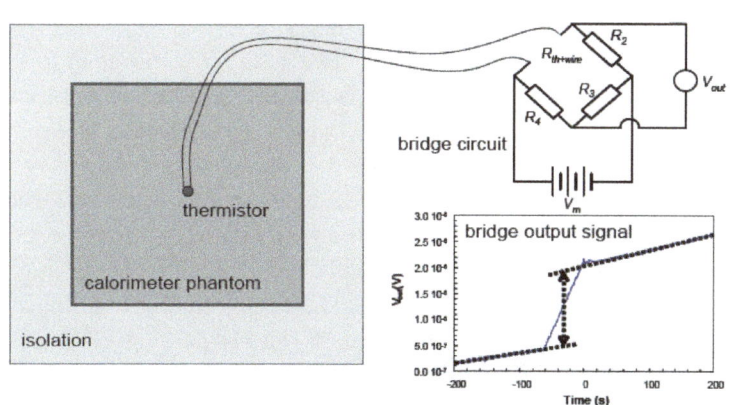

Calorimeter measurement principle
(Figures courtesy of Dr. Malcolm McEwen, NRC, Canada)

in such materials. This temperature rise is very minute and is usually measured using a thermistor embedded in the calorimeter, connecting it in a Wheatstone bridge and recoding the voltage signal.

Since the reference medium for clinical dosimetry is water, a water calorimeter is the ideal choice for absorbed dose standard. (The graphite calorimeter will include an additional conversion factor to convert graphite dose to water absorbed dose, which adds an additional uncertainty in the determination of absorbed dose in water.) The water calorimeter, on the other hand, involves a head defect correction, which adds to its uncertainty.

A calorimeter is a very complex device to set up since measuring a few millidegree rise against a few degrees ambient change is very difficult and requires many control circuits. (To measure with an uncertainty of 0.1%, as required in primary standards laboratories, a temperature difference of 10^{-5} deg. must be detectable.

41.   a. iii.   b. i. and iv.   c. ii.

42.   c, e   Both c and e are correct. Gray is just the name of the unit (e.g., the absorbed dose at a certain point in a water phantom is 100 J/kg or 1 Gy or 100 cGy).

43.   a, b, c   In clinical dosimetry one frequently measures the exposure or absorbed dose. In radiation protection one usually measures the ambient dose equivalent or personal dose equivalent. The radiation protection dosimeters are generally referred to as radiation monitors since monitoring implies measuring dosimetric quantities that involve poorer accuracy.

44.   b, c, d   Precision indicates how reproducible the measurements are while accuracy relates to whether the measurement represents the quantity we intend to measure. The later is related to the accuracy of calibration of the dosimeter. To give a simple example, if you measure the length of a table with a scale that is, say, short by 1 cm, the measurements will always be short by 10 mm even if the reproducibility is 100%. See the figure below: A) Good precision, poor accuracy; B) Good precision and accuracy; C) Good accuracy, poor precision.

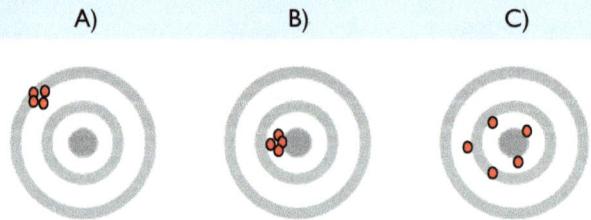

Illustration of the concepts of precision and accuracy

45.   a, b   As the previous question shows, the measurements have two basic uncertainties: namely, how reproducible are the measurements and how close is the measurement to the true value of the quantity one is trying to measure. These uncertainties were earlier referred to as random and systematic uncertainties, but for more than 10 years now the standards labs and the metrology bodies are using the terms Type A and Type B uncertainties. Type A is a statistical uncertainty and is expressed in terms of standard deviation. Type B uncertainties are guess estimates since the "true value" of a quantity is unknown in most of the measurements. There are methods of setting limits for these errors, and adding them usually in quadrature one arrives at a total uncertainty for the measurement with some confidence level (usually 95%). When we measure, for instance, the output of a linac beam using the TG-51 protocol, the measurement involves many factors, and each one of them has an uncertainty. So, one uses the propagation of uncertainties to evaluate the total uncertainty of such quantities. It is important to keep these ideas in mind while making any clinical dosimetric measurements.

46. e It is easy to understand why these properties are important. A dosimeter is calibrated at some energy (beam quality), at certain dose, and in some geometry, but the user conditions are frequently different from the calibration conditions, and so we must ensure that changes in dose or energy or orientation of the dosimeter will not significantly influence its response exhibited during calibration.

Directional dependence is particularly important for radiation protection monitors since the scatter radiation at the measurement location usually comes from all directions. There are other properties, too, that need to be considered depending on the clinical situation (e.g., spatial resolution, convenience of use, repeated use, etc.).

47. b, c, d Exposure is defined for air, and so the ionization chamber designed for exposure measurements must be preferably air equivalent. When radiation passes through a medium, it produces a spatial dose distribution, and at every point the dose will be different depending on the attenuation and scattering properties of the medium. So in principle, the dose is a point quantity, and the dosimeter must be of the smallest dimensions so that it will not integrate the dose from several depths when we try to measure at one depth. This is of particular importance in small field dosimetry. The dosimeter can be of any shape since all dosimeters are calibrated before being used in dosimetry. However, some geometries may suit some applications better (e.g., a plane parallel design for surface dose measurements, spherical design for isotropic response, etc.).

48. b

49. d The chambers used are usually of low Z and are of about the same equilibrium thickness (say about 500 mg/cm^2 for ^{60}Co measurements). They will not significantly influence the response, even if the two chambers are having different wall materials, but the volume is the single most important factor that determines the response. That is the reason why radiation protection survey meters need much larger volume chambers. A 1 cc chamber would give negligible response.

50. c The fundamental property of ionizing radiation is to cause ionization in the medium traversed. Only in gas-filled dosimeters can these ionizations be conveniently collected by applying a voltage before they recombine. So the most common dosimeter for detection of ionizing radiation is an ionization chamber. Other responses like temperature rise, chemical reactions, etc. occur in special types of materials and are also used for the purposes of dosimetry (e.g., a calorimeter, primary standard of absorbed dose measures a temperature increase, a chemical dosimeter—$Fe_2(SO)_4$ dosimeter—measures the yield of the reaction through absorbance, etc.).

51. a

52. b

53. b, c For high-energy electron beams, both PP and Farmer type chambers can be used, but for low-energy electron beams (<10 MeV) a PP chamber is preferable. For very low (< about 6 MeV) energies, a PP chamber is a must.

54. a, c Each point is a detector on a film, and hence it gives higher resolution. It is a 2D detector and gives dose at all points in the exposed area (in a single exposure). It is an integrating system, and the amount of blackening is a measure of dose and not dose rate.

55. d, e Si diode dosimeters and MOSFET dosimeters are commercially available and are most extensively used for *in vivo* dosimetry. Diode- and scintillation-based dosimeters are now available in the market which can be used for *in vivo* dosimetry.

56. e

57. c

58. b   There is no special significance for using this volume for radiotherapy dosimetry. The first chamber made by Farmer in the UK happened to have this volume, and this later became commercially available. Other vendors then marketed Farmer-type chambers for dosimetry. The in-water calibration method recommended for ^{60}Co later required a displacement correction which was evaluated and made available only for a chamber of this size, so the use of this chamber had become almost mandatory due to this reason. With modern protocols (e.g., TG-51), no displacement correction is required, so other sizes of chambers can also be used for dosimetry purposes and have been recommended in TG-51.

59. d   Some hospitals make use of in-air measurement at short distances using Farmer-type chambers, but this can cause large errors in dosimetry due to the short distances involved (≈ 10 cm) and hence are not recommended. The majority of people use the well chamber for the AKS measurements.

60. e   The electrometer also should have high sensitivity and high input impedance, which is usually the case. The leakage should be $<10^{-15}$A.

61. b, c, d, e

62. e   These are also the disadvantages with the use of conventional films in dosimetry, namely high energy dependence, the need for a darkroom, chemical processing, and QC.

63. a, b, c, d
          The radiochromic film is grainless and, hence, exhibits very high resolution. It is also dose rate independent.

64. a, b, c

65. e   The following figure helps to explain the concepts behind thermoluminescence.

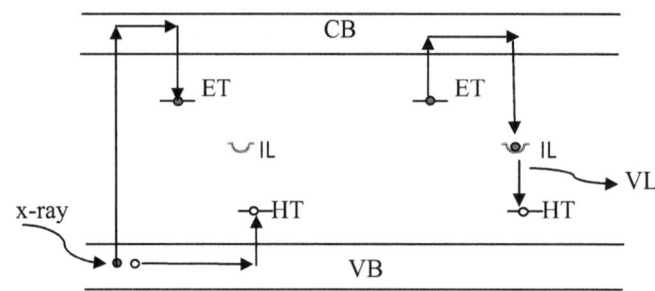

VB:	valance band
CB:	conduction band
ET:	electron trap
IL:	impurity level
HT:	hole trap
VL:	visible light

Concepts of thermoluminescence

The dopants must be chosen to produce deeper traps (not close to the conduction or valency bands) to prevent thermal excitation of traps by room temperatures. Otherwise there will be significant fading of the signal with time. If the stimulation is light instead of heat, it is known as optically stimulated luminescence (OSL).

66. a, b, d   The dosimeter usually exhibits a limited linearity of response, and so to what dose the response is linear must be established before using it for clinical dosimetry. If the dose range exceeds the linearity region, the response curve can be made use of for correcting for nonlinearity.

67. a, b, c   High-Z TLDs have high sensitivities and are useful for measuring low levels of radiation, such as in personal monitoring, but they would exhibit pronounced energy dependence in the keV region (see the figure below) and are not suitable for clinical dosimetry work.

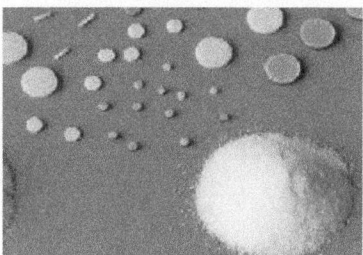

TLDs in various forms

Rel. Response = $[(\mu_{en})_{TLD} / (\mu_{en})_{air}]$

Energy response of TLD dosimeters

The energy dependence curve is simply the ratio $[(_m\mu_{en})_{Zeff} / (_m\mu_{en})_{air}]$ since the dosimeter response is proportional to the energy absorption in the sensitive volume which depends on $[(_m\mu_{en})_{Zeff}$ for the effective Z (Z_{eff}) of the dosimeter. The term $(_m\mu_{en})_{air}$ appears in the denominator because exposure (measured in roentgens) is defined for air, and so the relative exposure response depends on this term. This applies not only to TLDs, but to any clinical dosimeter because it depends only on the effective Z of the dosimeter.

68. a, e The following two figures illustrate the principle involved. The first figure shows a photomultiplier tube, which is basically a photocathode that converts TL light into electrons followed by a series of dynodes to amplify this signal to get a measurable current. In the second figure, the electrometer connected to the PMT measures the total charge (current integrated over a time period) which is proportional to the TL emitted.

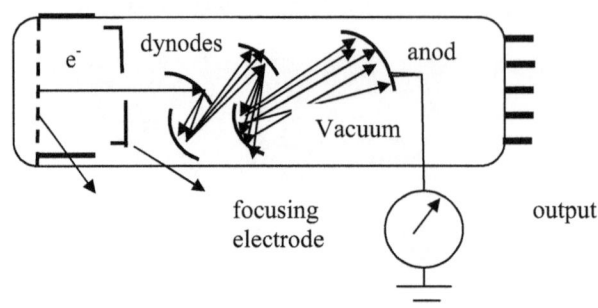

Photo multiplier tube (PMT)

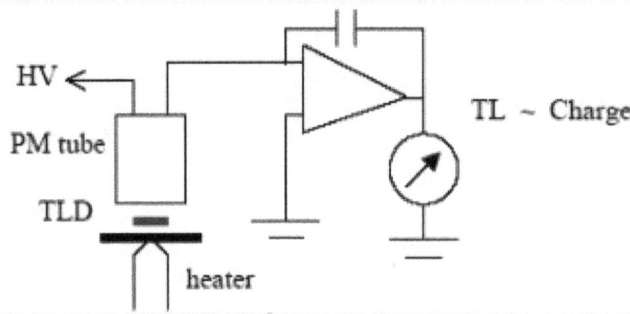

Electrometer coupled to PMT

69. a, b, c Post readout annealing vacates all the higher-temperature traps. The readout would have vacated only the traps in the reading temperature peak of the glow curve. A reproducible annealing procedure is very necessary for the reproducibility of the results.

70.   a, b, c, d

The concept behind the diode dosimeter is illustrated below. The figure shows the following: The positive and negative ions in p and n regions (free carriers not shown); the diffusion of majority carriers produces a depletion region; and the voltage barrier stops further recombining of majority carriers, causing an equilibrium. So, the majority carriers cannot all cross and neutralize the Si crystal. The diode dosimeter can be operated in a forward bias or a reverse bias by applying the proper bias voltage.

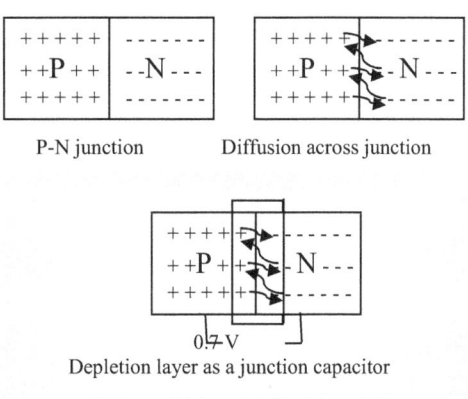

Concept of P-N junction diode

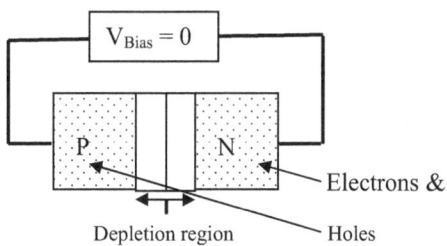

A diode dosimeter under zero external bias

71.   a, b, e   The intrinsic junction voltage prevents any diode conduction. A reverse bias will increase the junction voltage and the depletion depth and continue to prevent any conduction of majority carriers. The forward bias does not have much effect once the intrinsic junction voltage is neutralized. Only bulk resistance comes into play. The current is saturated. This region is not used for dosimetry.

With a reverse bias, there is only a minority carrier combination current across the reverse direction. Both majority and minority career currents depend on temperature since thermal excitations create electron hole pairs.

In the reverse bias or zero bias mode, the diode can function as a dosimeter. At zero bias, the leakage current is at a minimum and is therefore the preferred method for Si diode dosimetry. The high electric field across the n-p junction can sweep all the electrons and holes across the junction. This current is proportional to the dose rate being measured. The electron-hole pairs produced in the small region, bordering on the depletion layer, is the diffusion layer, and the carriers produced here can diffuse to the depletion layer and contribute to current. The diode dosimeter mechanism and the current/charge reader are shown in the figure below.

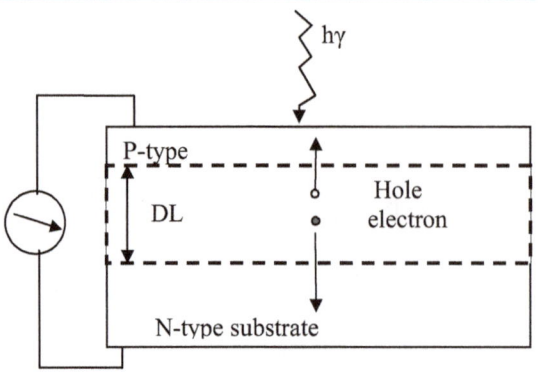

Ionization in diode depletion region

72. a, d   The air cavity has density about 10^3 times less compared to Si material. The average energy required to produce an ion pair is 34 eV in an ionization chamber compared to a Si diode, which is 3.4 eV. So a Si diode dosimeter is approximately 10^4 times more sensitive compared to an ion chamber.

The voltage across the depletion layer is small ($\approx$1 V). However, the electric field is very high ($\approx 10^3$ V/cm) since the depletion thickness is extremely small (in μM).

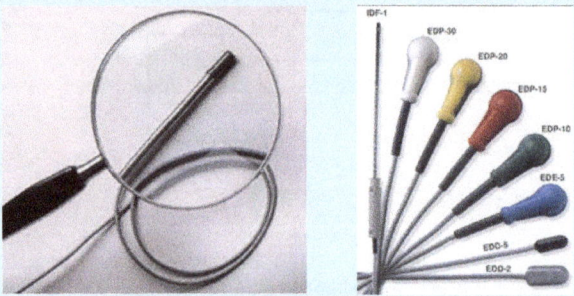

Si diode dosimeters (Courtesy of IBA Dosimetry and Scanditronix)

73. b, c, d, e

The diode sensitivity decreases with the accumulated dose. The diodes must be used with a buildup cap, similar to the use of a Farmer chamber in ^{60}Co beam calibration or for collimator scatter factors determination. The equilibrium thickness must be roughly (d_{max} / ρ) where d_{max} is the peak depth in water for the beam quality in question and ρ is the density of the buildup material used in the diode. This is the reason we have different photon detectors for different beam qualities. The higher beam quality diodes can be used with lower beam qualities, though they cause unnecessarily higher attenuation.

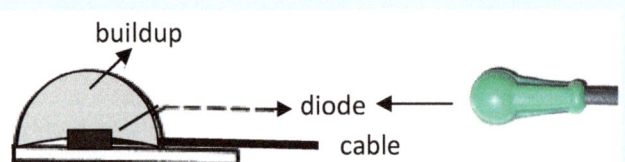

Diode dosimeter design (buildup integral part of the diode)

Buildup cap for a bare diode dosimeter (Courtesy of Radiation Products Design, Inc.)

74. d   At higher doses rates the recombination centers are "occupied" and not available, resulting in lesser recombination and hence higher diode response (signal). So any parameter that depends on dose rate (e.g., SSD, field size, introducing wedge in the beam, etc.) will alter dose rate and hence the diode response. This is more pronounced for n-type diodes compared to p-typed diodes.

75. b, c, d

76. e

77. a, b, c   Unbiased single MOSFETs are often used as single-use dosimeters. They have a short range of linearity and the response is not very stable. Like all single MOSFETs, they are temperature dependent. There may be 10 degree of temperature difference between the room temperature and body temperature, which will significantly affect the dosimeter response. In such cases it is advisable to keep the dosimeter on an insulating paper placed on the patient's skin, or one has to evaluate the response change and correct for it. They are, however, very useful as implantable dosimeters for daily dose verification.

Single bias, single MOSFETs also suffer from same problems, except that the linearity of response range improves (increases) with the bias voltage. A dual-bias, dual-MOSFET system uses two matched MOSFETS (0.2 mm x 0.2 mm) on a silicon die (1 mm x 1 mm). They exhibit excellent dosimetric characteristics, namely good temperature independence, better stability, and better reproducibility compared to single MOSFETs.

78. f

79. c   Here I will explain only the response of the system, omitting the physics part. When a sufficiently large negative bias is applied to the gate, a sufficient concentration of holes accumulate at the SiO_2/Si interface, allowing a channel current (I_{ds}) between the source and the drain. The voltage necessary for initiating this current is the threshold voltage V_{TH0}. On exposure to radiation, electron-hole pairs are generated, and ultimately holes move toward the SiO_2/Si interface, hopping through traps where they become trapped in sites that will lost for years. This affects the channel current and the MOSFET cannot be switched ON. The voltage must be further increased to establish the same channel current (I_{ds}). The new threshold voltage is V_{TH}. $(V_{TH} - V_{TH0}) = \Delta V_{TH}$, which is proportional to dose. The following figure illustrates this principle:

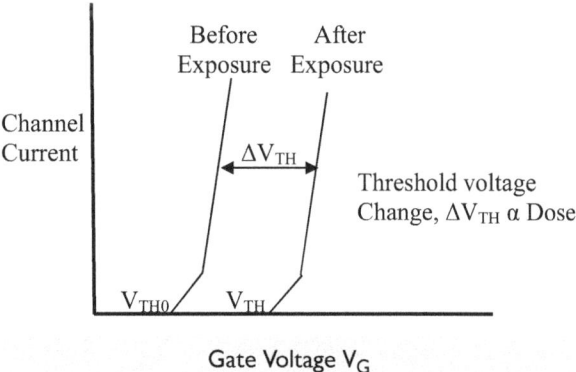

Gate Voltage V_G

This also incidentally explains the finite life of MOSFET dosimeters. Since the trapped charges can remain trapped for a long time, repeated exposures increase the trapping and, after a specific dose saturation occurs, the MOSFET dosimeter will stop working. The life depends on the thickness of the oxide layer. According to literature, MOSFET detectors may be used up to 20,000 mV threshold voltage change. The mechanism also implies that the sensitivity will change with accumulated dose.

80.   e   It is the smallest detector in the dosimetric field. Its axial anisotropy is ±2% for 360°.

81.   a, b, c, d

Though MOSFET dosimeters have excellent properties for clinical dosimetry use, they, like other solid state dosimeters, are not to be used for beam calibrations. An ionization chamber is the only recommended chamber for this purpose. Solid state dosimeter response may change with repeated use, and their constancy of response must be frequently checked depending on the stability.

82.   b, c, d

83.   a, d   The scintillation detector was perhaps the first radiation detector discovered, when Roentgen noticed some crystals fluorescing when the x-ray tube was activated. These detectors are widely used in x-ray cassettes, image intensifiers, gamma cameras, PET scanners, etc.

84.   a, b, c   Scintillation requires impurities to produce energy levels in the forbidden gap to cause light emission. So the dosimeters are not inherently scintillating. In NaI(Tl) scintillators, thallium is the impurity atom. Ionizing radiation produces electron hole pairs. The energy is transferred to the impurity when the electron decays to the impurity atom level before the emission of light (see the figure below).

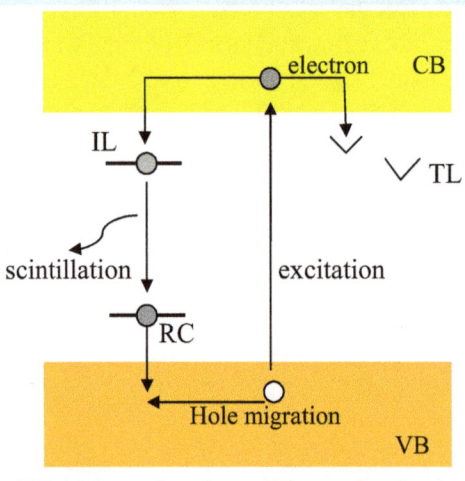

VB: valence band   CB: conduction band
IL: impurity level   TL: trap level
RC: (electron-hole) recombination

Scintillation production in inorganic scintillators

The Z or effective Z of a material determines the tissue equivalence of a dosimeter. Inorganic scintillators have high Z and are essentially non tissue equivalent and rarely used in clinical dosimetry. Organic scintillators (e.g., plastic scintillators) are of low Z and, hence, are used in radiation oncology.

85.   a   Since they emit light, the scintillator dosimeter is connected not to an electrical wire like copper, but light-carrying "wires" called optical fibers. Though this system has an advantage of not requiring voltage cables, the optical fiber has problems properly coupling to scintillators. The production of Cerenkov radiation in MV radiation also contributes to noise in the system. The plastic scintillation dosimeter has very little directional dependence and dose rate dependence, excellent energy independence, very little temperature dependence, and is less susceptible to radiation damage compared to diodes.

86.   a, c, d, e

The gel dosimeters are tissue equivalent.

II. Localization

A. Treatment Planning Concepts

Circle the right answer (Yes or No):

1. (Yes / No) The aim of treatment planning is to plan a treatment that gives the prescribed dose to the target volume of interest and no dose to the surrounding normal tissues.

2. (Yes / No) "Tumor localization" refers to the quantitative evaluation of the size, shape, and position of the target volume in a patient.

3. (Yes / No) Organs at risk (OAR) are the critical normal tissues or structures whose irradiation may significantly influence the treatment plan or dose prescription.

4. (Yes / No) Most of the tumors are treated using single fields.

5. (Yes / No) Dose-volume histograms for the target volumes and organs at risk should be reported for 3D CRT treatments.

6. (Yes / No) The image data storage format is not the same for all CTs.

7. (Yes / No) TPS systems can accept data from any model of CT.

8. (Yes / No) Electronic portal imaging can be used for the evaluation of setup errors.

9. (Yes / No) Treatment planning without CT can lead to inadequate coverage of PTV or larger coverage of PTV than necessary.

10. (Yes / No) 3D-CRT will not affect TCP or the cure rate compared to conventional treatment.

11. (Yes / No) Digitally reconstructed radiographs (DRRs) can be compared with simulator films to verify the accuracy of beam placement.

12. (Yes / No) DRRs can be compared with port films to ensure the accuracy of treatment delivery.

13. (Yes / No) From transverse CT slice images, the CT image for any arbitrary plane can be reconstructed.

14. (Yes / No) A beam's-eye view (BEV) helps the oncologist to see on the monitor that the beam adequately covers the target volume.

15. (Yes / No) The user can choose different calculation algorithms in the TPS depending on the accuracy required.

16. (Yes / No)   The accuracy of dose calculations performed using TPS depends on the accuracy of the input data provided by the user.

17. (Yes / No)   Patient immobilization is required only during treatment and not during simulation.

18. (Yes / No)   The patient support assembly (or the couch top) must be identical during CT, simulation, and treatment for accurate planning and dose delivery.

19. (Yes / No)   Production of a quality radiograph in a simulator depends only on the technique factors used in the simulator.

20. (Yes / No)   Since the image of a treatment portal can be viewed on a monitor, it is not necessary to have the radiography mode in simulators.

21. (Yes / No)   Diagnostic CT can be used for the purposes of CT planning.

22. (Yes / No)   DRRs are digitally reconstructed radiographs generated for any patient using the axial scan.

23. (Yes / No)   The patient must hold his breadth while taking CT slices for planning.

24. (Yes / No)   The positions of organs in the body in a CT slice are independent of patient couch top or the patient's breathing pattern.

25. (Yes / No)   PTV-CTV margin, when conservatively set, can lead to recurrence.

26. (Yes / No)   PTV-CTV margin must always be liberally set for optimal patient treatment.

27. (Yes / No)   Patient movements can be eliminated during CT scanning or treatment simply by following the instructions of the radiation oncologist or the physicist.

28. (Yes / No)   Can GTV be defined in all clinical cases.

29. (Yes / No)   GTV-CTV margin can be easily visualized on a CT image.

30. (Yes / No)   CTV can be a moving target in the patient.

31. (Yes / No)   The target (CTV + ITV) can be treated without any uncertainty during treatment delivery.

32. (Yes / No)   The GTV is not uniquely identified by CT.

33. (Yes / No)   The contouring of GTV is not dependent on the contouring physician.

34. (Yes / No)   Beam penumbra is NOT part of the margin included in the definition of PTV.

35. (Yes / No)   Due to the beam penumbra, the periphery of the PTV always receives less dose compared to the central regions.

36. (Yes / No)3   The prescription isodose must closely envelop PTV.

37. (Yes / No)   PTV localization can be done only with planning CT, and it cannot be verified on a daily basis.

38. (Yes / No)   In order to deliver the prescription dose to the tumor, it is enough to treat just the whole of GTV to the prescription dose.

39. (Yes / No)   The decay of M_{xy} signal and the recovery of M_z signal occur at the same rate.

40. (Yes / No) T_1 and T_2 properties of tissues can be exploited by MRI to aid in the diagnosis of cancer.

41. (Yes / No)   Among all the imaging modalities, MRI gives zero dose to the patient.

42. (Yes / No)   Any respiratory or cardiac motions or involuntary motions can cause motion artifacts in MRI imaging.

43. (Yes / No)   Verification is the means of ensuring that the patient is being treated as planned during localization procedures.

44. (Yes / No)   The definition of GTV and CTV were modified in ICRU 62 and ICRU 83 following the development of advanced treatment techniques like IMRT and IGRT.

45. (Yes / No)   Not all GTVs are very well defined in oncology.

46. (Yes / No)   The PTV-CTV margin around CTV is uniform.

47. (Yes / No)   Inadequate dose to irradiated volume, but outside the PTV, can lead to recurrence.

48. (Yes / No)   Ultrasound is particularly useful for imaging prostate.

49. (Yes / No)   The staging accuracy of PET/CT is significantly higher than that of PET or CT alone for a multitude of malignancies studied.

50. (Yes / No)   PET and MRI can often distinguish between benign and malignant lesions.

51. (Yes / No)   Pet imaging can identify hypoxic cells and proliferating cells in the tumor.

52. (Yes / No)   Geometric inaccuracy does not affect the tumor dose delivery accuracy.

53. (Yes / No)   PTV localization can be done only with planning CT, and any changes cannot be verified during treatment.

54. (Yes / No)   A safety margin around OAR has no physical significance.

55. (Yes / No)   If the daily setup error is greater than the treatment planning margin, the prescription dose to the target may not be achieved, or the tolerance dose to the normal tissues may be exceeded.

56. (Yes / No)   Setup uncertainties can arise only from intrafraction motion and not inter-fraction motion.

57. (Yes / No)   4D imaging refers to imaging patients' anatomy in 3D as a function of a time variant parameter.

58. (Yes / No)   Two important applications of imaging in radiation oncology are accurate localization of tumor and normal structures and accurate and precise delivery of treatment dose.

59. (Yes / No)   Pixel size resolution is the same as system resolution.

60. (Yes / No)   The organs or tissues that do not show up in imaging can be identified by implanting a radiopaque marker in them.

61. (Yes / No)   For imaging and treating moving targets, one needs 4D CT.

62. (Yes / No)   To adjust patient positioning error in 3D, we require at least one pair of orthogonal reference images and verification images.

63. (Yes / No)   Respiratory-gated radiotherapy decreases the treatment time.

64. (Yes / No)   Respiratory-correlated imaging should be recommended particularly for cancers of the lung and upper abdominal regions.

65. (Yes / No)   CT simulator software can calculate the geometric center of the treatment volume.

66. (Yes / No)   While treating moving targets like the lung lesion, tighter margins for CTV-PTV cannot be set.

67. (Yes / No)   The base plate used to secure head immobilization systems must be made of heavy metal to have enough strength.

68. (Yes / No)   The contrast in MRI imaging depends on the electron density differences.

69. (Yes / No)   CT can produce only axial images since it can only scan in axial planes.

B. Acquisition of Patient Data

Choose the right answer(s) (more than one may be correct):

1.   The CT number for any tissue is defined by _____.
   a. μ_{tissue}
   b. $\mu_{tissue} / \mu_{water}$
   c. $(\mu_{tissue} - \mu_{water}) / \mu_{water}$
   d. $\dfrac{(\mu_{tissue} - \mu_{water})}{\mu_{water}} \times 100$

2.   The CT number _____.
   a. gives the absolute value of attenuation coefficient of a pixel element
   b. can be related to electron density information by calibration
   c. is linearly proportional to electron density over a certain range
   d. is required for inhomogeneity corrections or dose computations

3.   CT numbers are expressed in terms of _____.
   a. cm^2/g
   b. cm^{-1}
   c. roentgen units (RU)
   d. Hounsfield units (HU)

4.   The CT number for air is _____.
   a. 1
   b. 0
   c. −1
   d. −1000

5.   The CT number for water is _____.
   a. 0
   b. 1
   c. 1000
   d. none of the above

6. The CT number for bone can be as high as _____.
   a. 10
   b. 100
   c. 1000
   d. 5000

7. The number of gray levels displayed on the CT monitor is _____.
   a. 32
   b. 256
   c. 1000
   d. 2000

8. The number of gray levels that the eye can distinguish is _____.
   a. 100
   b. 256
   c. 1000
   d. 2000

9. While decreasing the window width, the image contrast _____.
   a. decreases
   b. increases
   c. decreases and then increases
   d. remains the same

10. Increasing window width allows _____.
    a. increasing of image contrast
    b. viewing of structures with a larger CT number range to be viewed
    c. looking for differences in soft tissues
    d. none of the above

11. For radiation therapy treatment planning, one requires _____.
    a. the position of PTV
    b. the position of OAR
    c. their location with respect to external landmarks on the skin
    d. the electron density of the pixels in the CT image

12. Axial CT scans provide the important patient data of _____.
    a. tumor target volume
    b. volumes of critical organs
    c. external patient contour
    d. electron density data

13. CT scanners for radiation therapy treatment planning are required to have _____.
    a. large gantry aperture
    b. small scan field of view
    c. alignment lasers
    d. flat couch top with provision for attaching immobilization devices

14. Diagnostic CTs are not ideal for treatment planning because _____.
    a. the couch top is flat
    b. the aperture is not wide enough to simulate all kinds of treatments
    c. the slice thickness cannot be varied
    d. none of the above

15. While planning to acquire CT images for treatment planning, the following points are important.
    a. The CT table top should be flat.
    b. The immobilization/accessory devices should be accommodated during simulation and CT planning.
    c. The CT gantry should be fixed at 360 degrees.
    d. The thickness of CT cut does not matter.

16. Digitally reconstructed radiographs (DRRs) are generated from _____.
    a. the simulator's fluoroscopy images
    b. ultrasound images
    c. x-ray images
    d. CT images

17. Scout views from CT are not used for planning because _____.
    a. CT beam has divergence in both transverse and sagittal planes
    b. CT beam has divergence only in the transverse plane
    c. CT beam has no divergence
    d. none of the above

18. In the A/P radiograph of the pelvis, the prostate gland is found at the _____.
    a. level of sacroiliac joint
    b. mid pelvic canal
    c. level of symphysis pubis
    d. level of ischial tuberosity

19. The important requirements of CT for treatment planning are _____.
    a. the ability to scan thin slices
    b. a flat couch top similar to the treatment unit
    c. accommodating immobilization devices during scanning
    d. a large scanning time

20. CT slices are acquired in _____.
    a. the coronal plane
    b. the transverse plane
    c. the sagittal plane
    d. all planes

21. The desired dose uniformity in PTV as per ICRU-50 is _____.
    a. ±1%
    b. ±2%
    c. ±3%
    d. ±5%

22. A/an _____ is used for on-line treatment verification.
    a. simulation image
    b. electronic portal image
    c. DRR
    d. MRI image

23. _____ is superior to make soft tissue discrimination for treatment planning in radiotherapy.
    a. CT
    b. MRI
    c. x-rays
    d. SPECT

24. The radiation oncologist should mark on each CT slice _____.
    a. the gross tumor area
    b. the clinical target area
    c. the planning target area
    d. the irradiated area

25. The electron density information required for dose calculations can be obtained from _____.
    a. MRI images
    b. CT images
    c. PET images
    d. SPECT images

26. The clinical target volume is _____.
    a. always equal to the tumor volume
    b. larger than the tumor volume
    c. generally smaller than the tumor volume
    d. 90% of the tumor volume

27. Positron emission tomography (PET) is based on imaging with photons produced in _____.
    a. pair production
    b. annihilation of positrons by electrons
    c. fluorescence caused by positrons in certain screens
    d. Compton scattering of photons

28. Patient contouring can be obtained using _____.
    a. solder wire
    b. plaster strips
    c. mechanical contouring devices
    d. CT

29. Patient contours are required for _____.
    a. precise dose calculations in the plane of the contour
    b. checking the adequacy of treatment plan in the contour plane
    c. customizing patient treatment
    d. none of the above

30. CT gives _____.
    a. good delineation of the periphery of the tumor
    b. the best spatial accuracy
    c. good definition of bony structures
    d. electron density data

31. MRI gives _____.
    a. good delineation of soft tissue sarcomas
    b. information only in axial plane
    c. good definition of bony structures
    d. electron density data

32. MRI _____.
    a. can give a scan image in any plane
    b. can give a scan image only in the axial plane
    c. scans involve patient dose
    d. images can be fused with CT images

33. Ultrasound _____.
    a. can be used for localizing tumors and delineating patient contour
    b. scans involve radiation dose to patients
    c. is more expensive than a CT or MRI scan
    d. imaging will soon replace other imaging modalities

34. A CT simulator _____.
    a. is a CT scanner with simulation software
    b. can reconstruct images in any plane using the axial scan images
    c. produces 2D images in simulator mode
    d. is only a new type of simulator

35. CT simulators _____.
    a. have "virtual simulation" software to reconstruct images in any plane
    b. reconstruct images from axial scans
    c. provide a DRR of the treatment field that can be compared against a simulator film or port film for verification
    d. require the presence of the patient for image reconstruction.

36. The center of the clinical target volume (CTV) can be determined _____.
    a. from orthogonal radiographs
    b. by CT slices across the tumor volume
    c. by physical examination of the patient
    d. using calipers

37. A lateral radiograph shows the spinal cord to be at a measured depth of 9 cm. If the film magnification is 1.5, the actual depth of the cord in the patient is _____.
    a. 4 cm
    b. 6 cm
    c. 9 cm
    d. none of the above

38. While reporting patient dose, the radiation oncologist must specify _____.
    a. maximum target dose
    b. minimum target dose
    c. reference dose
    d. none of the above

39. Accurate patient data is required for _____.
    a. precise dose calculations to the target volume and other organs
    b. precise dose corrections for body inhomogeneities
    c. getting static landmarks for reproducible patient setup
    d. none of the above

40. The patient data required for treatment planning are _____.
    a. relevant patient contours in treatment position
    b. data on patient inhomogeneities
    c. patient height and weight
    d. electron densities of all relevant tissues
    e. delineation of PTV and OAR

41. In the treatment of the entire spinal cord, the lengths of the two adjacent fields have the sizes of 7 cm x 25 cm and 8 cm x 20 cm (the longer lengths are the abutting lengths). The fields need to be matched at the depth of 4 cm. What is the separation of the fields (in cm) on the skin surface when treated at an SSD of 100 cm?
    a. 1.10
    b. 0.9
    c. 1.0
    d. 1.2

42. In an isocentric treatment using a 100 cm SAD machine, the anterior SSD along the central axis is given by 88 cm. The tumor (center) depth (in cm) is _____.
    a. 8
    b. 10
    c. 12
    d. none of the above

43. In an isocentric treatment using a 100 cm SAD machine, the anterior SSD and the posterior SSD along the central axis were measured to be 86.5 cm and 85.5 cm respectively. The patient thickness (in cm) along CA is given by _____.
    a. 14.5
    b. 24
    c. 28
    d. none of the above

44. In clinical photon beam therapy planning, the PTV must be normally enclosed by _____.
    a. 99% isodose
    b. 95% isodose
    c. 80% isodose
    d. 50% isodose

45. Conformal therapy techniques _____.
    a. can reduce doses to organs bordering on the target volume and reduce normal tissue complications
    b. can escalate tumor dose and increase cure rate
    c. can cure all types of cancers
    d. none of the above

46. Errors in patient treatment can arise due to incorrect _____.
    a. transfer of parameters from simulator to treatment unit
    b. simulation parameters transferred to the treatment unit
    c. patient setup
    d. custom block size or position

47. An A/P radiograph _____.
    a. gives an image in the sagittal plane
    b. displays anatomical structures in the inferior/superior direction
    c. displays anatomical structures in lateral directions
    d. gives information for 3D planning

48. Disadvantages of a portal radiograph compared to a simulator film are _____.
    a. low contrast
    b. poor spatial resolution
    c. high sensitivity
    d. inability of digitization

49. Advantages of electronic portal imaging compared to portal radiography are _____.
    a. more convenient and time saving
    b. improved image quality
    c. image acquisition for any beam portal
    d. none of these reasons

50. The concept of target volume and uniform dose prescription was first described by _____.
    a. AAPM
    b. ICRP
    c. NCRP
    d. ICRU
    e. IAEA

51. The ICRU reports that deal with photon beam prescription and reporting are _____.
    a. ICRU 29
    b. ICRU 50
    c. ICRU 62
    d. ICRU 83
    e. all the above

52. The delineation accuracy of GTV can be improved by _____.
    a. CT imaging
    b. MRI
    c. US
    d. multi-modality imaging
    e. experienced technologists

53. Which of the following statements are true regarding ITV?
    a. It takes into account motion of CTV in a patient.
    b. The ITV = CTV + IM.
    c. It sets internal margin to CTV to define a moving CTV.
    d. It also includes setup uncertainties.
    e. All of the above are true.

54. Which of the following relationships are correct? PTV = _____.
    a. GTV + IM
    b. GTV + (IM + SM)
    c. CTV + IM
    d. CTV + (IM + SM)
    e. CTV + [(IM +SM) combined in a certain way]

55. Variations in the position of PTV due to internal movements (e.g, breathing) are accounted for in _____.
    a. GTV
    b. CTV
    c. ITV part of CTV
    d. ITV part of PTV
    e. SM part of PTV

56. A margin around the normal tissue (OAR) at risk defines the _____ volume.
    a. CTV
    b. PTV
    c. ITV
    d. PRV
    e. any of the above volumes

57. Some of the common causes of setup uncertainties are _____.
    a. technologist's error in setting up the patient
    b. variable filling of bladder or digestive organs
    c. weight gain/loss affecting PTV size and shape
    d. PTV delineation error
    e. patient's voluntary movement
    f. all of the above

58. The diameter of a spherical PTV target is 10 cm. If the CTV-PTV margin is reduced by 5 mm, the volume of the tissue irradiated reduces by about _____.
    a. 5%
    b. 10%
    c. 20%
    d. 30%
    e. 50%

59. Some of the methods for reducing the CTV-PTV margin are _____.
    a. skin marks and weekly port films
    b. daily setup to bony anatomy
    c. immobilization and gating
    d. planar and volumetric image guidance
    e. stereotactic delivery
    f. all of the above

60. The clinical cases where considerable differences between CTV and PTV can be expected are in the treatment of _____.
    a. brain
    b. head and neck
    c. rectum
    d. lung
    e. all the above

61. A hot spot is identified as the region that receives _____.
    a. maximum dose inside the PTV
    b. minimum dose outside the PTV
    c. dose larger than the tissue tolerance dose for that region
    d. larger than the prescription dose and lies outside PTV
    e. maximum dose in the irradiated volume

62. To be considered significant, a "hot spot" must have a minimum diameter (in cm) of _____.
    a. 3
    b. 2.5
    c. 1.5
    d. 1
    e. 0.5

63. Correction of setup errors _____.
    a. reduces CTV-PTV margin
    b. is done through in-room or on-board verification imaging
    c. involves complete modification of the treatment plan
    d. does not influence treatment accuracy
    e. on a daily basis improves geometric accuracy compared to weekly correction

64. _____ can be used for GTV determination.
    a. Clinical examination
    b. CT
    c. MRI
    d. PET
    e. US
    f. All of the above

65. If the treated volume is not wholly enclosing the PTV, then _____.
    a. it will not adversely affect the treatment outcome
    b. the TCP is reduced
    c. the case requires preplanning or the treatment aim needs to be reconsidered
    d. it will still be a good plan for stereotactic treatment of small volumes
    e. escalating the prescription dose would increase the tumor control probability

66. With modern technology, patient contouring and image acquisition can be achieved to an accuracy of about _____.
    a. 2 cm
    b. 1 cm
    c. 0.5 mm
    d. 0.1 mm
    e. 0.05 mm

67. ICRU 62 introduces the new concepts of _____ into radiotherapy treatment planning.
    a. PTV
    b. conformity index
    c. classification of organs at risk
    d. PRV
    e. all of the above

68. Conformity index is defined as _____.
    a. volume covered by prescription isodose / GTV
    b. PTV / GTV
    c. volume covered by 50% isodose / PTV
    d. irradiated volume / PTV
    e. none of the above

69. Conformity index must be _____.
    a. <<1
    b. >>1
    c. >1 and as close to 1 as possible
    d. exactly equal to 1
    e. near zero

70. Which of the following statements are true?
    a. It is difficult to draw the PTV contours of moving targets.
    b. If CT scanning speed << tumor motion speed, the target we get is a smeared tumor image.
    c. If CT scanning speed >> tumor motion speed, tumor position and shape are captured at an arbitrary breathing phase, which does not represent the realistic case.
    d. If CT scanning speed ~ tumor motion speed, tumor position and shape are heavily distorted (motion artifacts).
    e. All of the above are true.

71. The volume which will receive a dose that is considered significant in relation to normal tissue tolerance is called _____.
    a. significant volume
    b. normal tissue sparing volume
    c. planned target volume
    d. irradiated volume
    e. any of the above

72. The "irradiated volume" is often specified by choosing an isodose volume of _____.
    a. 95%
    b. 50%
    c. 20% or 30%
    d. 10%
    e. 5%

73. "Irradiated volume" refers to _____.
    a. PTV volume
    b. volume that encloses 95% isodose surface
    c. the whole volume of the patient that receives radiation
    d. volume that receives a dose that is considered significant in relation to normal tissue tolerance
    e. whole patient volume

74. _____ are the common methods of tumor and organs at risk localization.
    a. Querying the patient
    b. Imaging
    c. Clinical examination
    d. Surgery
    e. Consultancy with other physicians

75. The advantages of CT in radiotherapy are _____.
    a. 3D data set enabling treatment simulation in 3D
    b. enabling treatment planning
    c. 3D tumor localization
    d. 3D organs at risk localization
    e. electron density data for dose calculations
    f. all of the above

76. The capabilities of a CT scanner that are critical for accurate treatment planning include _____.
    a. a large aperture enabling patient immobilization
    b. a flat couch top or couch top insert
    c. accurate couch positioning
    d. axial scanning
    e. all of the above

77. Some of the important properties of modern CT scanners are _____.
    a. helical scan and data acquisition
    b. multi-slice scanning
    c. complex algorithms for image creation
    d. fast scanning
    e. gated scanning capability
    f. all of the above

78. CT is most widely used for patient simulation due to _____.
    a. excellent spatial localization of patient anatomy
    b. good differentiation between bone, soft tissue, air, and fat
    c. capability for 3D data
    d. minimal data acquisition time
    e. all the above reasons

79. Decreasing slice width leads to _____.
    a. better resolution in the scan direction (Z)
    b. more partial volume averaging
    c. increased image noise
    d. increasing dose (to keep the noise low)
    e. all of the above

80. The spatial resolution of a CT image depends on _____.
    a. focal spot size
    b. detector size
    c. field of view (FOV)
    d. algorithms used
    e. slice thickness
    f. all the above

81. Medical scanners usually exhibit a CT number range between −1024 and +3072. To illustrate the complete range of the HU on a volume, _____ bit voxels are required.
    a. 4
    b. 8
    c. 12
    d. 16
    e. 24

82. For a 512-by-512 matrix image displayed in 40 cm field of view (FOV), the pixel size is given by _____.
    a. 1.0
    b. 0.8
    c. 0.6
    d. 0.4
    e. 0.2

83. In the above problem, the pixel size resolution (in lp/cm) is _____.
    a. 16
    b. 12.8
    c. 6.3
    d. 3.2
    e. 2

84. Some of the factors contributing to artifacts in CT imaging are _____.
    a. beam hardening causing streaks while passing dense regions (e.g., bone)
    b. motion artifacts
    c. partial volume sampling
    d. tissue artifacts
    e. all of the above

85. The coils that produce gradient fields in an MRI scanner are called _____.
    a. magnetic coils
    b. field coils
    c. gradient coils
    d. circular coils
    e. any of the above names

86. In the following figure, match the coil with the respective gradient it produces.

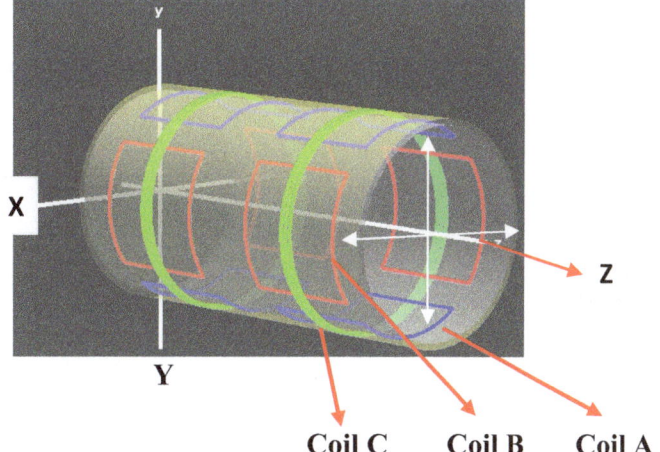

Coil		Gradient	
a. Coil A		_____	i. X gradient
b. Coil B		_____	ii. Z gradient
c. Coil C		_____	iii. Y gradient

87. MRI signals come from _____.
    a. water molecules
    b. oxygen nuclei
    c. carbon atoms
    d. hydrogen nuclei
    e. all of the above

88. The purpose of the gradient magnetic fields in an MRI system is _____.
    a. to identify the signal from each image voxel
    b. to flip the magnetization vector
    c. to identify the disease
    d. to make Larmor frequency to vanish
    e. none of the above

89. The purpose of an RF signal in MRI is _____.
    a. to create a field gradient
    b. to flip the magnetization vector by some angle usually by 90°
    c. to receive the signal
    d. to prolong the signal emission time

90. What are the consequences of flipping the spin vector M_Z by 90°?
    a. The magnetic vector M_{XY} dephases on loss of coherence and decays off in a short time.
    b. The longitudinal magnetization M_Z slowly builds up over a certain time.
    c. Decay of M_{XY} vector and the buildup of M_Z vector emit MR signals which results in image formation.
    d. The finite decay time of M_{XY} (T_2) and finite buildup time of M_Z (T_1) allow for manipulation of contrast between various tissue types because of the nature of its variation.
    e. All the above are true.

91. Which of the following statements are true?
    a. T_1 is the spin-lattice or longitudinal relaxation time, and T_2 is the spin-spin or transverse relaxation time.
    b. T_1 and T_2 are magnetization vectors which differ from one tissue to the next. They provide the contrast in MRI images.
    c. The T_1 and T_2 time constants determine the *shape* of the exponential recovery and decay curves of the longitudinal and transverse magnetization, respectively.
    d. T_2 is zero for many tissues, making imaging of these tissues difficult.
    e. All of the above are true.

92. Which of the following statements regarding T_1 are true?
    a. The T_1 relaxation constant is the time needed to recover 63% of the longitudinal magnetization, Mz.
    b. The T_1 relaxation constant is the time for the M_{xy} vector to dephase to 37% of its original value.
    c. T_1 and T_2 relaxations are entirely independent and occur simultaneously.
    d. T_1 and T_2 are very much dependent on the physical characteristics of the tissues.
    e. All of the above are true.

93. Which of the following statements are true?
    a. T_1 increases with higher field strengths.
    b. T_1 arises as a result of spin-lattice relaxation.
    c. Differences in T_1, T_2, and T_2* (along with proton density variations) provide high contrast in MRI.
    d. T_1 and T_2 decay constants are fundamental properties of tissues.
    e. All of the above are true.

94. The Larmor angular frequency $\omega_0 = \gamma B_0$. The frequency f_0 is given by _____.
    a. γB_0
    b. $2\pi B_0$
    c. $(\gamma/2\pi) B_0$
    d. $\gamma 2/ B_0$
    e. $(\gamma/2\pi B_0)$

95. _____ is the Larmor frequency (in MHz) of a 1.5 T MRI scanner (γ for proton = 42.57 MHz/T).
    a. 127.7
    b. 108.4
    c. 63.86
    d. 42.57
    e. 21.28

96. When a patient is placed on a 1 T MRI scanner, the scanner applies a magnetic field along his long axis. Which of the following consequences occur in this situation?
   a. All the protons in the body are aligned parallel to the magnetic field.
   b. The protons align parallel or anti-parallel to the scanner magnetic field.
   c. The parallel alignment is a bit more favored (about 6 parts per million for a 1 T unit).
   d. This creates a minute net magnetization along the scanner field direction.
   e. This net magnetization can be easily measured.

97. Which of the statements deduced from the following figure are true?

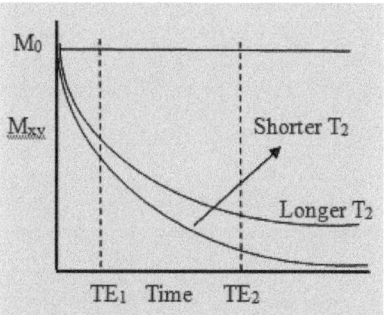

Dephasing signal decay

   a. Different tissues exhibit different T_2 values.
   b. The contrast depends on the timing of measuring the MR signals.
   c. At TE_2 the contrast is less compared to contrast measured at TE_1.
   d. The dephasing signal M_{xy} becomes negligible after a few time constants.
   e. All of the above are true.

98. Which of the statements deduced from the following figure are true?

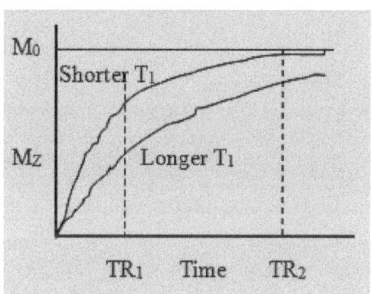

Recovery of MZ signal

   a. T_1 values are characteristic of the tissues.
   b. The contrast depends on the timing of measuring the MR signals.
   c. At TR_2, the contrast is less compared to contrast measured at TR_1.
   d. By proper choice of TR, the image contrast can be manipulated.
   e. For large TR, the signal strength is high but contrast is low.
   f. All of the above are true.

99. From the above two figures and the conditions on TR (repetition time) and TE (echo time) match the type of image to the values of TR and TE.

Image Type		TR /TE	
a. T_1 weighted	_____	i.	Long/Long
b. T_2 weighted	_____	ii.	Long/Short
c. Proton density (PD) weighted	_____	iii.	Short/Short

100. How slice selection is made in an MRI (use the figure given below).

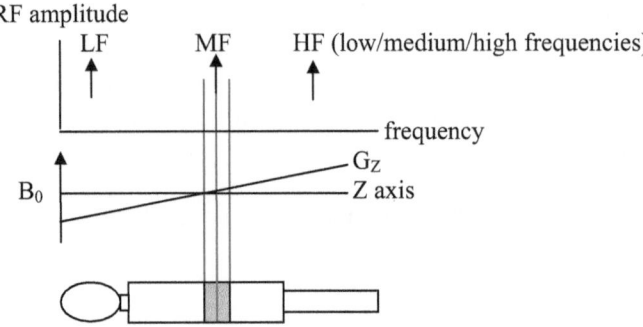

Gradient field G_Z varies effective field and, hence, the Larmor frequency, along Z _____.
a. by applying a homogeneous magnetic field
b. by applying a gradient field along the longitudinal axis
c. by applying both the fields simultaneously
d. by applying both fields and sending an excitation pulse simultaneously with the application of gradient field, the RF pulse frequency matching the slice (Larmor frequency) chosen
e. all the above

101. From the figure shown above, one can surmise that slices can be obtained in any arbitrary orientation with respect to the patient axis or the scanner longitudinal axis by _____.
a. changing the direction of the field B_0
b. changing the direction (orientation) of the gradient field
c. changing the orientation of both the fields
d. introducing a third gradient field in the transverse direction

102. From the figure given in problem 100 above, one can guess that slice thickness can be varied by _____.
a. varying the field gradient
b. varying the field B_0
c. varying the RF pulse bandwidth
d. varying the RF pulse frequency
e. all the above methods

103. How is the slice selection made in MRI imaging?
a. An RF pulse is directed at the section of interest in the patient.
b. The imaging slice location is moved to the center of the magnet bore.
c. A magnetic field gradient is temporarily applied along the patient axis.
d. The gradients are applied along X, Y, and Z directions.
e. Slice selection is made by all the above methods.

104. One can uniquely identify each pixel in the slice chosen by the Z gradient field and the RF pulse frequency by _____.
a. applying a phase gradient along a perpendicular direction, say along the X direction
b. applying a frequency gradient along the same direction
c. applying a phase gradient along X followed by a frequency gradient along the Y direction
d. there is no way to identify the signal from each pixel element
e. method b or c

105. The current role of MRI in treatment planning can be summarized as _____.
    a. tumor visualization
    b. target delineation
    c. anatomical structure definition in favorable cases
    d. soft tissue discrimination
    e. routine target and normal tissue dose calculations
    f. all of the above

106. Which of the following statements are true in MRI imaging?
    a. MRI contrast agents are used for improved visualization of abnormal structures and lesions in the body.
    b. Compounds containing gadolinum are used as contrast agents.
    c. Gadolinum contrast agents are not absolutely risk free.
    d. MRI can be used to study the functioning of the brain.
    e. All of the above are true.

107. MRI cannot be used exclusively for treatment planning because _____.
    a. MRIs are not extensively available in cancer centers
    b. it cannot give electron density information for each voxel that is required for dose calculations
    c. geometric distortions are a real problem in this imaging
    d. MRI lacks patient setup verification images comparable to CT DRR
    e. all of the above

108. Artifacts and geometric distortions in MRI imaging are caused by _____.
    a. any inhomogeneity in the magnetic, RF, or gradient fields
    b. sudden changes in magnetic susceptibility around tissue boundaries
    c. surgical clips or dental work
    d. eddy currents
    e. all the above

109. Which of the following statements regarding MRI are true?
    a. It has superior soft tissue resolution.
    b. It has the ability to assess neural and marrow infiltration.
    c. Imaging of metabolic activity through MR spectroscopy is possible.
    d. Imaging of tumor vasculature and blood supply is possible.
    e. All of the above are true.

110. _____ occurs when US is propagated through a medium.
    a. Reflection
    b. Refraction
    c. Scatter
    d. Attenuation
    e. Signal amplification
    f. All of the above

111. If υ if the frequency of the ultrasound (US) wave, the period or the time duration of one wave cycle is given by _____.
    a. $v \, \upsilon$ where v is the velocity of propagation of the wave
    b. v / υ
    c. $1 / \upsilon$
    d. υ

112. The US frequency used in medical imaging typically falls in the range _____.
    a. 2 kHz to 10 kHz
    b. 20 kHz to 200 kHz
    c. 1 kHz to 5 MHz
    d. 2 kHz to 10 MHz
    e. 20 kHz to 20 MHz

113. The velocity of sound in soft tissue (in m/sec) is _____.
    a. 150
    b. 1000
    c. 1250
    d. 1450
    e. 1540

114. Which of the following statements regarding the velocity of sound are correct. Sound velocity _____.
    a. depends on the material through which it traverses
    b. is very high in media like air
    c. depends on the elasticity of the medium
    d. is very low in materials like bone
    e. characteristics above are all true

115. Which of the following statements are true regarding sound propagation through a medium?
    a. Wavelength depends on the velocity of propagation in the medium.
    b. Frequency depends on the velocity of propagation in the medium.
    c. Depth of penetration depends on the velocity of sound in the medium.
    d. Contrast depends on sound velocity in the medium.
    e. All of the above are true.

116. The attenuation of ultrasound in soft tissue, in db/cm MHz, is approximately _____.
    a. 0.1
    b. 1.0
    c. 3
    d. 5
    e. 10

117. A 4 MHz ultrasound beam traverses 8 cm of soft tissue. The original intensity is reduced by _____ dB. (Attenuation of beam is given as 1 dB/cm MHz.)
    a. 64
    b. 48
    c. 32
    d. 16.4
    e. 12

118. The wavelength (in mm) of a 5 MHz sound wave in a tissue medium is _____. ($v = 1540$ m/sec.)
    a. 1.8
    b. 0.3
    c. 0.15
    d. 0.03
    e. 0. 005

119. The velocity of ultrasound in the transducer element is 4000 m/sec. The thickness of a transducer required (in mm) to produce a 5 MHz pulse is _____.
    a. 4
    b. 2.4
    c. 0.4
    d. 0.2
    e. 0.04

120. In the above problem, _____ is the transducer thickness (in mm) required for producing a 10 MHz pulse.
    a. 0.4
    b. 0.2
    c. 0.1
    d. 0.02
    e. 0.005

121. Which of the following are true? The ultrasound transducer _____.
    a. is made of a piezoelectric material
    b. can transmit and receive US pulses
    c. transmitted pulse length determines the axial resolution of the scan
    d. converts electrical pulse to mechanical vibrations
    e. converts mechanical vibrations to electrical pulses
    f. properties above are all correct

122. In US imaging, a gel is applied between the transducer and the skin _____.
    a. because good contact cannot be established between the transducer and the skin
    b. to prevent the transducer from hurting the skin by continuous rubbing
    c. because air film and the transducer have a large Z difference causing poor transmission (high reflection) of the US wave
    d. because air film and skin have large Z difference, causing poor transmission of US into the patient
    e. all the above

123. _____ is the spatial pulse length (SPL) in mm of the pulse shown in the figure below ($\lambda = 0.8$ mm).

λ

Pulse
length

An ultrasound pulse

    a. 3.2
    b. 2.4
    c. 1.6
    d   0.8
    e. 0.4

124. _____ is the axial resolution (in mm) of the image if the above (10 MHz) pulse is used for scanning.
    a. 2.4
    b. 1.6
    c. 0.8
    d. 0.4
    e. 0.2

125. Lateral resolution depends on the _____.
    a. number of scan lines
    b. beam width
    c. depth of scanning
    d. axial resolution
    e. all of the above

126. The best axial resolution is obtained with a _____ transducer.
    a. 3 MHz pulsed
    b. 3 MHz continuous
    c. 6 MHz pulsed
    d. 10 MHz continuous
    e. 10 MHz pulsed

127. The acoustic impedance of any tissue depends on _____.
    a. US frequency and tissue viscosity
    b. US wavelength and tissue density
    c. tissue elasticity and density
    d. US velocity and tissue density
    e. US frequency and velocity

128. The advantages of US imaging are _____.
    a. noninvasive
    b. real time
    c. low cost
    d. no patient harm
    e. no special room requirement
    f. all of the above

129. A sound pulse is sent through a tissue medium at time $t = 0$ and an echo from certain depth d is received at time $t = t_1$. The velocity of the sound pulse in the medium is V. The depth d is given by _____.
    a. V/t_1
    b. $V t_1$
    c. $V t_1 / 2$
    d. V^2 / t_1
    e. t_1 / V

130. The reflection coefficient (R) of an ultrasound pulse traveling from soft tissue to bone is 0.23. _____ will be the reflection coefficient when the pulse travels from bone to soft tissue. (The acoustic impedance of bone, Z_2, is about three times that of the soft tissue, Z_1.)
    a. 0.23
    b. 0.46
    c. 0.69
    d. 1
    e. 1.5

131. The fraction reflected at the air/muscle interface is _____. ($Z_{muscle} = 1.70$; $Z_{air} = 0.0004$)
     a. 1.7004
     b. 1.70
     c. 1.25
     d. nearly 1
     e. nearly 0

132. _____ is the fraction reflected at the liver/spleen interface. ($Z_{liver} = 1.65$; $Z_{spleen} = 1.64$)
     a. 3.29
     b. 1.645
     c. 1.24
     d. Nearly 1
     e. Nearly 0

133. Approximately _____ fraction of the incident beam is reflected from an interface between two media with Z values 1.65 and 1.55.
     a. 1
     b. 0.5
     c. 0.1
     d. 0.01
     e. 0.001

134. _____ is the acoustic impedance for water given $V_{water} = 1480$ m/sec.
     a. 14.8×10^3
     b. 1.48×10^6
     c. 1.480
     d. 2.8×10^{-3}
     e. 1.48×10^{-6}

135. Which of the following statements regarding the matching interface layer of the US transducer are true? The interface layer _____.
     a. thickness must be ($\lambda/4$) where λ is the wavelength of sound in the interface layer
     b. must have an impedance matching the impedance of soft tissue
     c. transmissions arising from multiple internal reflections are in phase and reinforce the transmitted pulse
     d. transmissions into the transducer are in opposite phase and cancel one another.
     e. all of the above

136. Factors that help to maximize the transmission of the US beam into the patient from the transducer include _____.
     a. a quarter wave matching layer
     b. acoustical gel between the patient and the transducer
     c. increased Q factor
     d. making the crystal thickness equal to twice the wavelength
     e. all the above

137. Which of the following statements are true regarding the image formation in ultrasound scanning?
     a. The US beam is intermittently transmitted with most of the time used for listening to echoes.
     b One pulse echo sequence on a scan line produces one line of image.
     c. Time delay between the transmission pulse and the return echo is directly related to the depth.
     d. The strength of the echo pulse is represented by a dot of a certain shade of gray.
     e. A 2D scanning gives rise to a 2D image of different shades of gray.
     f. All of the above are true.

138. The different modes of US scanning are _____.
    a. A mode
    b. B mode
    c. C mode
    d. M mode
    e. universal mode

139. The data acquisition time for a frame in US imaging varies between 25 msec to about 66 msec. _____ is the frame rate for real-time imaging.
    a. 25–66
    b. 20–50
    c. 15–40
    d. 10–35
    e. 1–15

140. The pulse repetition period (PRP) is 500 µsec. _____ is the pulse repetition frequency (PRF) in kHz.
    a. 0.5
    b. 2
    c. 5
    d. 10
    e. 14

141. Image quality in US imaging depends on _____.
    a. frame rate
    b. FOV
    c. number of lines
    d. penetration depth
    e. all of the above

142. Which of the following statements are true regarding the damping of transducer resonance vibrations?
    a. A damping material backing the transducer is used to reduce the oscillations to zero.
    b. A "light damping" slightly changes the emitted frequency, but the vibrations last longer (i.e., the pulse is much longer).
    c. A "heavy damping" arrests the vibrations quicker (i.e., smaller pulse) but changes the frequency of oscillations a great deal.
    d. Short pulses are required for imaging purposes.
    e. Blood velocity estimates require narrow bandwidth to preserve velocity information.
    f. All of the above are true.

143. _____ is the resonance frequency (in MHz) of a 1 mm thick transducer. (The velocity of propagation of the ultrasound wave is 4000 m/sec.)
    a. 2
    b. 3
    c. 5
    d. 7.5
    e. 10

144. _____ is the Q factor of a 3 MHz transducer with a frequency bandwidth of 1.5 MHz.
    a. 3.0
    b. 2.5
    c. 2.0
    d. 1.0
    e. 0.5

145. The most widely used radioisotope in PET imaging is _____.
    a. ^{14}C
    b. ^{32}P
    c. ^{18}F
    d. ^{131}I
    e. none of the above

146. Which of the following statements regarding PET are true? PET _____.
    a. imaging system detects the positrons emitted by the active biological molecules
    b. detectors are arranged in a circular array around the patient
    c. positron emitters are short-lived isotopes
    d. systems are available for incorporating the PET information into the CT image for more accurate tumor localization
    e. details given above are all true

147. As biological molecules, _____ are labeled with positron emitter in PET imaging.
    a. sugars
    b. amino acids
    c. nucleic acids
    d. receptor-binding ligands
    e. all of the above

148. Which of the following statements are true with respect to PET imaging in radiotherapy?
    a. PET/CT imaging improves target delineation.
    b. PET/CT improves the detection of metastasis, nodal involvement, etc.
    c. PET/CT improves dosimetric accuracy.
    d. PET/CT improves staging accuracy.
    e. PET/CT improves detection of recurrence.

149. Which of the following statements regarding PET/CT imaging are true?
    a. In a large number of cases, the CTV and PTV are modified by PET.
    b. The target volumes are always increased because of the PET information.
    c. Nodal delineation improves considerably as a result of PET imaging.
    d. Some cancers can be detected earlier because functional changes precede structural changes.
    e. PET imaging is not useful for follow-up studies.

C. Patient Positioning and Immobilization

Choose the right answer(s) (more than one may be correct)

1. Shoulder retractors are used for _____.
    a. pushing the shoulder into the treatment field
    b. pulling the shoulders out of the treatment field
    c. providing dose uniformity
    d. none of the above

2. A patient's head position should be made comfortable by _____.
    a. using the softest head support possible
    b. placing a head support of the right height under the head
    c. using a regular bed pillow under the head
    d. avoiding the use of head support

3. Patient movement during simulation or dose delivery is minimized by _____.
   a. giving instructions to patients
   b. observing the patient during treatment
   c. using immobilization devices
   d. remotely holding the patient

4. The patient's head position when treating a vertex field should be with the _____.
   a. chin extended as much as possible
   b. chin tucked down on the chest
   c. chin turned toward one side
   d. head in any position

5. To treat a vertex field, we may have to _____.
   a. keep the couch at 0° and turn the gantry to get the desired angle
   b. turn the couch 90° and rotate the gantry to get the desired angle
   c. turn only the collimator and couch
   d. turn only the gantry and collimator

6. When treating an extremity, the anatomical axis must be _____.
   a. parallel with treatment couch
   b. elevated until it is horizontal
   c. perpendicular to the beam

7. A belly board _____.
   a. promotes small bowel movement
   b. restricts bowel movement
   c. helps to keep all the flabby skin folds out of the treatment field
   d. helps to keep the small bowel out of the pelvis

8. A patient with rectal carcinoma is best treated in the _____.
   a. supine position using parallel opposed pair of fields
   b. supine position via posterior and opposed lateral fields that are flashing over the posterior surface
   c. prone position via only lateral fields that are flashing over the posterior surface
   d. prone position via posterior and opposed lateral fields that are flashing over the posterior surface

9. A vertex field is difficult to treat because _____.
   a. the gantry and collimator must be turned
   b. the couch and gantry must be turned
   c. the couch and gantry must be turned, and sometimes the collimator, too
   d. there is no reliable way of obtaining a port film because the beam exits along the patient's torso

10. To make patients more comfortable during treatment, it is good idea to _____.
    a. make them lie prone
    b. provide supports under head, knees, and arms
    c. place a soft pillow under the back
    d. ask the patient to adjust to a comfortable position

11. Immobilization devices are required to _____.
    a. prevent the patient from falling off the couch top
    b. improve the dose distribution
    c. prevent patient motion during treatment
    d. reproduce the treatment position during the full course of treatment

12. Immobilization devices help in _____.
    a. restricting patient movement
    b. reducing setup errors
    c. reducing treatment duration
    d. reducing PTV-CTV margin

13. The most popular and convenient immobilization device for the head and neck region is a _____.
    a. thermoplastic mask
    b. Vac-Lok™ vacuum cushion
    c. bite block system
    d. belly board

14. For pelvic treatments, _____.
    a. skin marks are sufficient for reproducible patient positioning
    b. there are many dose-limiting structures
    c. immobilization devices are necessary
    d. bony landmarks are available for accurate patient positioning

15. For the treatment of head and neck cases, _____.
    a. immobilization is critical
    b. immobilization is unnecessary
    c. good immobilization can be achieved to reduce movements to about 5 mm
    d. accurate and reproducible patient setup is possible because of bony landmarks

16. To minimize the beam divergence into the eye on the opposite side, _____ when treating a brain through opposed lateral fields.
    a. the collimator should be angled so there is no divergence
    b. the beam is centered just behind the eye
    c. the patient head is turned when each field is treated so that the beam exits behind the eyes
    d. the gantry can be angled so that the beam exits behind the eyes

17. In craniospinal irradiation, the spinal cord field length is 30 cm. The skull field length is 15 cm. To align the skull field with the divergent spinal field, _____ collimator rotation is required for the skull field.
    a. 10°
    b. 8.53°
    c. 9.53°
    d. 15°

18. In craniospinal irradiation, the spinal cord field length is 30 cm. The brain field length is 15 cm. To eliminate the divergence of the brain portal and spinal portal in the abutting plane, the table is rotated _____ through the angle.
    a. 8.3°
    b. 5.3°
    c. 4.3°
    d. 6.3°

19. When treating a left lateral field, the gantry and collimator angles are respectively 90° and 25°. If an opposed portal is planned with a gantry angle of 270°, then the collimator angle should be _____.
    a. 25°
    b. 335°
    c. 345°
    d. 0°

20. The volume of lung within an anterior supraclavicular field can be reduced when treating breast carcinoma by
    _____.
    a. elevating the arm
    b. treating the supraclavicular field only every other day
    c. trying to move the breast lower on the chest wall by placing an angled board under the chest
    d. moving the breast down on the chest wall and making the match line between the tangential fields and the supraclavicular field as high on the chest as possible

21. Partial kidney shielding is best accomplished by means of _____.
    a. 5 HVT in both the anterior and posterior fields throughout the course of treatment
    b. partial shielding of the kidneys during the first week of treatment only
    c. partial shielding of the kidneys in the anterior and the posterior fields until kidney tolerance is reached, followed by total removal of the blocks
    d. partial shielding of the kidneys in the anterior and the posterior fields until kidney tolerance is reached, followed by total shielding of the kidneys

22. When the entire spinal axis must be treated, the patient's posterior surface should be as flat as possible _____.
    a. to make field matching easier
    b. to make the dose more uniform
    c. so the position and matching of the adjacent field can be reproduced

23. Lateral brain fields and a posterior spinal field are best matched by turning the _____.
    a. collimator when treating the lateral fields and turning the couch when treating the posterior spinal field
    b. couch when treating the lateral fields and turning the collimator when treating the posterior spinal field
    c. collimator and the couch when treating the lateral fields
    d. couch and the collimator when treating the posterior spinal field

24. The size of the treatment field (at SAD = 100 cm) is 14 cm x 17 cm, and a template is to be made for shielding a rectangular corner of size 4 cm x 5 cm. The shielding tray is at a distance of 65 cm. For an isocentric treatment, the size of the template is _____.
    a. 10 cm x 12 cm
    b. 14 cm x 17 cm
    c. 9.1 cm x 11.05 cm
    d. none of the above

25. In a head and neck treatment where left lateral, right lateral, and anterior fields are commonly used, one of the following statements is not true.
    a. Matching of the borders is important so as to prevent underdose/overdose.
    b. The spinal cord is protected by shielding.
    c. Sometimes central lead shielding is used to protect the spinal cord and the upper area digestive tract.
    d. The field junction can be present in the area where the tumor is present.

26. Patient immobilization devices _____.
    a. can have any atomic number
    b. must not interfere with the treatment beam
    c. must also be separately designed for CT and simulator
    d. must be very flexible for accurate patient positioning

27. Immobilization devices _____.
    a. restrict external movement of the body or body parts
    b. restrict the motion of internal organs in the body
    c. can reduce patient setup errors
    d. do none of the above

28. Patient immobilization devices give rise to _____.
    a. reproducible patient positioning
    b. better patient comfort
    c. better accuracy in treatment
    d. none of the above

29. Field placement errors during patient setup arise as a result of _____.
    a. data transfer errors between TPS and delivery system
    b. incorrect design or use of treatment accessories (e.g., blocks)
    c. lack of reproducibility in patient setup
    d. none of the above

30. Beam positioning errors can be detected by _____.
    a. comparison of portal image with reference image (e.g., DRR)
    b. excessive skin reactions on the patient
    c. comparison of electronic portal images taken during successive treatment fractions
    d. none of the above

31. Field placement errors can be reduced significantly by _____.
    a. using thermoplastic masks for immobilization in head and neck treatments
    b. pelvic immobilization during treatment
    c. the use of lasers for patient setup
    d. none of the above methods

32. To fully exploit the potential of high-precision conformal radiotherapy we must _____.
    a. use frames and other immobilization devices
    b. detect setup errors by pretreatment imaging and making appropriate corrections to the patient couch
    c. arrest internal motions of the target or adapt to its motion through some means
    d. modify the prescription dose
    e. do all the above

33. With modern technology, patient positioning and immobilization can be achieved to an accuracy of about _____.
    a. 2 cm
    b. 1 cm
    c. 0.5 mm
    d. 0.1 mm
    e. 0.05 mm

34. Setup uncertainties are caused by _____.
    a. movement of CTV due to physiological reasons
    b. lack of reproducibility in patient positioning
    c. beam alignment error in treatment
    d. changes in patient physical condition (e.g., weight loss)
    e. tumor shrinking

35. An internal margin is considered to account for _____.
    a. CTV movement due to breathing, etc.
    b. variations in patient positioning
    c. light/radiation field misalignment
    d. human errors
    e. data transfer errors

36. Immobilization devices help _____.
    a. to reduce setup error
    b. to reduce patient movement
    c. to increase treatment accuracy
    d. to increase patient comfort
    e. in localization
    f. in all the above

37. Some of the ancillary treatment devices used for head and neck cases are _____.
    a. masks
    b. molds
    c. invasive frames
    d. non-invasive frames
    e. all of the above

38. Immobilization devices used in radiosurgery or Gamma Knife procedures should give positional accuracy of
    _____.
    a. ±5 mm
    b. ±3 mm
    c. ±1 mm
    d. ±0.5 mm
    e. ±0.05 mm

39. A breast board _____.
    a. is used for the treatment of breast cancer with parallel opposed pair of fields
    b. has several features to manipulate the positioning of arms, hands, wrists, heads, shoulders, etc.
    c. is generally constructed of low-density foam and carbon fiber
    d. helps in reproducible patient positioning
    e. is used for all the above

40. Which of the following statements are true regarding thermoplastics? Thermoplastics _____.
    a. are used for making masks for immobilization
    b. soften at 60°
    c. melt at 150°
    d. are available as plastic sheets of mesh in different thickness
    e. properties above are all true

D. Ancillary Treatment Devices

Choose the right answer(s) (more than one may be correct):

1. Pencil eye shields are used to shield the _____.
   a. cornea
   b. sclera
   c. eye lens
   d. retina

2. Hyperextension of the neck is achieved by using _____.
   a. head rests and neck rolls
   b. a thermoplastic mask
   c. a breast board
   d. none of the above

3. A bite block is used for _____.
   a. head immobilization
   b. chin extension
   c. moving the tongue out of the treatment field
   d. none of the above

4. Perforated thermoplastics are better than nonperforated thermoplastics because _____.
   a. they reduce average skin dose
   b. they increase average skin dose
   c. they are more sturdy
   d. reproducibility is good with them

5. All thermoplastic materials need some kind of base plate support on the couch top. It is preferable to have these plates made of _____.
   a. high-density materials
   b. non-tissue equivalent materials
   c. carbon fiber materials
   d. none of the above

6. A breast board is used to _____.
   a. bring the arm above the shoulders and out of the lateral field
   b. provide unobstructed access to the breast by the lateral field
   c. allow the patient to be positioned so that the chest wall is horizontal and thus avoids angulation of the collimator
   d. use gravity to pull the large breast down into a better treatment position

E. Treatment Simulation and Verification

Choose the right answer(s) (more than one may be correct):

1. Any error in patient marking in CT or simulator procedures _____.
   a. will act as a systematic error during patient treatment
   b. will always get canceled during patient setup for treatment
   c. can lead to excess dose to nearby critical structures
   d. can be detected from portal film

2. A simulator helps the radiation oncologist in _____.
   a. localizing PTV and OAR
   b. determining the exact volume of PTV
   c. choosing field placement
   d. treatment verification

3. Simulator films, compared to port films, have _____.
    a. poor quality
    b. better quality
    c. the same quality

4. The beam-limiting diaphragms (lead or tungsten) in a simulator have a thickness of _____.
    a. a few millimeters
    b. >5 mm
    c. >10 mm
    d. >15 mm

5. The x-ray field on the simulator radiograph must be _____.
    a. larger than the film size
    b. equal to the film size
    c. smaller than the film size

6. Increasing the object film distance in a simulator _____.
    a. magnifies the image
    b. demagnifies the image
    c. increases the optical density on the radiograph
    d. decreases the optical density of the radiograph

7. An image intensifier _____.
    a. converts an x-ray image into a light image at the input screen
    b. converts a light image into an electronic image at the input screen
    c. converts an electronic image into a light image at the output screen
    d. amplifies the image at the output screen

8. On a film, relatively lesser tissue density produces _____.
    a. a darker image
    b. a lighter image
    c. no change in gray level

9. Port films make use of a _____.
    a. phosphor screen
    b. metal screen
    c. metal screen to cut off electrons from reaching the film
    d. metal screen to increase film sensitivity

10. Metal screen films make use of _____.
    a. calcium tungstate as screen
    b. rare earth phosphor as screen
    c. copper or lead as screen
    d. none of the above

11. The patient structures seen on a simulator film are _____.
    a. magnified
    b. minified
    c. the same size

12. The focus-to-film distance in a simulator is 140 cm. A lead wire of 5 cm length kept on the patient gives an A/P image on the simulator film. The focus-to-skin distance is 100 cm. The image magnification factor is given by _____.
    a. 1
    b. 140/100
    c. 100/140
    d. none of the above

13. Shape distortion occurs in a simulator film if the _____.
    a. object and film or image plane are not parallel
    b. x-ray beam is not incident normally on the object, causing unequal magnification
    c. object is not spherical
    d. film is far away from the object

14. Scatter radiation reaching the film _____.
    a. reduces contrast
    b. affects image quality
    c. is significant for larger thicknesses (e.g., pelvis)
    d. can be reduced by using a grid

15. A grid used in a simulation procedure _____.
    a. filters scatter and improves contrast
    b. does not cut off primary radiation
    c. does not warrant any change in technique factors
    d. cuts off more scatter if a larger grid ratio is used

16. A port film check is essential _____.
    a. if any critical structure is close to the clinical target volume
    b. if skin marks are likely to move
    c. for 3D conformal treatments
    d. for "parallel opposed pair of fields" treatment

17. To get a consistently good radiographic image with a simulator, _____.
    a. exposure output must be reproducible
    b. exposure output must be linearly proportional to kVp set on the unit
    c. exposure output must be linearly proportional to mAs set on the unit
    d. none of the above is true

18. In a simulator, if the mAs and kVp are kept constant and mA is changed, _____.
    a. output will change
    b. it would result in a proportional change in exposure time
    c. it would result in a reciprocal change in exposure time
    d. mR/mAs will remain constant (within the acceptable level of tolerance)

19. In a simulator, _____ if the mA and kVp are kept constant and mAs is changed.
    a. output will change
    b. a proportional change in exposure time will occur
    c. a reciprocal change in exposure time will occur
    d. mR/mAs will remain constant (within the acceptable level of tolerance)

20. When treating the paraaortic lymph nodes and the splenic pedicle through parallel opposed isocentric fields, clips marking the splenic pedicle will always appear on port films at the same position with respect to the central axes of the beam if _____.
    a. the central axes is on the clips
    b. the clips are at the depth in the patient as the isocenter
    c. a and b are correct
    d. a and b are incorrect

21. If the treatment position is not exactly reproducible during the course of a treatment, it can lead to _____.
    a. missing part of the target volume
    b. treating a larger volume of normal tissues
    c. higher cure rate
    d. none of the above

22. During treatment with a planned field, a volume other than the clinical target volume may get irradiated due to _____.
    a. internal motion of organs
    b. change of SSD or collimator setting
    c. patient movement during treatment
    d. setup error

23. Patient data must be acquired for _____.
    a. correcting for patient skin curvature and body inhomogeneity
    b. computation of dose to tumor and critical structures
    c. determination of tumor volume
    d. none of the above

24. A transverse CT slice provides 2D information on the _____.
    a. patient contour
    b. extent of inhomogeneity in the patient
    c. target area to be treated
    d. none of the above.

25. Setup errors or reproducibility of setup can be detected by comparing _____.
    a. simulator film with port film or EPID image
    b. EPID images (of any treatment portal) during the course of treatment
    c. excessive skin reactions
    d. none of the above means

26. The field wires on the simulator defines _____.
    a. 80% isodose line on skin
    b. 50% isodose line on skin
    c. 20% isodose line on skin
    d. none of the above

27. CT image data are transferred to the TPS via _____.
    a. digital serial link
    b. magnetic tape
    c. disk
    d. none of the above

28. Match the CT image to the densities.

    CT Image   *Intensity*
    a. air cavity   _____ i.   white
    b. fat tissue   _____ ii. gray
    c. fluid   _____ iii. light gray
    d. bone   _____ iv. black

29. An MRI scan _____.
    a. provides functional information
    b. provides scan in any patient plane
    c. is inexpensive compared to CT
    d. aids in visualizing the boundaries of soft tissue tumors

30. Match the stages to the treatment planning processes in the correct sequence.

    Stage   *Various Processes*
    a. Stage 1   _____ i.   Virtual simulation
    b. Stage 2   _____ ii. CT data acquisition
    c. Stage 3   _____ iii. Patient position verification on the linac
    d. Stage 4   _____ iv. Treatment delivery
    e. Stage 5   _____ v.   Dose calculation

31. Some of the errors affecting treatment are _____.
    a. planning errors that propagate through the treatment
    b. patient moving between CT scanning and skin marking
    c. contouring errors
    d. treating a wrong site
    e. all of the above

32. Some important functions of a simulator are _____.
    a. patient reproducible positioning and setting isocenter
    b. creating films for patient position verification
    c. setting field boundaries
    d. treatment dose delivery
    e. all of the above

33. A simulator is used for _____.
    a. tumor localization
    b. critical organs localization
    c. treatment simulation
    d. treatment verification (after plan completion)
    e. treatment monitoring
    f. all of the above

34. Originally the main functions of a simulator were to _____.
    a. check if the treatment setup planned is possible
    b. verify if beams were correctly chosen to cover the tumor and properly directed
    c. mark the fields and isocenter on the skin
    d. include target motion in the beam
    e. realize all the above

35. A simulator can be used to _____.
    a. plan the treatment fields and directly treat patients in palliative cases
    b. verify that planned fields are actually treatable on a linac
    c. ensure field coverage in the case of moving tumors (or dynamic targets)
    d. treat patients in emergencies
    e. for all the above

36. Simulation can be accomplished with _____.
    a. only a therapy simulator
    b. CT
    c. MRI
    d. PET
    e. ultrasound
    f. all the above

37. The degrees of freedom of a simulator are _____.
    a. one
    b. two
    c. three
    d. many
    e. depends on the make and model

38. Which of the following statements regarding CT simulation are true? CT has _____.
    a. a flat couch top
    b. a bore diameter 70 to 85 cm
    c. a scanning field of view (SFOV) 70 to 85 cm
    d. multi slice and thinner slice capabilities
    e. gating capabilities
    f. all of the above

39. The advantages of CT simulator over conventional simulator are _____.
    a. 3D visualization
    b. patient not required for simulation or verification simulation
    c. image in any plane
    d. no limitation on simulation geometry
    e. use of software for generating verification film
    f. all of the above

40. The advantages of multislice CT in CT simulation are _____.
    a. thin slices giving clearer DRR
    b. increased resolution
    c. decreased acquisition time
    d. decrease in scanning volume
    e. all of the above

41. DRRs are used for _____.
    a. comparison with portal image
    b. setup verification
    c. beam arrangement
    d. patient dose calculations
    e. all of the above

42. Which of the following statements regarding virtual simulation (VS) are true?
    a. It is based on software and patient data without the presence of an actual patient.
    b. It is always based on CT data.
    c. It can recreate treatment machine capabilities.
    d. It must allow import of data from any data acquisition unit.
    e. It can be integrated with CT.

43. Which of the following statements are true about lasers?
    a. They are used in simulation and treatment rooms for accurate and reproducible patient positioning.
    b. They help accurately determine the isocenter of the treatment delivery equipment.
    c. They are used to mark the treatment field size on the patient.
    d. They are used for patient marking (with permanent tattoos).
    e. They must be regularly checked for quality assurance.

44. Which of the following statements are true with regard to errors in patient treatment?
    a. The treatment delivery error can be beam related or patient related.
    b. Errors in field size, field placement, MLC error, etc. are beam related.
    c. Changes in PTV size, planning CT error, etc. are patient-related errors.
    d. It is possible to distinguish between beam- and patient-related errors.
    e. All of the above are true.

45. An AP MV image of fiducial in prostate at 12 cm depth was taken delivering 3 MUs. Given the following beam data:
    $$TMR(12, 15 \times 15) = 0.752; OF = 1.071$$
    $$Dcal(100, 10 \times 10, d_{max}) = 1 \ cGy / Mu$$
    The dose delivered for taking the verification image is _____.
    a. 2.4
    b. 1.075
    c. 1.0
    d. 0.75
    e. 0.4

46. The advantages of kV CBCT over MV portal radiograph are _____.
    a. lower radiation dose
    b. better identification of soft-tissue targets
    c. superior image quality
    d. 3D versus 2D images
    e. kV inherently better contrast than MV imaging
    f. all of the above

47. Which of the statements regarding kV CBCT treatment verification imaging are true?
    a. Dose delivered in kV CBCT imaging is usually considered to be negligible.
    b. Organ doses from kV CBCT systems are typically <3 cGy/fraction.
    c. With daily imaging, doses can approach 1 Gy.
    d. MV imaging delivers less dose compared to kV imaging.
    e. All of the above are true.

48. The advantages of on-board volumetric imaging over fiducial localization are _____.
    a. the ability to detect prostate deformation
    b. real-time replanning
    c. visualization of adjacent normal critical structures
    d. yielding functional information
    e. all of the above

49. Important considerations in CT simulation are _____.
    a. treatment position reproducibility
    b. accurate tumor localization
    c. CT number accuracy
    d. patient dose
    e. treatment site

50. Some of the important features of a simulator are _____.
    a. use of 3D image data set to generate verification DRRs
    b. external lasers for patient marking
    c. flat table top
    d. virtual simulation software
    e. scanning capability to generate patient data
    f. all of the above

51. Match the positioning lasers to the defining planes.
    Laser   *Defining Plane*
    a. vertical wall laser   _____ i.   sagittal plane
    b. horizontal wall laser   _____ ii. transverse plane
    c. ceiling laser   _____ iii. coronal plane

52. DCR _____.
    a. refers to a digitally composited radiograph
    b. is another name of DRR
    c. means transmission radiograph
    d. is similar to a DRR but certain tissue types are suppressed or intentionally enhanced to create an image
    e. is a digitized image created by a CR system

53. The CT simulation software calculates the geometric center of the tumor. _____ cross-sectional images are required to identify the (x,y,z) coordinates of the tumor center.
    a. Only X-Y (transverse) plane images
    b. Only X-Z (coronal) plane images
    c. X-Y and X-Z plane images
    d. Only Y-Z (sagittal) plane images
    e. (X-Y) and (Y-Z) plane images

54. A moving target _____.
    a. introduces motion artifacts in CT imaging
    b. increases treatment volume, thus irradiating a larger volume of normal tissue
    c. needs 4D imaging and gated therapy for optimal imaging and treatment delivery
    d. occurs in the treatment of lung lesions
    e. can be made static by immobilization devices
    f. makes all the above true

55. The methods used to manage motion in radiation therapy are _____.
    a. ignoring the motion and setting CTV-PTV margin liberally to account for the target motion
    b. restricting motion by holding breath, by compressing abdomen, by fixing body, etc.
    c. tracking target motion and adapting it for treatment dose delivery
    d. 4D imaging and treatment delivery using gating techniques
    e. all of the above

56. 4D imaging refers to _____.
    a. imaging a patient with respect to four space coordinates
    b. imaging a patient with respect to two space coordinates and two time coordinates
    c. 3D body imaging at two different times
    d. 3D body imaging with the inclusion of temporal changes into patient anatomy
    e. all of the above

57. A moving target (e.g., respiratory motion of a lung lesion) may be imaged by _____.
    a. fluoroscopy observed directly or inferred from the motion of a surrogate
    b. radiography
    c. ultrasound
    d. CT
    e. all the above

58. In a respiratory motion of lung tumor observed in fluoroscopy, the greatest amplitude is in the _____.
    a. superior-inferior (cranio-caudal) direction
    b. medial (medio-lateral) direction
    c. dorsal (dorso-ventral) direction
    d. same direction for all patients
    e. superior-inferior direction, as long as the patient is standing while breathing

59. State which of the following statements are true regarding the tumor influenced by respiratory motion?
    a. The scanning speed of CT would influence the image of the tumor.
    b. A slow CT scan (<< tumor motion) results in a smeared tumor image.
    c. A fast scan (>> tumor motion) results in a frozen image of the tumor motion at the existing breathing phase.
    d. A medium speed scan ($\approx$ tumor motion) results in heavily distorted tumor position and shape.
    e. A free breathing scan will be greatly distorted and displaced.
    f. All of the above are true.

60. The Varian Trilogy linac is equipped with a gantry-mounted imaging system known as on-board imager (OBI). This system has following capabilities:
    a. It can do dynamic targeting in image-guided radiotherapy (IGRT).
    b. The OBI has a kV x-ray source and a flat panel detector (FPD).
    c. It is an MV portal imaging device.
    d. The kV imaging and MV imaging systems are parallel to each other.
    e. All of the above are true.

61. Which of the following statements are true regarding the Trilogy system?
    a. The kV and MV planar radiographic imaging is possible for patient repositioning.
    b. The kV cone beam CT (CBCT) 3D imaging is possible for patient repositioning.
    c. The kV planar fluoroscopic imaging is possible for gating verification.
    d. Its MRI imaging checks for localization errors.
    e. All of the above are true.

62. The advantages of on-board kV imaging over MV imaging are _____.
    a. better bone/soft tissue contrast
    b. less radiation dose
    c. real time imaging
    d. fluoroscopic imaging capability
    e. all of he above

63. The important advantages of electronic portal imaging devices are _____.
    a. verification of patient setup
    b. verification of individual field placements
    c. portal dosimetry
    d. treatment MU calculations
    e. all of the above

64. Match the stages to the IGRT treatment process.

Stage		*Treatment Process*
a. Stage 1	_____	i. image reconstruction
b. Stage 2	_____	ii. patient alignment
c. Stage 3	_____	iii. image acquisition
d. Stage 4	_____	iv. image registration

65. In planar image-based IGRT, _____.
    a. kV planar image or an MV planar image is compared with planning CT DRR
    b. misalignment parallel to the imaging plane is identified
    c. misalignment in the direction perpendicular to the imaging plane is identified
    d. superposition of anatomy in the projected image obscures the anatomy shift
    e. all the above are true

66. Which of the following statements are true regarding CT-based IGRT?
    a. CT-based imaging provides 3D patient data, enabling patient position adjustment in 3D.
    b. There is no superposition of anatomy, as in planar imaging, providing better spatial resolution.
    c. CT can be a stand-alone CT in the treatment room or it can be linac mounted.
    d. Both kV and CBCT verification images are used for correction of intrafraction errors.
    e. All of the above are true.

67. Respiratory-gated radiotherapy has the potential for _____.
    a. dose escalation to the target
    b. increased normal tissue sparing by decreasing the CTV-PTV margin
    c. treating patients having large respiratory motion amplitude
    d. treating patients who respond to breath coaching
    e. all the above

68. The American Association of Physicists in Medicine (AAPM) Report No. 91 recommends that respiratory management techniques be considered if _____.
    a. the GTV-CTV margin is greater than 7 mm
    b. the treatment delivery system has CBCT imaging facility
    c. >5 mm range of tumor motion is observed in any direction
    d. a significant normal tissue sparing as determined by the clinic can be gained through the use of a respiration management technique
    e. all of the above

69. A CT window of (0,500) CT numbers are displayed on a CT window. Which of the following statements are true in the display of these numbers?
    a. The display range covers a width of 500 HU.
    b. The display is centered at 250 HU.
    c. Values in the range from –250 to +250 HU are mapped into gray shades.
    d. Values below –250 HU are displayed in black.
    e. Values above +250 HU are displayed in white.
    f. All the above are true.

70. A virtual simulation system can _____.
    a. plan the treatment fields for a virtual patient using the scanned patient data
    b. display the room's view
    c. display the beam's eye view
    d. show the PTV and the organs at risk in 3D, suppressing other unnecessary bone/tissue data
    e. deliver the treatment as planned
    f. do all the above

71. Which of the following statements regarding treatment verification are true?
    a. A reference image acquired during the planning stage is required to compare with the verification image.
    b. The verification image is obtained in the treatment room prior to treatment or during the treatment fraction.
    c. The purpose of verification is to adjust the treatment geometry if the discrepancy between planning and execution exceeds an action level.
    d. The verification can be 2D or 3D systems to give different degrees of translational or rotational setup adjustment.
    e. The verification can be on-line or off-line.
    f. All of the above are true.

72. Methods for reducing setup uncertainties are _____.
    a. IMRT treatment
    b. Monte Carlo-based dose calculation algorithms in TPS
    c. patient immobilization
    d. image fusion
    e. in-room image guidance

73. During treatment delivery, the treatment field can be slightly offset with respect to the PTV due to _____.
    a. movement of skin marks made on the patient
    b. movement of PTV due to physiological reasons
    c. error in setting up the patient
    d. field offset
    e. all of the above

74. Which of the following statements are true?
    a. Random errors blur dose distribution around the CTV.
    b. Systematic errors shift the dose distribution away from CTV.
    c. Random errors can be reduced to zero.
    d. Systematic errors can be reduced to zero.
    e. The uncertainty margins can be minimized by immobilization and organ motion tracking.

II. Localization ANSWERS

A. Treatment Planning Concepts

1. No

2. Yes

3. Yes

4. No

5. Yes

6. Yes

7. No   The two should be DICOM compatible.

8. Yes   Day-to-day images for the same treatment field can be compared.

9. Yes

10. No   It does. Reduction in normal tissue dose will help in escalating tumor dose.

11. Yes   See Khan (2010).

12. Yes

13. Yes

14. Yes

15. No

16. Yes

17. No   Simulation must simulate patient treatment in every respect.

18. Yes

19. No   Film processor is the main factor influencing the quality of a radiograph in many centers. So the film processor, too, must be quality controlled by film sensitometry.

20. No   It is very useful for comparing with a port radiograph or DRR for treatment verification. Many centers keep a record of treatment fields using portal films. They also form part of the patient record and will come in handy in case of any litigation.

21. No

22. Yes

23. No   The patient does not hold his/her breath during treatment.

24.   No   They are very much dependent on these factors.

25.   Yes   The margins cannot be set too conservatively.

26.   No   The margins must be optimally set.

27.   No   Immobilization devices are often necessary.

28.   No   For instance, there is no GTV when the tumor has been removed by surgery.

29.   No   This margin includes suspected subclinical tumor spread that is contoured by the radiation oncologist based on his or her clinical experience.

30.   Yes   Targets like prostate, lung, etc. are subject to internal physiological movements and are likely to move during treatment. Any internal margin (IM) must be set for this movement and the whole volume (CTV + ITV) must be adequately treated in radiotherapy. See the figure below from a Paul Keall presentation on the topic.

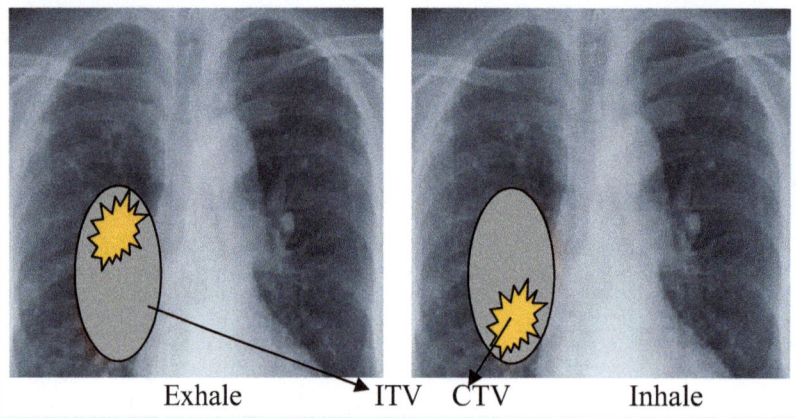

Generation of ITV

31.   No   The patient cannot be positioned identically every time with zero uncertainty. The laser alignment could be off with respect to the isocenter, so to treat (CTV + ITV), an additional setup margin, SM, that would take into account these uncertainties, must be included, leading to the concept of PTV. If we can treat PTV, the CTV would receive the prescription dose.

32.   Yes

33.   No   Unfortunately, there is some variability among radiation oncologists when they are asked to contour the same tumor. However, experienced radiation oncologists show good agreement in contouring a tumor. While planning, the physicist or the dosimetrist must assume that the given GTV is the "true value" for which the planning needs to be done.

34.   Yes   It is not included while defining the PTV. A margin is set around PTV to define the beam aperture, and this margin would account for the beam penumbra or the dose gradient region not influencing the PTV.

35.   No   A separate margin is set for the penumbra so that the dose gradient does not influence dose uniformity in the target volume.

36.   Yes   The prescription dose must tightly enclose PTV so that surrounding normal tissues do not receive high dose.

37. No   Modern linacs come equipped with kV x-ray imaging system and FPDs for daily image guidance. Linacs with CBCT attachment are now available while providing in-room 3D imaging.

38. No   We must deliver the prescription dose to the CTV and not just GTV (to avoid recurrence). This can be realized in the face of treatment uncertainties only when the PTV is subjected to the prescription dose.

39. No   Dephasing of M_{xy} signal depends on the spin-spin interactions, while the recovery of M_z signal depends on spin-lattice interactions. T_2 is typically much smaller than T_1.

40. Yes

41. Yes   Because no ionizing radiation is involved in MRI imaging.

42. Yes

43. Yes   This is very important for the success of radiotherapy. The whole planning goes to waste, leading to recurrence or complications, if the target is missed or normal tissues are irradiated for not conforming to the plan. The uncertainty in the dose delivery must be within the limits set by the uncertainty margin allowed in the treatment plan.

44. No   They are oncologic concepts and remain the same.

45. Yes   There are many ill-defined tumors due to diffuse infiltration or lack of sufficient tissue contrast between tumor and normal tissue.

46. No   It can be different in different directions. It depends on the nature of movement of the CTV.

47. No   Any in-field recurrence is due to inadequate dose to PTV and not inadequate volume coverage, assuming that PTV has taken into account all margins.

48. Yes

49. Yes   This has been demonstrated by J. Czernin et al. in *J. Nucl. Med.* 2007 48 Suppl 1:78S-88S.

50. No   PET can, but CT and MRI can't.

51. Yes   Only PET has this capability since it can image the behavior of cells at molecular level. In the future, this can lead to more targeted treatment of tumors.

52. No   It does. It can lead to the under-dosing of the tumor periphery or the overdosing of normal tissues near tumor periphery. So, for the best dose accuracy, the best geometric accuracy is essential. Instead of skin marks or bony structure-based verification, soft tissue-based alignment (like PET/CT) must be adopted to match planned dose to the tumor.

53. No   With on-board CBCT imaging systems, daily image guidance is available in the treatment room, if required.

54. No   It has. It takes into consideration the uncertainties in its location.

55. Yes

56. No   Both contribute to setup uncertainty. Errors in reproducible positioning of the patient lead to inter-fraction error.

57.   Yes   In radiotherapy, it commonly refers to imaging the patient in 3D in different phases of a breathing cycle. By restricting the treatment to certain phases of the breathing cycle, one can avoid irradiating larger volumes of normal tissues. This comes at the cost of more complex treatment technology and longer treatment duration.

58.   Yes

59.   No   Pixel size is determined by the FOV and the matrix size. System limiting spatial resolution depends mainly on the detector size.

60.   Yes   They can be used as surrogates for the organs or tissues and can be used for identifying setup errors by comparing against reference images. However, it requires an invasive procedure to implant the markers in the tissues of interest.

61.   Yes

62.   Yes

63.   No   It increases treatment time since only part of the breathing cycle is utilized for treatment.

64.   Yes   The lungs, esophagus, liver, pancreas, breast, prostate, and kidneys, among other organs, are known to move with breathing, and respiratory-correlated treatments are highly recommended while treating these organs.

65.   Yes   Tighter margins can be set provided we use gating techniques to treat such lesions.

66.   No

67.   No   Though the material is expected to be strong, it should also produce minimum beam attenuation, so they are usually made of low-Z materials like acrylic or carbon fiber.

68.   No   In MRI, the contrast depends on magnetic properties and not on electron density differences, as in CT. Therefore, by changing the magnetic properties, the contrast can be manipulated.

69.   No   Since the image data is 3D, the image for any plane can be reconstructed from the data by the software.

B. Acquisition of Patient Data

1.   d

2.   b, c, d

3.   d

4.   c

5.   a

6.   c

7.   b   This corresponds to an 8-bit depth. Modern CTs can reconstruct a 512 x 512 image matrix with 12 or 16 bits depth, but eye capacity to distinguish gray levels is very limited. The eye can distinguish structures only over a limited range of CT numbers, so to take advantage of the contrast resolution offered by the imaging capability of CT, the images for display have to be windowed. For instance, CT numbers 0 to +500 can display some soft tissue and bone. CT numbers −100 to +100 will display soft and adipose tissue. Most soft tissues in the human body lie in this range. For instance, to examine the liver in a CT image, the radiation oncologist usually windows the image with center at 80 HU with the black and white gray level set at 30 and 130 HU.

8.   a

9.   b

10.   b

11.   a, b, c, d

12.   a, b, c, d

13.   a, c, d

14.   b

15.   a, b

16.   d

17.   b

18.   c

19.   a, b, c

20.   b

21.   d

22.   b

23.   b

24.   c

25.   b

26.   b

27.   b

28.   a, c, d

29.   a, b, c

30.   c, d

31.   a

32. a, d

33. a

34. a, b

35. a, b, c

36. a, b

37. b

38. a, b, c   The reference dose is the dose at ICRU-defined reference point. See Hendee and Ibbott (1996).

39. a, b, c

40. a, b, d, e

41. b

42. c

43. c

44. b

45. a, b

46. a, b, c, d

47. b, c

48. a, b

49. a, b, c

50. d

51. e

52. d

53. a, b, c

54. e   The two margins are not added linearly, but generally in quadrature to give a more realistic margin. With the introduction of high-precision radiotherapy techniques, an idealistic margin would defeat the whole purpose of conformal therapy objectives.

55. d

56. d

57. f

58. d   This problem emphasizes the importance of reducing CTV-PTV margin as much as possible to reduce normal tissue complications, which will help to escalate tumor dose.

59. f

60. d

61. d

62. c   If the hot spot is less than this size, it is not clinically significant. However, if the organ itself is very small (e.g., eye, optic nerve, etc.) a size smaller than 15 mm is also considered significant.

63. a, b, e

64. f

65. b, c

66. c

67. b, c, d   The PTV concept was already there in the earlier protocols.

68. a

69. c

70. e

71. d

72. c

73. d

74. b, c

75. f

76. a, b, c

77. f

78. e

79. a, c, d   The mixing of tissue types occurs for larger voxels, hence less slice thickness leads to less partial volume effect.

80. f   Limiting FOV increases pixel size and hence the resolution.

81. c   For 4096, HU numbers with 12 bits are required. $2^{12} = 4096$.

82. b   Pixel size = 400 / 512 ≈ 0.8.

83. c   Each pixel can accommodate one line. An lp will occupy 1.6 mm. In 10 mm one can have 10/1.6 ≈ 6.3 lp or 0.63 lp/mm.

84. a, b, c

85. c

86.   a. Coil A iii.   b. Coil B i.   c. Coil C ii.

87.   d   From the nuclear magnetic resonance of protons in the applied magnetic field.

88.   a

89.   b

90.   e

91.   a, b, c

92.   e

93.   e

94.   c $\omega_0 = 2\pi f_0$

95.   c $1.5 \times 42.57 = 63.86$. This is called the center frequency of the scanner system.

96.   b, c, d   The magnetization is difficult to measure since this field is very minute compared to the scanner field. However, it becomes measurable if we flip this magnetization away from the direction of the scanner magnetic field, the main functioning principle of MRI.

97.   a, b, d   The contrast at TE_2 is larger compared to contrast at TE_1, as is clear from the figure.

98.   f

99.   a. iii.   b. i.   c. ii.

100. d   When the field is homogeneous, all the protons are having the same Larmor frequency. The gradient G_Z changes the effective field along Z, and hence the frequencies become a function of Z. Now the slice in question fixes the effective field and the Larmor frequency of the slice, and hence the RF pulse frequency and width that need to be applied for excitation of this slice. See the figure given in the problem.

101. b

102. a, c   It is easy to see that reducing pulse width reduces slice thickness. If you double the field gradient, the frequency range also doubles. So for the same pulse width, the slice becomes half. For example, if the gradient is 5 mT/m, a 10 mm slice will have a frequency range of 42, and 58 MHz/T $\times$ 5 mT/m = 2. 13 kHz. A pulse of this width can excite all protons in this slice. If the gradient is 10 mT/m, the range of pulse frequencies is 4.3 kHz, so the pulse of 2.13 kHz will excite protons in 5 mm slice thickness. Slice thickness is proportional to the bandwidth and inversely proportional to applied field gradient.

103. c   An RF pulse applied is not directed at any slice. It excites protons throughout the scanner bore. The field gradient G_S varies the Larmor frequency throughout the length of the patient. The RF will excite the protons in the slice whose frequencies are in resonance with the RF pulse frequencies.

104. c   The figure below explains several features of slice selection in MRI imaging. (The figures are adapted from a presentation by Xiaojuan Li, Dept. of Radiology, UCSF, USA.)

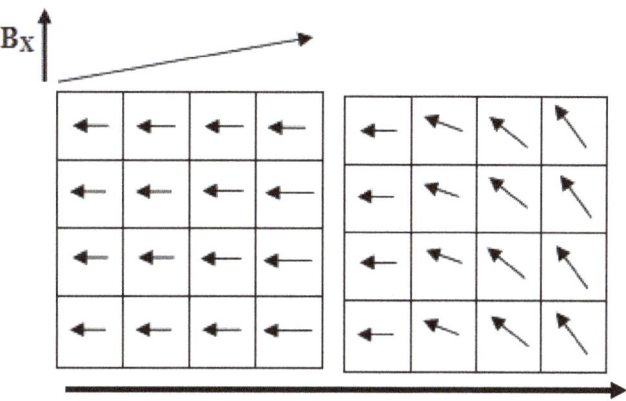

Phase gradient applied for a time ΔT

a) The slide has finite thickness (or width, say ΔZ) and so contains a range of frequencies centered on a certain frequency. This pulse can excite all the protons in this slice. for a given field gradient (say G_Z), reducing the pulse width of δf reduces the slice thickness. ($\Delta Z \propto \delta f$)

b) It can also be noticed that by keeping δf constant, the slice thickness can be varied by varying the field gradient. For instance, by doubling the field gradient, the slice thickness becomes half (see the figure below).

c) If the central frequency is 58 MHz/T, for the G_Z field, $\delta f = 58$ MHz/T $\times$ 5 mT/m = 2.13 kHz. If the gradient is 10 mT/m, the range of pulse frequencies is 4.3 kHz, so a pulse of 2.13 kHz will excite protons in 5 mm slice thickness. ($\Delta Z \propto 1/G_Z$).

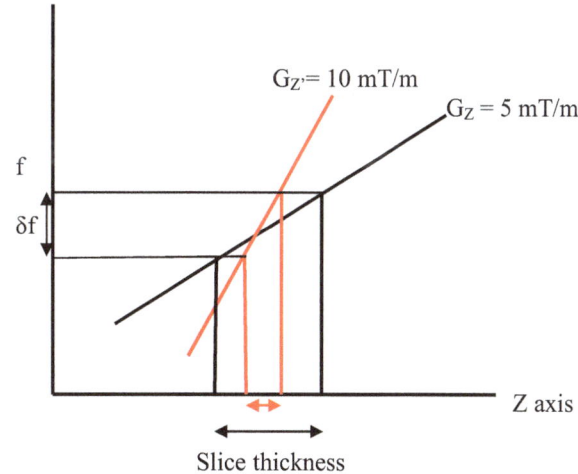

The data is acquired (signal recorded) while the frequency gradient is ON.

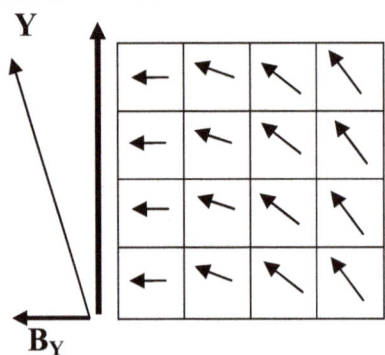

Application of frequency gradient along Y direction

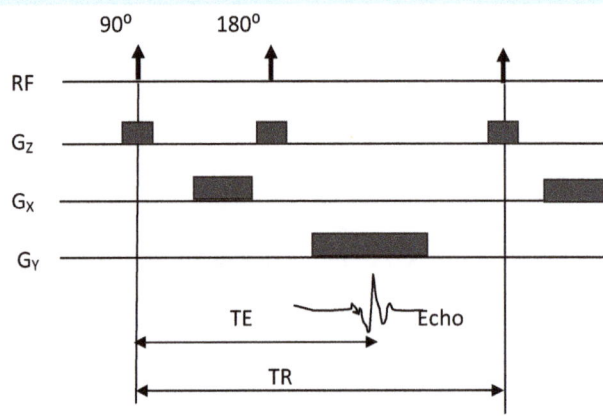

Pulse sequence applied for MRI imaging

A 90° RF pulse produces the flip of the magnetization vector onto the XY plane. The resulting dephasing signal decays fast. 180° pulse applied at TE/2 rephases the dephasing signal and gives us more time for signal encoding before the rephase signal can be measured. (Following figure taken from Yu Cao and Lili Chan presentation on MRI, Fox Chase Cancer Center, USA.)

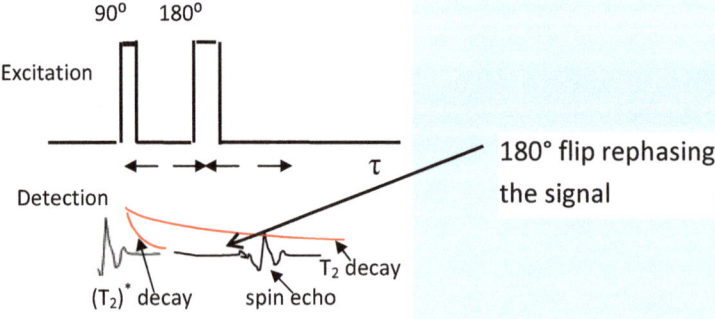

Phase gradient and frequency gradients applied and the data acquired during frequency decoding duration. The pulse sequence is repeated to read data of each line, before decoding pixel data.

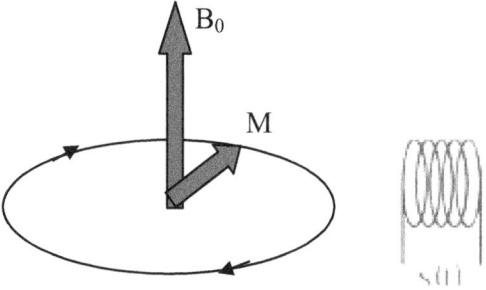

Receiver coil reading the dephasing signal data

If the image is a 256-by-256 matrix, 256 repeated pulse sequences will read 256 rows and the FT will yield the 256-by-256 pixel intensities.

105. a, b, c, d

106. e   Some studies have indicated that certain gadolinium agents have an element of risk to advanced kidney impaired patients. There are some recommendations to discontinue or restrict the use of these specific compounds, replacing them with more appropriate compounds recommended for patients with kidney diseases.

The MR signal strength depends on the oxygenation level. The blood and brain consume 20% of the oxygen that a person intakes, though the blood and brain constitute less than a few percent of the total body mass, so the neural activity can be imaged using MRI. This is called functional MRI or fMRI.

107. b, c, d   Though electron density information is not available, bulk density values can be used for treatment dose calculations for homogeneous regions (e.g., prostate, brain, etc.) with fairly accurate results (to about 2%) with MRI planning alone. For patient setup verification, insufficient bony structure information is available for verification.

108. e

109. e   Imaging of tumor vasculature and blood supply is performed by a new technique called dynamic contrast-enhanced MRI.

110. a, b, c, d

111. c

112. d

113. e

114. a, b   Sound can move faster in a medium where the particles (atoms or molecules) are closer together (denser) and stronger bound. This transfers momentum faster and also resists deformation. This is so in solids compared to liquids and gases, so sound would travel faster in bone than in air. In an easily deformable medium like rubber, sound would travel slowly.

115. a, d   v = υ λ. Velocity determines the λ. Frequency depends on the source and not the medium. Excitation frequency determines the vibration and hence the signal frequency. Penetration depends on the medium's attenuation characteristics. Velocity differences give differences in acoustic impedance and hence give signal contrast.

116. b

117. c   Attenuation of 4 MHz beam = 4 dB/cm or 32 in 8 cm.

118. b $\lambda = v / υ$. Express v in mm and υ in cycles/sec. (1 MHz = 10^6 cycles/sec)

119. c $\lambda = v / υ$ (both are given). Thickness for resonant frequency = T = λ / 2.

120. b   T α (1/υ). Doubling the frequency reduces thickness by half.

121. f

122. c, d   So, a matching element of required Z (gel) is used as an interface.

123. c   SPL = number of waves x λ

124. c   Half the spatial pulse length is the theoretical axial resolution. For example, in the first figure below (for a 5 MHz pulse) the axial resolution is sufficient to distinguish the 2 target objects as separate because the incident wave hits the first target before hitting the second.

In the second figure (2.5 MHz pulse, pulse length = 1.8 mm; 3 cycles), the axial resolution is no longer adequate. Because both target are hit by the same wave, they cannot be resolved and will appear as one. (Graphics courtesy of http://usra.ca/imageresolution.php.)

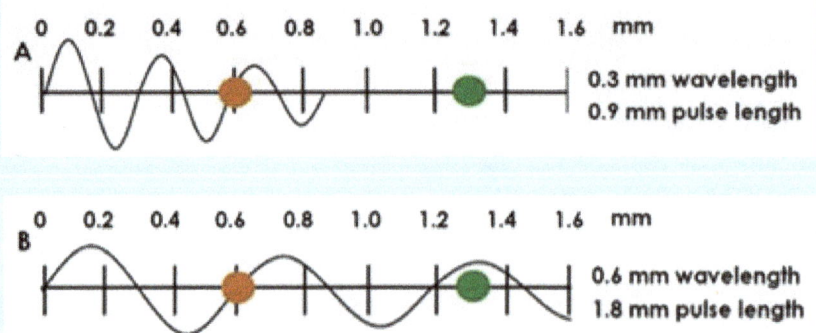

Pulse length and target resolution

The following figure shows the concept that when the pulses meet at FWHM point, the pulses will be just distinguishable as two maxima, i.e., λ/2.

Axial resolution concept

125. a, b, c   The figure below illustrates the concept of lateral resolution. The nature of the US beam shows the frequency and depth dependence of beam width (i.e., lateral resolution) and that the best resolution occurs at the near field and far field interface. One can get the best lateral resolution if the scan depth is adjusted to be the depth of the interface. (Graphics courtesy of http://usra.ca/imageresolution.php.)

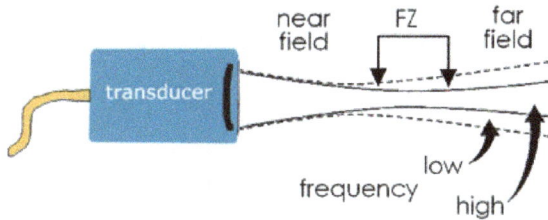

Concept of lateral resolution

126. e   For the same number of cycles, the largest frequency will have minimum special length.

127. d

128. f

129. c   Time to reach depth d = echo time / 2 = t_1 / 2. d = velocity $\times$ (t_1/2).

130. a   R of the pulse depends only on the difference in acoustic impedance and hence will be the same. $R = (Z_2 - Z_1)^2 / (Z_2 + Z_1)^2$

131. d Z_{air} is negligible. $(Z_{bone} - Z_{air})^2 / (Z_{bone} + Z_{air})^2 \approx 1$

132. e $Z_{liver} \approx Z_{spleen}$. $(Z_{liver} - Z_{spleen})^2 / (Z_{liver} + Z_{spleen})^2 \approx 0$

133. e $R = (0.01)^2 / (1.65 + 1.55)^2 \approx 1/1000$

134. b $Z = (\rho V) = 1.48 \times 10^6$. Use the same units (SI) for both quantities.

135. a, c, d   The impedance of intermediate layer must be $[Z_{transducer} \times Z_{tissue}]^{1/2}$.

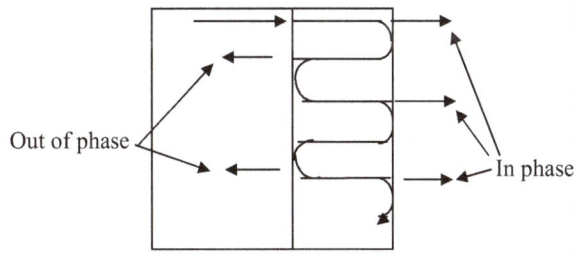

Out of phase   In phase

Interface layer transmitting US in phase

136. e

137. f

138. a, b, d

139. c   Frame rate = (1 / acquisition time per frame). The frame rate is between 15 to 40 frames/sec, which will permit motion to be followed.

140. b   PRF = 1/PRP = 2×10^3 / sec = 2 kHz.

141. e   Higher frame rates improve temporal resolution. Line density (scan lines per FOV) improves resolution in the scan plane. The S/N ratio decreases with depth, affecting image quality.

142. f   The S/N ratio decreases with depth affecting image quality. As shown in figures below, the damping of oscillations determine the pulse length and the bandwidth. Ideally one would desire a short pulse of resonance frequency which is not practical to achieve due to damping of resonance that is required.

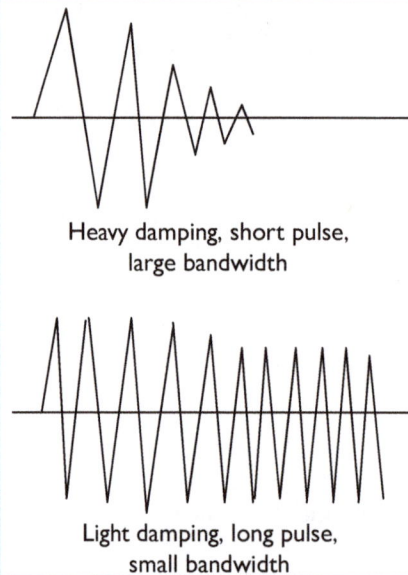

Heavy damping, short pulse,
large bandwidth

Light damping, long pulse,
small bandwidth

143. a   If v is the velocity of propagation in the crystal, $v = \upsilon \lambda$, and for resonance crystal thickness $T = \lambda/2$, i.e., $\upsilon = v/2T$. (Express the numerator and denominator in the same units as v on the LHS.)

144. c $Q = v_0 / \Delta v$

145. c

146. b, c, d   PET cannot detect positrons since they have negligible range in tissue. PET relies on detection of photons liberated by positron annihilation with electrons. Positrons coming to rest at the end of the range have zero momentum, so the annihilation photons created must have net momentum of zero, i.e., photons are emitted in opposite directions at 180° angle and simultaneously. Detection of this pair and subsequent mapping of the event of origin images the system. The principle is illustrated in the following figure.

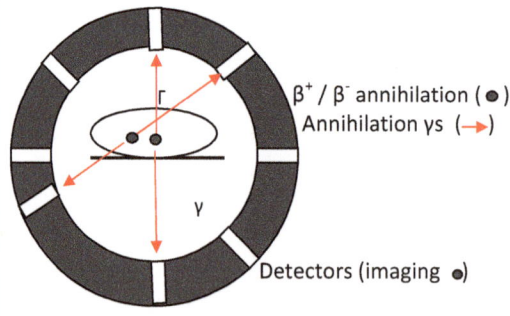

β^+ / β^- annihilation (●)
Annihilation γs (→)

Detectors (imaging ●)

Principle of PET scanning

147. e   By their rates of uptake, we can identify malignancies and other physiological functions occurring in various regions.

148. a, b, d, e
Without PET, there is larger uncertainty in marking the GTV-CTV margin. Distant metasatsis, nodal involvement, etc. not showing up on CT are easily identified on PET due to increased activity uptake in these regions. See the figures below (from A. Martinez et al., MEDICAMUNDI 54/2 2010, with kind permission).

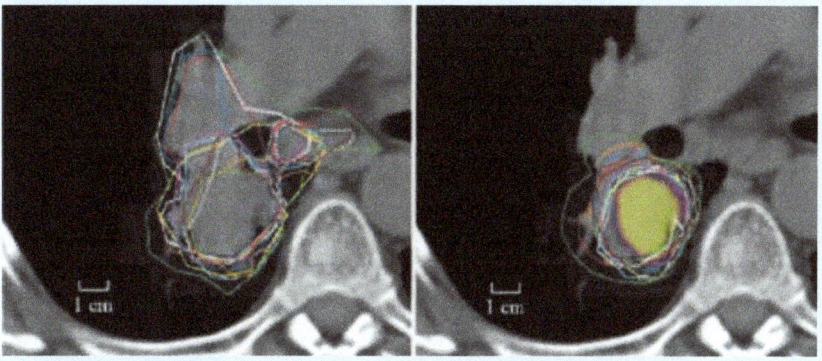

Contouring based on CT data   Contouring based on PET and CT data

Image segmentation

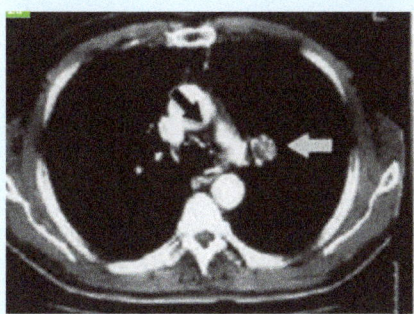 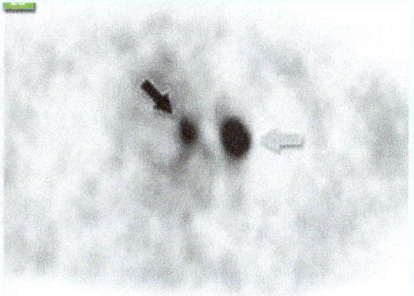

CT imaging shows no nodes   Nodal involvement (block arrow), near primary tumor (white arrow).

Biological information based on PET imaging

149. a, c, d   The target volume can also decrease from PET information. PET is useful for follow-up studies.

C. Patient Positioning and Immobilization

1.   b

2.   b

3.   c

4.   b

5.   b

6.   c

7.   d

8.   c

9.   c

10. b

11. c, d

12. a, b, d

13. a

14. d

15. a, c

16. b

17. b

18. c

19. b

20. d

21. d

22. b

23. c

24. c

25. d

26. b

27. a, c

28. a, c

29. a, b, c

30. a, c

31. a, b, c

32. a, b, c

33. c

34. b, c   Answers a, d, and e are accounted for in internal margin.

35. a   The rest are setup errors.

36. a, b, c

37. e

38. c

39. e

40. e

D. Ancillary Treatment Devices

1. c

2. a

3. a, c

4. a

5. c

6. a, b, c

E. Treatment Simulation and Verification

1. a, c, d

2. a, c, d

3. b   X-rays, being in the diagnostic range, give better contrast and visibility compared to higher-energy treatment beams.

4. a

5. c

6. a, d

7. a, b, c, d

8. a

9. b, d

10. b

11. a

12. b

13. a, b   For further study, see volume 1 of Washington and Leaver (1996).

14. a, b, c, d

15. a, d   The grid does cut off some primary radiation and, of course, the scatter. Grid absorption warrants an increase in technique factors, so c is wrong.

16. a, b, c

17. a, c

18. c, d

19. a, b, d

20. c

21. a, b

22. a, c, d

23. a, b, c

24. a, b, c

25. a, b

26. a

27. a, b

28. a. iv.   b. ii.   c. iii.   d. i.

29. b, d

30. a. ii.   b. i.   c. v.   d. iii.   e. iv.

31. e

32. a, b, c

33. f

34. e

35. e

36. b, c, d   Simulation can be accomplished with a radiotherapy simulator, but not only with that device. Ultrasound is generally used for target localization, and only for certain sites.

37. d

38. a, b, d, e
    The SFOV is in the range 48 to 60 cm.

39. f

40. a, b, c   Since one rotation covers multiple slices, the volume scanned per rotation increases.

41. a, b, c

42. a, c, d, e

   VS need not be CT based. It must be only capable of receiving data from any data acquisition unit and manipulate and display images for planning the fields.

43. a, b, d, e

44. e

45. a

46. f

47. a, b, c   See http://www.aapm.org/meetings/amos2/pdf/60-15015-86048-515.pdf.

48. a, b, c

49. a, b, c

50. a, b, c, d

51. a. ii.   b. iii.   c. i.

52. a, d

53. e

54. a, b, c, d

55. e

56. d

57. a, d

58. a

59. f

60. a, b, c   The kV and MV systems are orthogonal to each other, and three robotic arms independently position the two imaging systems (kV source, kV imager and MV imager), whenever imaging is required. Here we have given the Varian machine as an example. Other linac makes also have such facilities for delivering IGRT treatments.

61. a, b, c

62. a, b, d   The kV imaging system is orthogonal to the MV imaging system, so it cannot do real-time imaging.

63. a, b, c   EPID cannot be used for MU calculations in treatment delivery. However, since it is capable of portal dosimetry, it can be used for the verification of IMRT fluence or MUs delivered.

64. a iii.   b i.   c iv.   d ii.

65. a, b, d   In planar imaging, shift parallel to the imaging plane only can be identified. To identify a shift in a direction perpendicular to the imaging plane, we need an image parallel to this plane (i.e., we need an orthogonal image).

66.   a, b, c

67.   e   The figure below explains the concept of respiratory gated therapy.

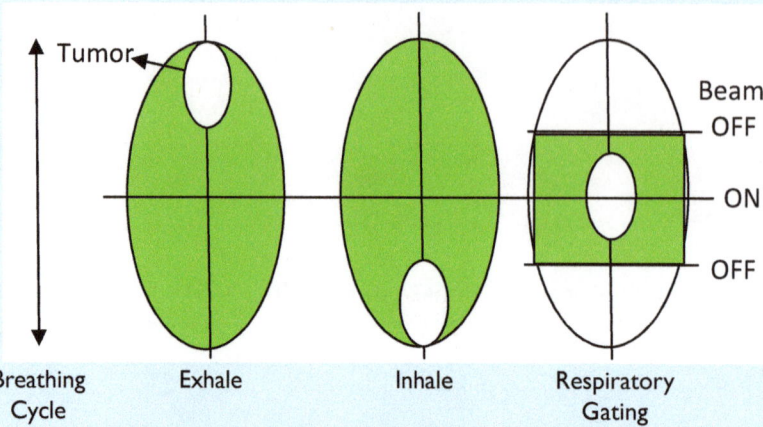

Reducing CTV-ITV margin by respiratory gating

68.   c, d   The 5 mm limit may be further reduced in case of special procedures like SBRT.

69.   f

70.   a, b, c, d

71.   f

72.   c, e

73.   e

74.   a, b, e

III. Treatment Planning, Techniques, and Delivery

Introduction

Treatment planning plays a very important role in radiation therapy. It is a very broad term, and one can include practically all steps in radiotherapy except treatment delivery under treatment planning. The more restricted meaning is, of course, the use of a treatment planning system (TPS) for treatment calculations. In this section we cover some basic aspects of treatment planning, various advanced treatment techniques, and some treatment delivery aspects. The treatment methodology in radiotherapy is based on radiobiology, so questions on this topic are also added in this section.

A. Treatment Planning Concepts

Circle the right answer (Yes or No):

1. (Yes / No)   The aim of treatment planning is to plan a treatment that gives the prescribed dose to the target volume of interest and no dose to the surrounding normal tissues.

2. (Yes / No)   "Tumor localization" refers to the quantitative evaluation of the size, shape, and position of the target volume in a patient.

3. (Yes / No)   Organs at risk (OAR) are the critical normal tissues or structures whose irradiation may significantly influence the treatment plan or dose prescription.

4. (Yes / No)   Most of the tumors are treated using single fields.

5. (Yes / No)   Dose-volume histograms for the target volumes and organs at risk should be reported for 3D CRT treatments.

6. (Yes / No)   The image data storage format is not the same for all CTs.

7. (Yes / No)   TPS systems can accept data from any model of CT.

8. (Yes / No)   Electronic portal imaging can be used for the evaluation of setup errors.

9. (Yes / No)   Treatment planning without CT can lead to inadequate coverage of PTV or larger coverage of PTV than necessary.

10. (Yes / No)   3D-CRT will not affect TCP or the cure rate compared to conventional treatment.

11. (Yes / No)   Digitally reconstructed radiographs (DRRs) can be compared with simulator films to verify the accuracy of beam placement.

12. (Yes / No)   DRRs can be compared with port films to ensure the accuracy of treatment delivery.

13. (Yes / No)   From transverse CT slice images, the CT image for any arbitrary plane can be reconstructed.

14. (Yes / No)   A beam's-eye view (BEV) helps the oncologist to see on the monitor that the beam adequately covers the target volume.

15. (Yes / No)   The user can choose different calculation algorithms in the TPS depending on the accuracy required.

16. (Yes / No)   The accuracy of dose calculations performed using TPS depends on the accuracy of the input data provided by the user.

17. (Yes / No)   Patient immobilization is required only during treatment and not during simulation.

18. (Yes / No)   The patient support assembly (or the couch top) must be identical during CT, simulation, and treatment for accurate planning and dose delivery.

19. (Yes / No)   Production of a quality radiograph in a simulator depends only on the technique factors used in the simulator.

20. (Yes / No)   Since the image of a treatment portal can be viewed on a monitor, it is not necessary to have the radiography mode in simulators.

21. (Yes / No)   Diagnostic CT can be used for the purposes of CT planning.

22. (Yes / No)   DRRs are digitally reconstructed radiographs generated for any patient using the axial scan data.

23. (Yes / No)   The patient must hold his breadth while taking CT slices for planning.

24. (Yes / No)   The positions of organs in the body, in a CT slice, are independent of patient couch top or the patient breathing pattern.

25. (Yes / No)   PTV-CTV margin, when conservatively set, can lead to recurrence.

26. (Yes / No)   PTV-CTV margin must always be liberally set for optimal patient treatment.

27. (Yes / No)   Patient movements can be eliminated during CT scanning or treatment simply by following the instructions of the radiation oncologist or the physicist.

28. (Yes / No)   Daily verification of patient positioning is not necessary in precision radiotherapy.

29. (Yes / No)   A systematic error is one which can be eliminated by repeating the measurement.

30. (Yes / No)   Patient positioning error is a random error.

31. (Yes / No)   Forgetting to insert the wedge in a wedge treatment is a systematic error.

32. (Yes / No)   Comparison of "Simulator-EPID" images and EPID-EPID images will reveal errors in planning stage to treatment stage and reproducibility of patient treatments.

33. (Yes / No)   If a patient cannot be positioned for treatment as planned on the simulator, the patient must be treated in a geometry that is nearest to the simulation geometry.

34. (Yes / No) One verification image obtained prior to the start of radiotherapy is enough to correct for setup uncertainties by comparing with the simulator image.

35. (Yes / No) The estimation of systematic and random errors can help to reduce the CTV-PTV margin.

36. (Yes / No) Portal imaging study and onboard imaging study can provide data to set the limits of random and systematic components of error.

37. (Yes / No) The changes in PDD characteristics become less pronounced as obliquity angle increases.

38. (Yes / No) Any material can be used as bolus in electron beam therapy.

39. (Yes / No) Bolus must be placed close to the patient surface in electron beam therapy.

40. (Yes / No) Internal shields are used to protect structures beyond the target volume.

41. (Yes / No) Electron arc therapy is most suited for treating superficial tissue layers lying under curved surfaces like chest wall.

42. (Yes / No) The PDD and output factors are not field size dependent in electron beams.

43. (Yes / No) The convenient method of measuring dose distributions of electron beams is by using an ion chamber in a water phantom.

44. (Yes / No) The tail of an electron PDD curve is caused by stray electrons in the medium.

45. (Yes / No) The electron beam can produce Bremsstrahlung only in the linac head and not in the patient.

B. Photon Beam Isodose Curve Parameters

Choose the right answer(s) (more than one may be correct):

1. The depth of maximum absorbed dose, d_{max}, along the beam central axis, is _____.
   a. the reference depth for treatment time or monitor unit calculation
   b. the reference depth for the calibration of clinical photon and electron beam
   c. independent of field size
   d. independent of beam quality

2. PDD is normalized at _____.
   a. d_{max} for SSD technique
   b. target center for isocentric techniques
   c. a reference depth of 10 cm
   d. none of the above

3. PDD _____.
   a. is normalized to 100% at d_{max}
   b. increases with depth, beyond the buildup region
   c. increases with field size
   d. is independent of SSD

4. The field size dependence of PDD occurs due to _____.
   a. increased collimator photon scatter that reduces the effective energy of the beam
   b. increased electron scatter that influences the dose on skin and in buildup depths
   c. increased phantom scatter
   d. none of the above

5. PDD depends on _____.
   a. the source or machine output
   b. patient size
   c. source decay in case of ^{60}Co machines
   d. none of the above

6. Surface dose _____.
   a. increases with field size
   b. increases with beam quality
   c. is more for electron beams compared to photon beams
   d. can be measured using a Farmer chamber

7. Surface dose for clean clinical photon beams is _____.
   a. about 10% to 50%
   b. about 5% to 10%
   c. zero
   d. more than 70%

8. Electron contamination _____.
   a. increases the surface dose
   b. increases dose in the buildup region
   c. decreases for increasing diaphragm skin (or surface) distance
   d. increases with field size

9. Dose buildup (in the buildup region) _____.
   a. is not linear
   b. is very steep in the initial portion of buildup
   c. is more pronounced for lower beam qualities
   d. leads to skin sparing

10. From isodose curves measured, e.g., in the principal plane, one can determine _____.
    a. central axis depth dose values
    b. cross beam profiles at any depth
    c. beam penumbra
    d. beam profile in any plane of interest

11. Standard isodose distributions are measured _____.
    a. in a water phantom.
    b. in a water-equivalent phantom
    c. in an anthropomorphic phantom
    d. for normal incidence of beams on a flat entry surface of the phantom

12. The shape of an isodose curve is influenced by _____.
    a. phantom scatter
    b. variation of beam quality at off-axis points
    c. side scatter at beam edges
    d. SSD

13. Isodose curves are normalized at _____.
    a. phantom surface
    b. peak depth for SSD type of treatments
    c. isocenter for SAD treatments
    d. any depth of convenience

14. Dose beyond the beam edge is influenced by _____.
    a. photon side scatter
    b. electron transport
    c. collimator transmission
    d. inverse square law variation

15. Isodose distribution depends on _____.
    a. beam quality
    b. SSD
    c. field size
    d. machine type

16. Isodose distributions, for commissioning purposes, are usually measured using _____.
    a. an RFA analyzer
    b. a miniature ionization chamber or semiconductor detectors
    c. films
    d. none of the above

17. By comparing isodose charts of different beam qualities, one can see _____.
    a. dose outside beam edges is greater for kV x-ray beams because of higher penetration through x-ray diaphragms
    b. dose inside beam edges and at depths of clinical interest is more uniform for linac photon beams compared to kV x-ray beams
    c. dose flatness is independent of depth
    d. dose fall with depth is more rapid for kV photon beams

18. The penumbra of the beam at any depth in phantom _____.
    a. can be determined from the cross-beam profile at that depth
    b. is usually defined as $P_{80/20}$ or the distance between 80% and 20% dose values
    c. depends on source size
    d. is simply the geometric penumbra

19. The exit dose of the patient calculated from the standard depth dose data, compared to the actual dose, is _____.
    a. less
    b. more
    c. the same

20. Dose to the patient's skin and buildup region _____.
    a. increases with increasing field size
    b. decreases with increasing angle of incidence
    c. increases with the introduction of a blocking tray
    d. decreases with decreasing skin-tray distance

21. For megavoltage energies, the skin dose can be decreased and buildup dose profile improved by _____.
    a. keeping suitable filters (e.g., lead glass, lead foil, or copper) in the beam path, below the blocking tray
    b. keeping filters close to the skin
    c. using filters of about 1 to 2 mm thickness
    d. using filters of 1 or 2 cm thickness

22. The depth dose curves of different accelerators with the same nominal energy (MV) are not the same because of _____.
    a. variation in energies of electrons striking the target
    b. the type of target and flattening filter used
    c. the materials and design of the collimator used
    d. the type of microwave power sources (klystron or magnetron) used

C. Photon Beam Isodose Distributions

Choose the right answer(s) (more than one may be correct):

1. While measuring isodose distributions using a water phantom, the water phantom must extend at least _____ cm (laterally) beyond the field or the depth of measurement.
    a. 1 cm
    b. 5 cm
    c. 10 cm
    d. none of the above

2. In parallel opposed fields, treatment _____.
    a. target dose (midline dose) is more than the dose to surrounding normal tissues
    b. dose to peripheral normal tissues can be significantly higher than the midline dose
    c. dose uniformity along central axis (CA) improves with increasing patient thickness
    d. dose uniformity along CA improves with increasing beam quality

3. While treating pelvis with parallel opposed fields, _____.
    a. a 10 MV beam is preferable to a ^{60}Co beam
    b. both give rise to the same dose distribution
    c. a peripheral dose is larger for a ^{60}Co beam
    d. none of the above is true

4. In a parallel opposed equally weighted constant SSD field treatment, the dose at d_{max} depth due to each field is _____.
    a. 100%
    b. <100%
    c. >100%
    d. none of the above

5. In a treatment of the abdomen using a set of POP fields, _____.
    a. full isodose distribution in the treatment plane is required
    b. isodose distribution in more than one plane is required
    c. dose to the midplane or midpoint of patient is sufficient
    d. dose at several points in the treatment plane is necessary

6. In a constant SSD treatment involving a set of POP fields in A/P and P/A directions, PDD in the patient midplane due to a single field is 70.5%. The total PDD at the tumor center is _____.
    a. 0%
    b. 70.5%
    c. 141%
    d. none of the above

7. In a treatment involving a set of POP fields in A/P and P/A directions, the PDD at the exit point due to the P/A field is 47%. The total PDD at d_{max} in the anterior direction is _____.
    a. 47%
    b. 53%
    c. 100%
    d. 147%

8. In a treatment involving a set of POP fields, the total PDD at tumor center is 150%. If the dose rate at the input ports (d_{max} depths for the respective fields) is 1.02 cGy/MU, the dose rate at the tumor center is _____ cGy/MU.
    a. 0.77
    b. 1.02
    c. 1.53
    d. 2.04

9. The advantages of multiple field treatments are _____.
    a. maximization of target dose
    b. minimization of normal tissue dose
    c. applicability in all clinical situations
    d. none of the above

10. The figure below shows a three-field treatment utilizing SSD technique. The target dose to be delivered in one fraction is 200 cGy. All three fields are weighted equally (at d_{max}).

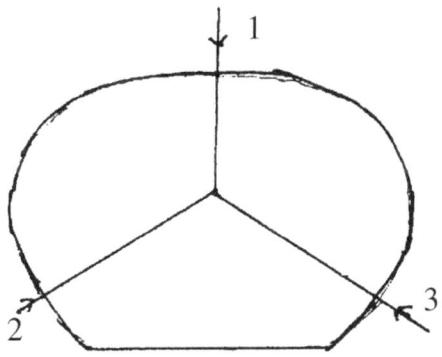

Three-field isocentric treatment

    The PDDs at the target center for the three fields are 65%, 50%, and 50% respectively. The dose at the input port (in cGy) is given by _____.
    a. 200
    b. 200×0.5
    c. 200×0.65
    d. $200 / 1.65$

11. In the above problem, the fields are weighted as follows:
Field 1 : Field 2 : Field 3 = 1 : 0.5 : 0.5.
To deliver 200 cGy to the target center, the dose at the three input ports is respectively given by _____.
   a. 200 / 1.15 $(200 \times 0.5) / 1.15$ $(200 \times 0.5) / 1.15$
   b. 200   200 / 0.5   200 / 0.5
   c. 200 200×0.5 200×0.5
   d. 200×1.15 200 / 0.5   200 / 0.5

12. In the above problem, the dose to the target center is normalized to 100%.
   i.   The renormalized PDDs at the target center for the three fields are given by, respectively, _____.
   a. $100 \times 65 / 115$ $100 \times 25 / 115$ $100 \times 25 / 115$
   b. 100 / 65   100 / 25   100 / 25
   c. $100 \times 115 / 65$ $100 \times 115 / 25$ $100 \times 115 / 25$
   d. none of the above

   ii. The renormalized PDDs at d_{max}, for the three fields, are respectively given by _____.
   a. 100 / 115 $100 / (115 \times 50)$ $100 / (115 \times 50)$
   b. $100 \times 100 / 115$ $100 \times 50 / 115$ $100 \times 50 / 115$
   c. none of the above

13. In problem 11 above, all fields contribute equally to the tumor. The dose at d_{max} for the three fields is respectively given by _____.
   a. $200 \times 100 / 65$ $200 \times 100 / 50$ $200 \times 100 / 50$
   b. 200×100 200×50 200×50
   c. $200 / 3 \times 100 / 65$ $200 / 3 \times 100 / 25$ $200 / 3 \times 100 / 25$
   d. $200 \times 100 / 165$ $200 \times 50 / 165$ $200 \times 50 / 165$

14. In a three-field isocentric setup using a ^{60}Co beam, a dose of 200 cGy is delivered at the target center. The TAR for the three fields is given by 0.707, 0.671, and 0.785. The "given dose" for the three fields, assuming equal doses, is given by _____.
   a. 200 / 2.163 $200 \times 0.671 / 2.163$ $200 \times 0.785 / 2.163$
   b. 200 / 2.163   200 / 2.163   200 / 2.163
   c. 200×0.707 200×0.671 200×0.785
   d. none of the above.

15. In the above question, if each field contributes equally to the target dose, the "given dose" for the three fields is respectively given by _____.
   a. 200 / 0.707   200 / 0.671   200 / 0.785
   b. $200 \times 0.707 / 2.163$ $200 \times 0.671 / 2.163$ $200 \times 0.785 / 2.163$
   c. $(200 / 3) / 0.707$ $(200/3) / 0.671$ $(200/3) / 0.785$
   d. none of the above

D. Electron Beam Dose Distributions

Choose the right answer(s) (more than one may be correct):

1. Electron beam dose distributions can be measured using _____.
   a. an ionization chamber
   b. a semiconductor diode
   c. film
   d. $FeSO_4$ dosimeters

2. _____ dosimeters can be used for the calibration of electron beams.
    a. Ionization chamber
    b. Diode
    c. Film
    d. TLD

3. The therapeutic depth (along beam central axis) for dose prescription is given by _____.
    a. depth of 90% of isodose level
    b. depth of 50% of isodose level
    c. peak depth
    d. none of the above

4. The skin dose in the case of electron beams is about _____.
    a. 80% to 90%
    b. 30% to 40%
    c. 100%
    d. none of the above

5. The therapeutic depth (depth of the 90% depth dose) in cm for a clinical electron beam of most probable energy E (MeV) is approximately given by _____.
    a. one-fourth of E
    b. one-half of E
    c. d_{max}
    d. none of these values

6. The depth of the 80% isodose line for a clinical electron beam of energy E (MeV) is approximately given by _____.
    a. one-fourth of E
    b. one-third of E
    c. one-half of E
    d. none of these values

7. A lesion at shallow depth has to be treated so that the 90% isodose line enclosing the lesion is at 3 cm depth. The suitable electron energy for treating this lesion is _____.
    a. 3 MeV
    b. 6 MeV
    c. 9 MeV
    d. 12 MeV

8. Skin dose is defined as the dose below the skin at a depth of _____ mm.
    a. 0
    b. 0.5
    c. 5
    d. 10

9. The parameters that are of importance in central axis depth dose distribution are _____.
    a. surface dose
    b. depth of peak dose, d_{max}
    c. therapeutic depth d_{90}
    d. x-ray contamination

10. A small increase (e.g., 10 cm or less) in nominal SSD will significantly affect _____.
    a. central axis depth dose
    b. cross-beam profile
    c. beam output
    d. beam penumbra

11. The thickness of Cerrobend™ material required for shielding can be determined by multiplying the lead thickness by a factor of _____.
    a. 1
    b. 1.1
    c. 1.2
    d. 1.5

12. The approximate thickness of lead required to cut off (stop) a 10 MeV electron beam is _____.
    a. 1 mm
    b. 5 mm
    c. 5 cm
    d. none of the above

13. A cutout is used in a 20 cm x 20 cm applicator to treat a smaller cross-sectional area, using a 12 MeV electron beam. The output and the depth dose will be significantly affected if the blocked field is less than _____ cm.
    a. 20 x 20
    b. 15 x 15
    c. 12 x 12
    d. none of these field sizes

14. A patient is to be treated with a single field of a 16 MeV electron beam. The depth of the 90% isodose line along the central axis is _____ cm.
    a. 2
    b. 4
    c. 8
    d. none of the above

15. A patient is to be treated with an 8 MeV electron beam. A lead cutout is used inside the applicator to shape the field. A rough thickness of _____ is required to reduce the dose below the cutout region to less than 5% of the useful beam.
    a. 1 mm
    b. 4 mm
    c. 8 mm
    d. none of the above

16. A patient is to be treated with a 15 MeV electron beam. The width of the lesion is 5 cm. A field size with a cone width of _____ is used for treating this lesion.
    a. 5 cm
    b. 4 cm
    c. 6 cm
    d. none of the above

17. The dose perturbation caused behind any inhomogeneity (e.g., lung or bone) depends on its _____.
    a. size
    b. shape
    c. electron density
    d. effective atomic number

18. The advantages of an electron beam in treating a considerable thickness of tissue (e.g., chest wall) are _____.
    a. near uniform dose from the skin
    b. drastic fall in dose beyond 80% dose level
    c. accompanying x-rays to give additional dose
    d. none of the above

19. Field flatness _____.
    a. is defined by "uniformity index" at a defined depth
    b. can be evaluated from the measured cross-beam profile
    c. does not change with depth
    d. cannot be evaluated for electron beams

20. Field symmetry _____.
    a. is defined for a pair of symmetric points on either side of the central axis
    b. is evaluated from the measured cross-beam profile
    c. must be better than 2%
    d. is not defined for electron beams

21. The most important characteristics of electron beam isodose curves that influence treatment planning are _____.
    a. constriction of 80% to 90% isodose curves
    b. ballooning (or lateral spread) of 20% to 50% isodose curves
    c. therapeutic depth
    d. none of the above

22. While treating the chest wall with an electron beam, part of the wall may be of lesser thickness. In order to deliver a uniform dose to target volume in such cases, _____.
    a. bolus may be used to make the thickness uniform
    b. the target may be treated by two adjacent fields of different electron energies with proper abutment of fields
    c. the curvature of the patient's surface will not affect planning
    d. none of the above is true

23. Which of the following properties of an electron beam are true?
    a. The electron beam diverges from the focus.
    b. The electron beam diverges from a virtual position in space which is closer than the SAD.
    c. The SSD is actually the distance from the virtual position to the edge of the electron cone applicator.
    d. The inverse square law correction for small air gaps between applicator and skin surface can be applied with good accuracy using nominal SSDs and air gap distances.
    e. All of the properties mentioned above are true.

24. The electron beam _____.
    a. does not exhibit a buildup region like in photon beams
    b. surface dose is much lower than for that for high-energy photon beams
    c. depth dose profile very much depends on the machine design and accessories used
    d. dose in the falloff region falls more sharply compared to the photon beams
    e. exhibits all the above characteristics

25. The electron PDD depends on _____.
    a. electron energy
    b. depth in the phantom
    c. field size
    d. prescription dose at this depth
    e. all the above factor

26. The PDD of electron beams can be measured using _____.
    a. a cylindrical ionization chamber in a water phantom
    b. a plane parallel ion chamber in a water phantom
    c. a diode detector in a water phantom
    d. film in a solid water phantom
    e. a re-entrant type chamber in a solid water phantom

27. If an ionization chamber is used to measure the depth of the ionization curve of an electron beam, to convert the curve into depth dose curve we must multiply ionization by _____.
    a. no factor, because depth ionization directly gives depth dose curve
    b. mass energy absorption coefficient ratio, water to air
    c. density ratio, water to air
    d. stopping power ratio, water to air
    e. efficiency of Bremsstrahlung production

28. If a film is used for electron dose measurements, to convert film density into dose we must _____.
    a. multiply optical density (OD) by the effective Z of film
    b. multiply OD by stopping power ratio, water to air
    c. multiply OD by stopping power ratio, film material to water
    d. use no conversion factor, because film density itself is numerically equivalent to dose
    e. use the H and D curve to get the dose

29. If a Si diode is used for electron dose measurements, to convert ionization signal into dose we must multiply ionization, i(d), by _____.
    a. effective Z of Si
    b. stopping power ratio, water to air
    c. stopping power ratio, film Si to water
    d. no factor, since stopping power ratio, water to Si, is independent of energy (and hence depth) and would get canceled while dividing i(d) by i(d_{max})
    e. mass energy absorption coefficient of Si

30. Which of the statements regarding oblique incidence of electron beams on a patient are true?
    a. As obliquity (angle of incidence) increases, the d_{max} moves away from the surface.
    b. The depth dose distribution is significantly altered.
    c. For large angles of obliquity (>60°), dose at d_{max} increases significantly compared to peak depth at normal incidence.
    d. Oblique incidence is encountered while treating chest wall, scalp. etc.
    e. All of the above are true.

31. The figure below shows the isodose curve of an 18 MeV electron beam containing the central axis of the beam. An inhomogeneity (bone) exists in the treatment field. Which of the statements regarding the curve is true? (The following two figures have been adapted from an excellent lecture presentation by John A. Antolak and Kenneth R. Hogstrom. Students should read all the presentations of K. R. Hogstrom to get a clear understanding of electron beam therapy and treatment planning.)

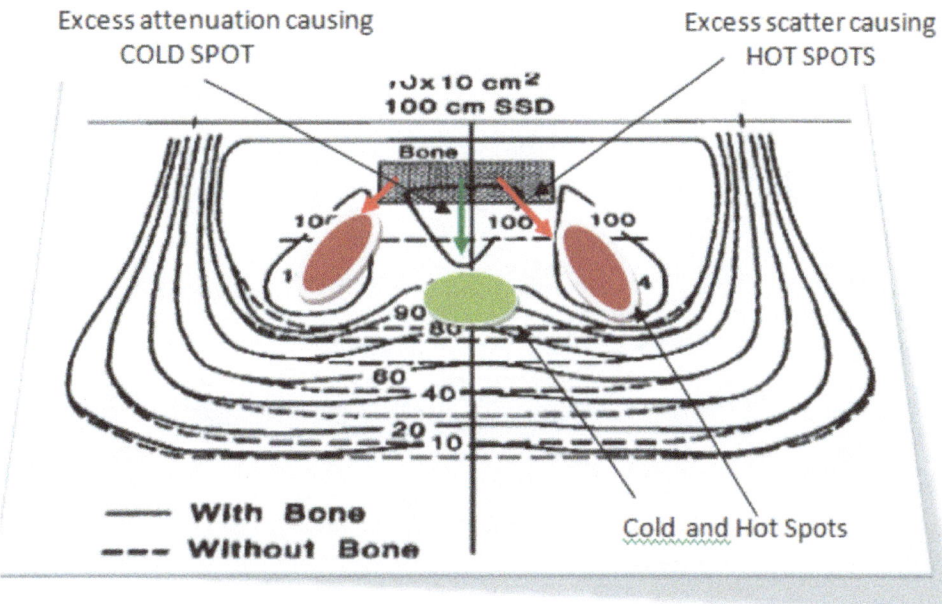

Electron isodose curve perturbation by bone heterogeneity

a. The therapeutic (90%) isodose curve shifts toward the surface below the bone due to higher stopping power of bone compared to tissue.
b. The hot spot lateral to the bone occurs because of larger bone scattering in that direction compared to tissue.
c. The small increase in dose to tissue above bone is due to larger back scatter from the bone compared to same amount of tissue.
d. The small increase in dose to the bone compared to tissue is due to larger multiple Coulomb scattering.
e. All of the above are true.

32. The following figure shows the isodose curve of an 18 MeV electron beam containing the central axis of the beam. An inhomogeneity (air cavity) exists in the treatment field. Which of the following statements regarding the curve are true?

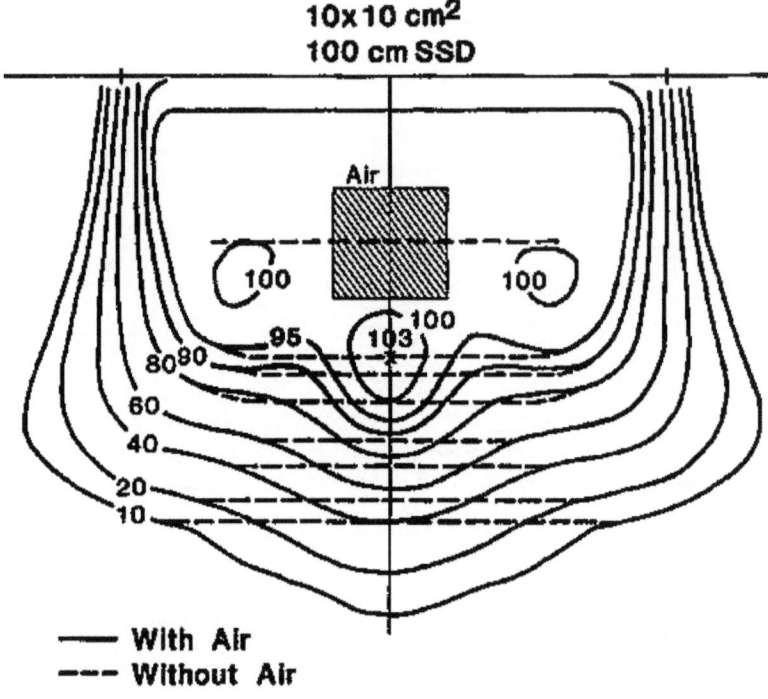

Electron Isodose curve perturbation due to air cavity heterogeneity

a. The therapeutic (90%) isodose curve shifts downstream from the original depth below the cavity due to lower stopping power of air compared to tissue.
b. The spots lateral to the cavity receive less scatter and push the curve upstream (cold spots).
c. The small decrease in dose to tissue above the cavity is due to lesser back scatter from the cavity compared to same amount of tissue.
d. The increase in dose below the cavity is due to lower stopping power of air compared to tissue (hot spots).
e. All of the above are true.

33. The significant effect of patient anatomy (heterogeneity and surface irregularities) on the electron beam dose distributions are _____.
a. isodose displacement due to bone, air cavity, or lung
b. hot and cold spots due to differences in stopping power and scattering strengths of inhomogeneities compared to water
c. oblique incidence of electron beam on the skin decreasing surface dose
d. an increase in peak dose compared to normal incidence
e. changes in penumbra due to two field edges being at different distances from the source

34. The effect of patient anatomy on the dose distributions must be properly accounted for in the treatment planning to ensure _____.
a. proper choice of electron energy to ensure that the isodose constriction or d_{90} shift will not miss any part of the CTV
b. adequate dose uniformity avoiding significant hot or cold spots inside the CTV
c. minimal dose to critical structures bordering on PTV
d. there is no underdosing of the tumor target in the buildup region
e. fields much larger than the target size are chosen to ensure 100% accuracy in patient treatment

35. SSD dependence of PDD for electron beams is not significant because the _____.
    a. change in SSD is no more than 10 cm, unlike the photon beam treatments that involve much larger SSDs
    b. therapeutic depths involved are much smaller compared to photon beams that involve much larger depths
    c. x-ray contamination neutralizes the SSD dependence of electron beam PDD
    d. inverse square law cannot be applied for electron beams
    e. all of the above

36. Electron dose distribution in water is different from the distribution in a patient due to _____.
    a. oblique incidence of electron beam
    b. surface irregularities
    c. bone heterogeneity
    d. lung heterogeneity
    e. cavities
    f. all of the above

37. How does the penumbra of an electron beam get affected when incident on a sloping patient surface?
    a. There is no change in penumbra.
    b. The field edge farther from the source exhibits larger penumbra.
    c. The field edge closer to the surface exhibits larger penumbra.
    d. The field edge farther from the source exhibits smaller penumbra.
    e. The field edge closer to the surface exhibits smaller penumbra.

38. Oblique incidence of electron beam on a patient's surface leads to decreased _____.
    a. surface dose
    b. peak dose (dose at d_{max} or R_{100})
    c. R_{100}
    d. R_{90}
    e. R_p

39. Major influencing factors for bone and air cavities in the patient are _____. (See the figures in questions 31 and 32.)
    a. the pulling of isodose curves (upstream) toward bone due to excess attenuation in bone compared to tissue
    b. the pushing of isodose curves away from cavity due to reduced attenuation in an air cavity compared to tissue
    c. the creation of hot and cold spots around the heterogeneity due to lack of side scatter equilibrium
    d. an increase in dose in tissues lying above bone due to electron back scatter
    e. all of the above

40. The influence of bone heterogeneity on dose distributions is _____.
    a. under dosing of any PTV existing beyond bone compared to the same volume of tissue
    b. under dosing of tissues in front of bone compared to the same volume of tissue
    c. increased dose to bone
    d. increase in electron range
    e. all of the above

41. Which of the following statements is true?
    a. Electron arc therapy increases the surface dose compared to a fixed-beam irradiation.
    b. Any protrusion (like the dose) creates hot spots lateral to the structure and cold spots below the structure.
    c. Any depression (like the ear canal) creates cold spots lateral to the structure and hot spots below the structure.
    d. Field abutment for treating extended field length is not possible with electron beams.
    e. All of the above are true.

42. A PTV having a 5 cm diameter and extending to a maximum depth of 4 cm is to be treated with an electron beam. The distal 90% isodose curve encloses the PTV. Which of the following is the appropriate beam energy and field size to choose (in MeV, cm diameter). (Use $E_{p,0} = 3.5 \, d_{90}$ and a field margin of 1 cm for the choice of field size.)
    a. 5, 5
    b. 6, 6
    c. 8, 6
    d. 10, 8
    e. 14, 7

E. Treatment Planning Systems

Choose the right answer(s) (more than one may be correct):

1. The aims of treatment planning are _____.
    a. uniform dose within target volume ($\pm 5\%$ desirable)
    b. conformal coverage of PTV with 95% of prescribed dose
    c. conformal avoidance of normal tissue with doses less than tissue tolerance dose
    d. maximum dose, preferably not exceeding the prescribed dose and contained in the target volume
    e. all of he above

2. The hardware components of a TPS system are _____.
    a. CPU
    b. graphics display
    c. calculation algorithms
    d. input/output devices
    e. archiving and network communication devices
    f. all of the above

3. The objectives of treatment planning are to _____.
    a. optimize "target dose/surrounding normal tissue dose"
    b. obtain more uniform dose-to-target volume
    c. exceed the tolerance dose of normal tissues
    d. spare critical organs

4. A TPS requires the user to input _____.
    a. beam data
    b. patient-specific data
    c. calculation algorithms
    d. none of the above

5. In order to use a TPS for accurate treatment planning, any TPS installed in a radiation therapy department must be _____,
    a. acceptance tested before commissioning
    b. provided with machine beam data input
    c. provided with patient-specific data input
    d. verified for the accuracy of its dose computation

6. The accuracy of the dose computed by the TPS depends on the _____.
    a. accuracy of the beam calibration
    b. dose calculation algorithm
    c. patient setup and treatment
    d. none of the above

7. The advantages of 3D TP are _____.
    a. escalation of tumor dose without increasing normal tissue complications
    b. more accurate inhomogeneity and patient contour corrections
    c. the possibility of non-coplanar TP
    d. better plan optimization using DVHs

8. Treatment planning errors _____.
    a. cannot have serious consequences
    b. can affect only individual patients and not a group of patients
    c. can be due to the operator or the planning system
    d. occur very often

9. _____ is/are patient dependent.
    a. CT data
    b. Beam data
    c. Simulation
    d. Field placement for dose planning

10. Acquisition of patient data is the responsibility of the _____.
    a. medical radiation dosimetrist
    b. therapy technologist (radiation therapist)
    c. physicist
    d. machine vendor

11. A 3D TPS _____.
    a. accepts patient data as input
    b. accepts machine data as input
    c. computes dose only to PTV
    d. optimizes the treatment plan

12. TPS incorporates _____.
    a. dose calculation algorithms
    b. software that implements the calculation algorithms
    c. provisions for modifying the algorithms
    d. self checks of the accuracy of its computation

13. Errors can arise in the dose computed by the TPS because of _____.
    a. inaccuracies in algorithms
    b. inaccuracies in the software implementation of algorithms
    c. operators inputting incorrect parameters (e.g., beam data)
    d. limitations of the algorithms while applying in a specific case

14. Acceptance testing and commissioning of a TPS is _____.
    a. very important and is the responsibility of the medical physicist
    b. not important since radiation safety is not involved
    c. the responsibility of the radiation oncologist
    d. the responsibility of the vendor

15. What are the differences between 2D and 3D treatment planning (2DP and 3DP)?
    a. In 2DP, regular fields or blocked fields are used in treatment while in 3DP, more complex fields are designed based on beam's eye view (BEV).
    b. 2D TP is based on a therapy simulator, while 3D TP is based on 3D localization of PTV and OARs with CT, MRI, and other modalities.
    c. 2D TP plan calculation, evaluation, and approval are based on single slice, while 3D TP is more sophisticated in both 2D and 3D and DVH analysis.
    d. Both 2D and 3D TP have extensive QA procedures.
    e. All of the above are true.

16. Which of the following statements regarding the TPS are true?
    a. Quality assurance is not as important as the QA of the treatment delivery systems.
    b. The TPS does not require any commissioning since it has been thoroughly tested and validated by the manufacturer.
    c. The TPS requires some basic input data.
    d. TPS systems in the market are all identical.
    e. All of the above are true.

17. _____ are brands of commercial treatment planning systems.
    a. Xio
    b. Plato
    c. Prowess
    d. Venus
    e. All of the above

18. The different types of algorithms used in dose calculations are _____.
    a. semi-empirically based
    b. model based
    c. Monte Carlo based
    d. hybrids (partly MC based)
    e. all of the above

19. Semi-empirical methods make use of _____.
    a. in-phantom dosimetry concepts
    b. measured relative dose quantities
    c. corrections for patient-phantom differences
    d. Monte Carlo algorithms in some cases
    e. all the above

20. Why do we need algorithms for dose calculations?
    a. We don't actually need algorithms.
    b. Measurements are phantom based and not patient based.
    c. Measurements are for specific geometric conditions.
    d. Measurements are for a homogeneous medium.
    e. Algorithms give true dose values in patients and require no verification.

21. Model-based dose calculation methods require _____.
    a. modeling the linac head and all components that come in contact with the beam to realistically characterize the beam incident on the patient
    b. adjusting model parameters to calculate the dose to required accuracy
    c. direct calculation of dose in patients, taking into account primary and scatter photon interactions
    d. no measured data
    e. all the above

22. Given the photon fluence Φ and the energy E of a monoenergetic photon beam, the energy fluence ψ is given by _____.
   a. Φ / E
   b. E / Φ
   c. ΦE
   d. ΦE^2
   e. Φ / E^2

23. The energy fluence ψ of a linac photon beam is given by _____.
   a. ΦE
   b. $\Phi_E E$
   c. Φ_E / E
   d. $\Phi_E E^2$
   e. none of the above

24. _____ quantities must be known to calculate the dose at a point in a homogeneous medium for a monoenergetic photon beam.
   a. Φ and E
   b. Φ and $[\mu_{tr}(E)]$
   c. Φ, E and $[\mu_{en}(E)/\rho]$
   d. Φ, E and $\mu_{en}(E)$
   e. ψ, Φ and $[\mu_{tr}(E)/\rho]$

25. The physical quantity $\psi [\mu(E)/\rho]$ is called _____.
   a. kerma
   b. collision kerma
   c. total kerma
   d. terma
   e. photon kerma

26. Dose deposited in a medium can be obtained from _____.
   a. terma
   b. terma and ψ
   c. terma and $\mu_{tr}(E)/\rho$
   d. terma and energy deposition kernel
   e. ψ

27. The various types of convolution kernels used in the TPS systems are _____.
   a. pencil beam convolution (PBC)
   b. collapsed-cone convolution (CCC)
   c. collapsed-cone convolution/superposition (CCCS)
   d. analytical anisotropic algorithm (AAA)
   e. slab kernel algorithm
   f. any of the above

28. Some of the MC codes used in dose calculations are _____.
   a. EGS4
   b. EGSnrc
   c. GEANT
   d. MCNP
   e. FLUKE
   f. all the above

29. Which of the following statements regarding TPS algorithms are true?
    a. The functioning and quality of a TPS system does not depend on the algorithm used.
    b. The only algorithm used in TPS is the dose calculation algorithm.
    c. Some basic ideas on the TPS algorithms will help the user know their capabilities and limitations.
    d. The user can fine tune the algorithms to improve agreement with measurements.
    e. All of the above are true.

30. Uncertainties in dose calculation in conventional TPS systems are introduced by _____.
    a. accuracy of input data provided by the user
    b. TPS calculations beyond the range of the data provided
    c. accuracy of beam output provided by the user
    d. modeling accuracy of dose calculation algorithms
    e. all of the above

31. Regarding Monte Carlo-based treatment planning systems, the uncertainty of dose calculation in MC-based systems arises from _____.
    a. imperfect modeling of the linac beam
    b. uncertainty of cross-section data
    c. standard deviation due to the finite number of histories simulated
    d. uncertainty in electron density derived from CT data
    e. treatment setup
    f. all of the above

F. Radiobiology and Clinical Oncology

Circle the right answer (yes or no):

1. (Yes / No)   A cell has the same inherent radiosensitivity throughout the cell cycle.

2. (Yes / No)   All cell types exhibit the same inherent radiosensitivity.

3. (Yes / No)   A tumor cell is more radiosensitive compared to a normal cell.

4. (Yes / No)   Both normal cells and tumor cells can repair cell damage.

5. (Yes / No)   Different tumors exhibit different radiosensitivity.

6. (Yes / No)   When a colony of cells is exposed to a high dose of radiation, all the cells are killed.

7. (Yes / No)   Radiosensitive tumors are radioresponsive.

8. (Yes / No)   Radioresponsive tumors are radiocurable.

9. (Yes / No)   The response of cells to radiation is linear.

10. (Yes / No)   Cell survival curves are enough to determine the tumor lethal dose (TLD) in radiation therapy.

11. (Yes / No)   Radiosensitive tumors are completely curable by radiation therapy.

12. (Yes / No)   Therapeutic ratio is the ratio of the tissue tolerance dose (TTD) for a certain level of complications to TLD for the tumor in question.

13. (Yes / No)   The oxygen enhancement ratio (OER) is the ratio of doses under hypoxic and aerated conditions for a given biological end point.

14. (Yes / No)   The tolerance dose for an organ decreases if only a portion of the organ is irradiated.

15. (Yes / No)   Fractionation increases the tolerance of normal tissues.

Choose the right answers:

16. Cell death refers to the _____.
    a. cell losing its capacity to divide in the case of proliferating cells
    b. cell losing some specific function in the case of nonproliferating cells
    c. cell transforming to another cell by mutation
    d. none of the above

17. Cell death occurs in an irradiated cell as a result of _____.
    a. lethal damage to DNA
    b. lethal damage to other critical sites
    c. damage to any site in the cell
    d. damage to any element of the cell

18. When cells are exposed to photons or electrons, damage to critical targets occurs as a result of _____.
    a. only direct action in cells
    b. only indirect action in cells
    c. direct action in the majority of interactions
    d. indirect action in the majority of interactions

19. Biological damage to a cell can be _____.
    a. only lethal
    b. lethal
    c. sublethal
    d. potentially lethal

20. The radiosensitivity of cells _____.
    a. can be only increased
    b. can be only decreased
    c. can be increased or decreased
    d. cannot be altered

21. When a colony of cells is exposed to a dose D, _____.
    a. all cells die
    b. a fraction of the cell population dies
    c. any number of cells can die
    d. no cells die

22. When radiation interactions occur in a cell, the cell death occurs because of _____.
    a. direct action with a critical target (e.g., DNA) causing irreparable lethal damage
    b. indirect action with a critical target causing irreparable lethal damage
    c. accumulated sublethal damages causing cell death
    d. none of the above

23. A colony of 100,000 cells was exposed to a certain dose of gamma rays resulting in the death of 50,000 cells. If a colony of 50,000 cells is exposed to the same dose, the number of cells killed will be about _____.
    a. 50,000
    b. 25,000
    c. 0
    d. any number

24. The size of a tumor that is barely detectable has a dimension of about 1 cm^3. Assuming a cell size of 0.01 mm x 0.01 mm x 0.01 mm, the number of cells in a tumor of size 1 cm^3 is given by _____.
    a. 10^3
    b. 10^6
    c. 10^9
    d. none of the above

25. The cell doubling time for a tumor is 60 days. The time taken by a single cell to reach a tumor size of one cm^3 is about _____.
    a. 6 months
    b. 1 year
    c. 5 years
    d. 10 years

26. Given a cell survival curve is characterized by quasi-threshold dose, D_q, extrapolation number, n, and mean lethal dose, D_0, then _____.
    a. n gives the number of critical targets in the cell
    b. D_q is a measure of cell's ability to repair damage
    c. D_0 is a measure of cell sensitivity

27. The shoulder portion of the cell survival curve _____.
    a. indicates repair of sublethal damage
    b. indicates no damage to cells due to small radiation doses
    c. is the same for all types of radiation
    d. is absent for high-LET radiation (e.g., alpha particles)

28. The mean lethal dose (MLD) is defined as dose to _____.
    a. reduce the survival fraction to 37% of the original value (in the exponential region)
    b. kill the whole cell population irradiated
    c. reduce the survival fraction to 10% of the original value
    d. kill all cancer cells

29. Tumor lethal dose (TLD) is _____.
    a. the same as MLD
    b. the dose that is detrimental to 100% of the irradiated cells
    c. the dose that causes complete destruction of the tumor

30. Benign tumors _____.
    a. are localized
    b. do not occur in human beings
    c. can never be lethal
    d. always turn malignant

31. Malignant tumors _____.
     a. are usually rapid growing
     b. tend to spread to other parts of the body
     c. can always be eliminated by surgery or destroyed by radiation therapy
     d. can turn into benign tumors

32. Tissue tolerance dose (TTD), $TD_{5/5}$, is defined as _____.
     a. maximum dose that can be tolerated by the tumor
     b. maximum dose that can be tolerated by normal tissues
     c. dose received by normal tissues in a patient treatment that would cause 5% complication rates in
         5 years' time
     d. none of the above

33. Tissue tolerance dose, $TD_{5/5}$, depends on _____.
     a. dose fraction
     b. volume of organ irradiated
     c. position of the organ in the body
     d. tumor dose

34. Therapeutic ratio _____.
     a. gives the ratio of tissue tolerance dose to tumor lethal dose
     b. is less than 1 for resistant tumors
     c. is larger than 1 for sensitive tumors
     d. can be modified by radiosensitizing tumors or radioprotecting normal tissue

35. The figure below shows the tumor control probability (TCP) and normal tissue complications probability
     (NTCP) plotted as a function of dose for the tumor and the adjacent critical organs.

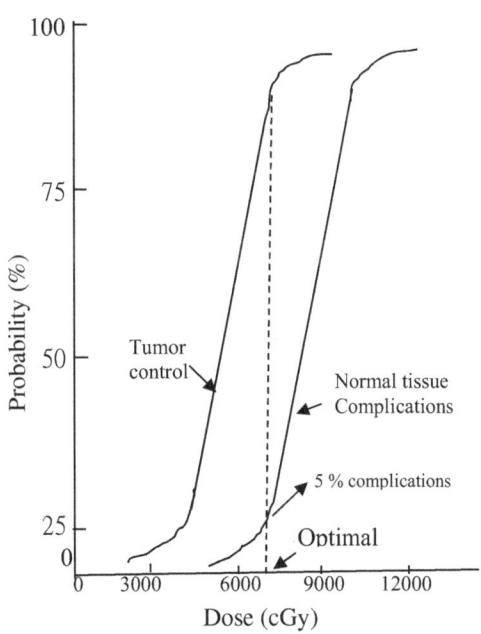

Dose response curves for tumor and normal tissues

     To increase the tumor dose (i.e., to increase TCP), for a given level of tissue complications, _____.
     a. the TCP curve must move toward the right
     b. the TCP curve must move toward the left
     c. the radiosensitivity of the tumor must be increased using radiosensitizers
     d. the radioresistance of the tumor must be increased using radioprotectors

36. The figure above shows the TCP and NTCP plotted as a function of dose for the tumor and the adjacent critical organ. To reduce normal tissue complications for the given tumor dose prescription, _____.
    a. the NTCP curve must move toward the right
    b. the NTCP curve must move toward the left
    c. the radiosensitivity of normal critical structures must be increased using radiosensitizers
    d. the radioresistance of critical normal tissue structures must be increased using radioprotectors

37. Multiple fractionations _____.
    a. for a given survival fraction decreases the total dose delivered to the tumor compared to the dose delivered as a single fraction
    b. allows time for recovery of normal tissues between fractions
    c. causes relatively more damage to the tumor compared to the organ at risk
    d. reduces setup errors compared to a single-fraction treatment

38. A single dose of 800 cGy delivered to a tumor, compared to 4 fractions of 200 cGy each, produces _____.
    a. the same biological damage
    b. more biological damage
    c. less biological damage
    d. no biological damage

39. Fractionated radiation therapy is successful because _____.
    a. normal tissues are not damaged by radiation
    b. normal tissues exhibit greater capacity for recovery compared to tumor mass
    c. fractionation increases tumor damage through reoxygenation and redistribution
    d. tumor cells cannot recover between fractions

40. The OER for photon or electron therapy is around _____.
    a. 1
    b. 2.5 to 3
    c. 10
    d. none of the above

41. The D_0 for an oxygenated tumor is 100 cGy. If OER for the tumor is 3, D_0 of the tumor under hypoxic conditions is _____.
    a. 100
    b. 300
    c. 600
    d. none of the above

42. In radiation therapy, failure of local control can occur due to _____.
    a. underdosing of tumor
    b. nonuniform irradiation of tumors
    c. error in localization
    d. error in patient setup

43. The PTV-CTV margin is _____.
    a. equal on the right-left axis of the patient
    b. equal on A/P-P/A direction of the patient
    c. equal on the cranio-caudal axis of the patient
    d. non-uniform in all directions

44. According to ICRU-50, _____ is correct.
    a. CTV = PTV
    b. treatment volume < PTV
    c. treatment volume > PTV
    d. treatment volume < irradiated volume

45. Which of the following statements regarding CTV and PTV are correct?
    a. PTV − CTV = margins for setup errors and patient/organ movements.
    b. The margin increases the volume of irradiated normal tissue.
    c. CTV is always smaller than PTV.
    d. Patient immobilization systems help us reduce the PTV-CTV margin.

46. Match the different volumes to their definitions.
    a. Gross tumor volume (GTV)
    b. Clinical target volume (CTV)
    c. Planning target volume (PTV)
    d. Treatment volume (TV)
    _____ i. Volume enclosing tumor and subclinical spread
    _____ ii. Tumor volume as determined by the radiation oncologist
    _____ iii. Volume enclosing tumor, its subclinical spread, and a margin to include patient (or organ) movements and setup error
    _____ iv. Defined by a specified isodose surface

47. A dose-volume histogram (in integral form) for PTV illustrates _____.
    a. what percentage of PTV receives at least the prescribed tumor dose
    b. spatial distribution of dose in the target volume
    c. amount of dose delivered outside the target volume
    d. none of the above

48. A dose-volume histogram (in integral form) for organs at risk (OARs) gives _____.
    a. what percentage of OAR receives at least the prescribed organ dose
    b. spatial distribution of dose in the organ
    c. amount of dose delivered outside the organ volume
    d. none of the above

49. A dose-volume histogram _____.
    a. helps in optimizing the treatment plan
    b. is the only method of optimizing treatment plans
    c. is the best method for optimizing treatment plans
    d. is a totally objective tool for optimization

50. The aim of curative radiotherapy is _____.
    a. to kill all tumor cells
    b. kill all tumor cells and no normal cells
    c. to cure all cancers
    d. to achieve local tumor control with normal tissue damage within acceptable limits
    e. to achieve local tumor control, normal tissue is of no concern

51. The four phases of a cell cycle are _____.
    a. G_1 phase
    b. S phase
    c. G_2 phase
    d. G_3 phase
    e. M phase

52. _____ is an inherent factor that affects the radiosensitivity of a cell.
    a. Mitotic rate
    b. Degree of differentiation
    c. Cell cycle phase
    d. LET
    e. Fractionation

53. _____ is an external factor that affects the radiosensitivity of a cell.
    a. Dose rate
    b. LET
    c. Fractionation
    d. Oxygenation
    e. Proliferating capacity
    f. all of the above

54. According to the law of Bergonie and Tribondeau, cell radiosensitivity depends on _____.
    a. oxygen content of the cells
    b. degree of differentiation
    c. fractionation of the radiation dose
    d. degree of mitotic activity

55. Which of the following statements regarding RBE are true?
    a. Equal doses of different types of radiation have the same biologic effect>
    b. The RBE is used to compare the biological efficiency of different types of radiation.
    c. In the definition of RBE, the reference radiation is 6 MV x-rays.
    d. The RBE of a radiation r is equal to D_{ref}/D_X, where D_{ref} and D_X are the doses of the reference radiation and the radiation in question, for the same biologic effect.
    e. RBE depends on dose and dose rate for low-LET radiation.
    f. All of the above are true.

56. If a dose of 400 cGy ^{60}Co gamma rays is required to kill 50% of the cells in a culture, and the same culture requires a dose of 80 cGy 200 kV x-rays for the same end point, the RBE of 200 kV x-rays relative to ^{60}Co gamma rays is _____.
    a. 0
    b. 1
    c. 5
    d. 20
    e. none of the above

57. If the LD_{50} for 250 kV x-rays is 6 Gy and the LD_{50} of a neutron beam is 4 Gy, the RBE of the neutron beam is equal to _____.
    a. 6×4
    b. $6 / 4$
    c. $4 / 6$
    d. $6 / 4^2$
    e. $4 / 6^2$

58. Which of the following statements regarding LET are true?
     a. LET stands for linear energy transfer.
     b. LET is defined as dE/dx in units of keV/μm.
     c. LET is a measure of the density of ionization along the particle track.
     d. In the case of low-LET radiation, OH radicals produced in water by radiation are responsible for much of the cell damage (indirect effect).
     e. All of the above are true.

59. Arrange alphas, neutrons, and electrons of 5 MeV in their increasing order LET.
     a. neutrons, alphas, electrons
     b. alphas, electrons, neutrons
     c. electrons, neutrons, alphas
     d. neutrons, electrons, alphas
     e. electrons, alphas, neutrons

60. _____ are sparsely ionizing radiations.
     a. X-rays
     b. Gamma rays
     c. Electrons
     d. Protons
     e. Neutrons
     f. All the above

61. Match the organ to radiation damage sensitivity. (The most sensitive organ is ranked 5 and the least sensitive is ranked 1.)

Radiosensitivity		*Organ*		
a. 1		_____	i.	Bone marrow
b. 2		_____	ii.	CNS
c. 3		_____	iii.	Eyes
d. 4		_____	iv.	Genital organs
e. 5		_____	v.	GI

62. The normal tissues can be classified as _____.
     a. acutely reacting or early responding
     b. late responding
     c. normal reacting
     d. not reacting
     e. all of the above

63. Normal tissue tolerance dose _____.
     a. is the dose below which NTCP (normal tissue complications probability) is acceptable
     b. is the same for all tissues
     c. depends on total dose
     d. depends on fractionation schedule
     e. is the volume of normal tissue irradiated
     f. depends on all the above

64. Match the organ to the complications.

Organ	_Complications_
a. brain	_____ i. myelitis necrosis
b. spinal cord	_____ ii. clinical stricture/perforation
c. lung	_____ iii. necrosis infarction
d. esophagus	_____ iv. clinical nephritis
e. kidney	_____ v. pneumonitis

65. Match the organ to the complications.

Organ	_Complications_
a. stomach	_____ i. obstruction perforation/ulceration/fistula
b. small intestine	_____ ii. ulceration, perforation
c. colon	_____ iii. obstruction perforation/fistula
d. liver	_____ iv. severe proctitis/necrosis/fistula
e. rectum	_____ v. liver failure

66. Which of the following statements regarding the tissue tolerance dose (TTD) are true?
    a. TTD depends on the tissue type.
    b. In some tissues, the severity of the tissue damage depends on the volume of the organ irradiated.
    b. In some organs, even a small-volume irradiation can cause serious damage.
    c. The time gap between irradiation and damage can vary widely depending on the tissue type.
    d. The proliferation capacity of cells have nothing to do with the damage response time of the tissues.
    e. All of the above are true.

67. What properties relating to OER are true?
    a. As the tumor grows, the central core tends to become oxygen deficient.
    b. OER property is used to increase the effect of radiation therapy in oncology treatments.
    c. Oxygen has no effect in high-LET radiations.
    d. Beyond about 80% oxygen tension, there is not significant change in oxygen enhancement.
    e. All of the above are true.

68. OER for high-LET radiation is _____.
    a. 0
    b. 0.1
    c. 0.5
    d. 1
    e. 5

69. OER for low-LET radiation and a fully oxygenated tumor is _____.
    a. −1
    b. 0
    c. 0.5
    d. 1
    e. 10

70. _____ is the OER for sparsely ionizing radiation.
    a. 0
    b. <1
    c. 2.5–3.5
    d. 10
    e. any of the above depending on the energy of the radiation

71. Which of the following statements regarding chronic hypoxia are true?
    a. The centers of large tumors are necrotic and surrounded by intact tumor cells.
    b. As the tumor grows, the necrotic center remains unchanged.
    c. Both large and small tumors have a necrotic center.
    d. The necrosis observed in the center of large tumors is not caused by the absence of oxygen.
    e. All of the above are true.

72. The percent of hypoxic cells in human tumors has been estimated to be around _____.
    a. 1%
    b. 5%
    c. 10% to 15%
    d. 20% to 25%
    e. 50%

73. The fraction of hypoxic cells in a tumor immediately following a single dose of irradiation _____.
    a. is zero
    b. is the same as before the irradiation
    c. is more than before
    d. is less than before
    e. depend on dose fraction

74. The advantages of high-LET radiation in radiotherapy are _____.
    a. much more affordable
    b. increased RBE for cell killing
    c. lower OER for dealing with hypoxic conditions in tumors
    d. improved dose distributions for critical tissues
    e. all of the above

75. Which of the following statements regarding LET and RBE are true?
    a. There is no relationship between LET and RBE.
    b. RBE is constant with respect to LET.
    c. RBE is directly proportional to LET.
    d. RBE is inversely proportional to LET.
    e. RBE increases with LET up to certain LET and then starts decreasing.

76. Reoxygenation in radiobiology refers to _____.
    a. hypoxic cells are turning into normoxic cells
    b. saturating tissues with oxygen in an oxygen chamber
    c. diffusion of oxygen to hypoxic cells
    d. tumors with hypoxic cells getting oxygenated in a couple of days following the first dose of radiation
    e. removal of oxygen from cells

77. If the cell survival fraction is 0.1, _____ is the percentage of cells killed.
    a. 0.1%
    b. 1%
    c. 10%
    d. 90%
    e. 100%

78. _____ is the dose required for 10% survival in terms of mean lethal dose D_0.
    a. $1/D_0$
    b. D_0
    c. $2.30 \times D_0$
    d. $2 \times (2.3 \times D_0)$
    e. $(2 D_0)^2$

79. In radiotherapy, the influence of four Rs depends on the total treatment time (T) and time between fractions (t). Match the Rs to their Ts or ts.

    Various Rs
    a. **R**eoxygenation
    b. **R**edistribution
    c. **R**epair
    d. **R**epopulation (or **R**egeneration)

    Influence on T or t
    _____ i. need minimum t for normal tissues
    _____ ii. need to reduce T for tumor
    _____ iii. need minimum t
    _____ iv. need minimum T

80. The figure below shows the SF curves for a radiosensitive cell population and a radioresistant cell population.

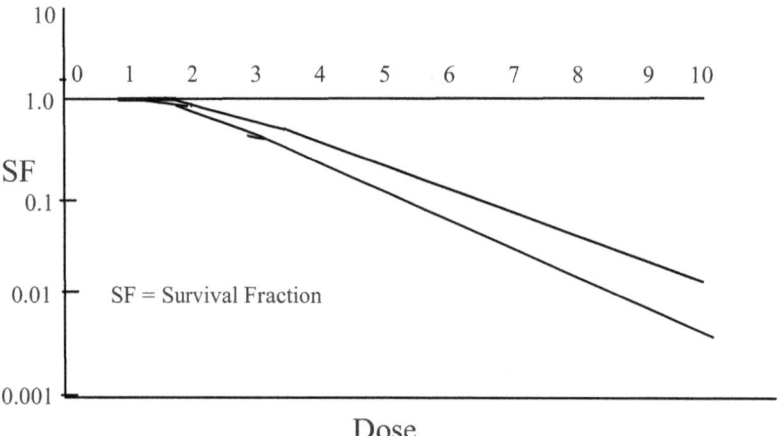

    The dose required to kill 90% of the cells from the radioresistant cell population is _____.
    a. 1 Gy
    b. 7 Gy
    c. 10 Gy
    d. 15 Gy
    e. 20 Gy

81. What conclusions can be drawn from the survival curve shown in the figure below for x-rays for the normal cells?

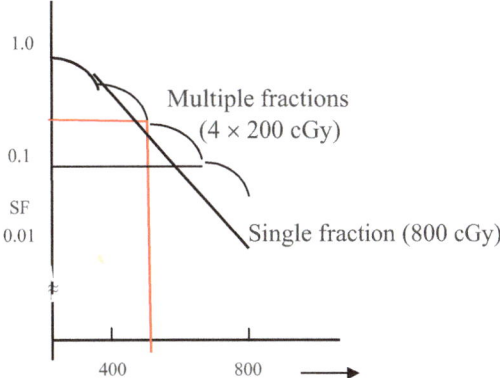

Cell survival curve for normal tissue for fractioned radiotherapy

a. It represents the survival curve for fractionated radiotherapy.
b. For a small dose, there is damage repair taking place.
c. Fractionation increases cell survival compared to a single large dose.
d. For the same SF, fractionation requires a larger dose compared to single dose.
e. High-dose treatment is preferable since more cells are damaged.
f. All of the above are true.

82. The cell survival curve for normal cell irradiation with x-rays is shown below.

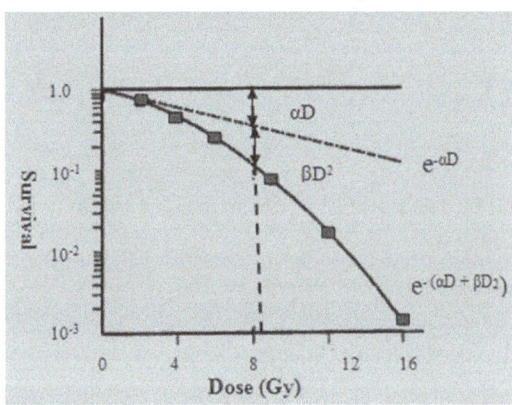

Survival curve in the Linear Quadratic (L-Q) model

This curve can be represented by a linear quadratic equation: $SF = \exp - (\alpha D + \beta D^2)$. What statements regarding the factors α and β in the equation are true?

a. The α component is responsible for the shoulder representing repairable sublethal damage to the target (DNA).
b. The shape of the curve at higher doses is governed by β which represents irreparable damage.
c. The dose range over which damage repair dominates is determined by α.
d. α/β is about the same for all normal tissues.
e. All of the above are true.

83. Factors affecting tumor growth include _____.
a. cell cycle time
b. growth fraction
c. cell loss fraction
d. tumor oxygenation
e. all of the above

84. Which of the following properties regarding cell cycle growth are true?
    a. Cell cycle time varies widely within a tumor.
    b. Malignant tumor cells cannot be slow cycling.
    c. Tumors of the same type may have different average cycle times.
    d. Benign tumors are generally slow cycling compared to the malignant types.
    e. All of the above are true.

85. The 4 Rs of radiobiology are _____.
    a. repair
    b. reassortment
    c. remodification
    d. reoxygenation
    e. repopulation

86. Which of the following parameters regarding BED are true?
    a. BED stands for biologically equal dose.
    b. BED can be used to differentiate between fractionation regimes.
    c. According the L-Q model, the BED can be shown to be D [1 + (D/n) / (α/β)].
    d. BED is the same for late-responding and early-responding tissues.
    e. All of the above are true.

87. A standard treatment scheme is 2 Gy/fraction, 30 fractions. If the dose per fraction has to be increased to 3 Gy, _____ fractions must be delivered for the same biological effect. The tissue is late-responding tissue. (Assume α/β = 3.)
    a. 30
    b. 24.2
    c. 16.7
    d. 15
    e. 10

88. In the above problem, _____ will be the number of fractions for the new regime if the tissues are early-responding tissues. (Assume α/β = 10.)
    a. 26
    b. 18.4
    c. 15
    d. 12.8
    e. 4

G. Treatment Techniques and Treatment Delivery

Choose the right answer(s) (more than one may be correct):

1. _____ are the IMRT plan objectives.
    a. Minimum dose to PTV
    b. Maximum dose to PTV
    c. Maximum dose to OAR
    d. Dose-volume constraints for OAR
    e. None of the above

2. IMRT _____.
   a. can be used to treat any site treated with conventional radiotherapy
   b. can deliver dose conformally to a concave target
   c. can escalate the tumor dose, sparing the OARS bordering on the PTV
   d. can reduce dose to critical structures without escalating tumor dose
   e. in the form of Rapid Arc or volumetric modulated arc therapy, can deliver highly conformal treatments with much-reduced treatment times
   f. can achieve all of the above

3. 3D CRT implies _____.
   a. acquisition of 3D image data of the patient region of interest (ROI)
   b. treatment planning in 3D
   c. design of treatment fields individually shaped for conformal irradiation
   d. treatment delivery in 3D conformal fashion
   e. all the above

4. IMRT _____.
   a. is an advanced form of 3D CRT
   b. confines high dose to PTV volume and spares normal tissues bordering on PTV
   c. replaces the use of wedges and compensators
   d. planning cannot be done with Sim-CT
   e. features listed above are all true

5. 3D volume information allows the use of noncoplanar beam geometries. The advantage(s) this treatment might have over traditional axial coplanar geometry include _____.
   a. reduced volume of normal tissues irradiated
   b. less CT image data required for planning
   c. more options to avoid critical structures (OARs)
   d. reduced treatment time
   e. all of the above

6. Comparing IMRT with 3D CRT, _____.
   a. IMRT allows for smaller margins and is better for concave-shaped targets
   b. IMRT is more dependent on beam energy than 3D
   c. beam directions through critical structures are not allowed with IMRT
   d. IMRT does not benefit from non-coplanar beams
   e. IMRT requires more QA

7. A conformal arc is defined as an arc treatment where the MLC conforms to the shape of the PTV at each gantry angle. The advantage(s) this treatment might have over a conventional 3D CRT include _____.
   a. reduced volume of normal tissue irradiated
   b. reduced volume of normal tissue irradiated to high dose levels (e.g., >80% Rx)
   c. reduced treatment time
   d. avoidance of radiation passing through critical structures
   e. optimized beam weighting with gantry angle

8. Compared to the four-field box technique, the total number of MU required for an IMRT plan using four fields is _____.
   a. approximately the same
   b. much less
   c. much greater
   d. greater or lesser depending on the target depth and field sizes
   e. depends on beam energy

9. The difference between an IMRT and 3D CRT delivery typically includes _____.
    a. nonuniform (modulated) beam intensities
    b. patient-specific beam shaping
    c. inverse planning for dose optimization
    d. dosimetric or biological objectives with relative weights
    e. significantly more complex dose calculation algorithm

10. The disadvantages of IMRT are _____.
    a. higher cost of equipment
    b. often longer treatment times
    c. increased normal tissue complications
    d. more complex quality assurance procedures
    e. all of the above

11. Four types of IMRT delivery methods are _____.
    a. compensator based
    b. helical tomotherapy
    c. segmental MLC
    d. dynamic MLC
    e. 3D CRT
    f. all of the above

12. The advantages of MLC in IMRT are _____.
    a. the replacement of Cerrobend for beam shaping
    b. automatic beam shaping for multiple fields
    c. dynamic MLC delivery
    d. modification of dose distribution within the field
    e. all of the above

13. Two concerns in MLC beam shaping are _____.
    a. leaf transmission
    b. leaf scattering
    c. penumbra
    d. interleaf transmission leakage
    e. leaf extension for large for large fields

14. Which of the following statements are true regarding 3D CRT and IMRT treatment planning?
    a. 3D CRT involves forward planning.
    b. 3D CRT planning is manually optimized.
    c. Optimal dose distributions are explicitly defined by the user in 3D CRT.
    d. IMRT involves inverse planning.
    e. Optimal intensities are computed by the TPS from the user-defined dose distributions.
    f. All of the above are true.

15. The term "step-and-shoot" is sometimes used to describe the _____ IMRT delivery technique.
    a. helical tomotherapy
    b. serial tomotherapy
    c. IMAT
    d. segmental MLC-IMRT
    e. dynamic MLC-IMRT

16. Delivery of step-and-shoot IMRT involves _____.
    a. segmenting of each field
    b. delivery of all segments of a field using each segmental delivery
    c. repeating for other fields (angles)
    d. controlling leaf motion during delivery
    e. all of the above

17. Match the IMRT delivery methods to their working principle.

    <u>Delivery Methods</u>   <u>Working Principle</u>
    a. compensator based   _____ i.   sliding window
    b. binary tomotherapy   _____ ii. physical modulator
    c. segmental MLC   _____ iii. Nomos Peacock system
    d. dynamic MLC   _____ iv. step-and-shoot

18. For a step-and-shoot IMRT treatment delivery, an MLC controller system introduces a 50 ms delay between the monitor chamber signal reaching a control point and beam termination. If the initial segment of a field is set to receive 2 MU, _____ percent error is introduced by this delay if the linac's output is set to 600 MU/min.
    a. <1
    b. 5
    c. 10
    d. 25
    e. 250

19. Which of the following statements are true in IMRT treatment delivery technique?
    a. In step-and-shoot IMRT, the collimator's shape is constant during irradiation and changes while the beam is "off."
    b. In "sliding window," the collimator's shape changes during irradiation.
    c. In "sliding window," the leaves move with same velocity.
    d. The "tongue-and-groove" effect is caused by the rounded MLC leaf ends.
    e. Leakage between the ends of a closed leaf pair may cause a small overdose region of 10% to 20%.

20. Which of the following statements regarding the tongue-and-groove MLC are true?
    a. The "tongue-and-groove" effect results in an increased transmission of ~2% between the sides (interleaf) of two MLC leaves (a maximum of about 5%).
    b. The interleaf transmission cannot be measured and must be calculated theoretically.
    c. The effects of MLC tongue-and-grove/leakage are greater for H&N IMRT than for prostate IMRT.
    d. The effects of MLC tongue-and-grove/leakage are greater for prostate VMAT than for prostate IMRT.
    e. All of the above are true.

21. Which of the following statements regarding IMRT are true?
    a. Intensity-based optimization includes delivery limitations (such as the tongue-and-groove effect for MLCs) during the optimization process.
    b. Aperture-based optimization includes delivery limitations (such as the tongue-and-groove effect for MLCs) during the optimization process.
    c. Ideal fluence maps are created during intensity-based optimization.
    d. In Eclipse, IMRT optimization is intensity-based, while VMAT optimization is aperture-based.
    e. All of the above are true.

22. Match the statement to the technique.

    Statement   *Technique*
    a. _____ MLCs move while the beam is ON   i. static MLC
    b. _____ MLCs do not move while the beam is ON   ii. dynamic MLC
    c. _____ this MLC delivery involves more MU
    d. _____ continuously changing shape, leaf speed, and dose rate
    e. _____ longer treatment times

23. Beamlet-based inverse planning involves _____.
    a. specification of beam weights for each beam
    b. specification of gantry speed
    c. fluence map optimization
    d. leaf sequencing
    e. all of the above

24. Intensity-modulated treatment fields of IMRT are _____.
    a. specified by the radiation oncologist
    b. specified by medical physicist
    c. determined by the medical physicist by calculations
    d. generated through an inverse planning treatment system
    e. not determined by any of the above methods

25. Which of the following statements are true regarding dMLC-IMRT?
    a. It has faster beam delivery compared to static IMRT methods.
    b. It is not suitable for complex dose distributions.
    c. Its leaf patterns are difficult to verify.
    d. It can produce dosimetric errors for very small field sizes.
    e. Inverse treatment planning is necessary.
    f. All of the above are true.

26. The advantages of the SMLC-IMRT method are _____.
    a. simpler treatment delivery technology
    b. forward planning is possible
    c. short treatment times
    d. few segments per field suffice for any complex treatment
    e. all of the above

27. IMRT will be very useful _____.
    a. for treating irregular targets
    b. for critical OARS bordering on the PTV
    c. where concave dose distributions are required (concave targets)
    d. for escalating the tumor dose or to reduce normal tissue complication probability
    e. in all the above cases

28. IMRT delivery techniques are commonly known as _____.
    a. static MLC technique (SMLC)
    b. dynamic MLC technique (DMLC)
    c. intensity-modulated arc therapy (IMAT)
    d. intensity-controlled radiotherapy (ICRP)
    e. volumetric-modulated arc therapy (VMAT)
    f. all of the above

29.   The main sources of errors in IMRT treatment are errors committed in _____.
   a.   commissioning
   b.   planning
   c.   transferring data to other systems
   d.   delivering planned treatment
   e.   execution by staff
   f.   all of the above

30.   Immobilization _____.
   a.   is critical for IMRT
   b.   helps in arresting patient movement
   c.   helps in setup reproducibility
   d.   gives comfort to the patient
   e.   achieves all the above

31.   IMAT _____.
   a.   stands for intensity-modulated arc therapy
   b.   was developed as an alternative to tomotherapy
   c.   combines gantry rotation with a DMLC system
   d.   delivery involves MLC dynamically stepping through a sequence of field shapes while the gantry rotates around the patient
   e.   the items above are all true

32.   _____ are the three factors that are varied for VMAT treatment delivery.
   a.   dose rate
   b.   gantry speed
   c.   MLC aperture shape
   d.   couch position

33.   In VMAT treatment delivery, the dose per degree is varied by _____.
   a.   varying dose rate
   b.   varying gantry speed
   c.   switching the beam OFF during rotation
   d.   none of the above methods

34.   The advantages of RapidArc over IMRT are _____.
   a.   reduced treatment time
   b.   less MUs
   c.   higher conformality
   d.   less time for QA
   e.   all of the above

35.   RapidArc treatment delivery is managed by _____.
   a.   a clinac controller
   b.   an MLC controller
   c.   a dose controller
   d.   a gantry controller
   e.   all of the above

36. The maximum gantry speed for RapidArc execution is 5.5°/sec, so _____ is the minimum possible seconds for one complete rotation of the gantry.
    a. 78
    b. 65
    c. 48
    d. 35
    e. 12

37. _____ is the maximum leaf speed for the Varian RapidArc treatment delivery.
    a. 20–60 MU/degree
    b. <3 cm/s (~0.5 cm/degree)
    c. 0.1 MU/degree
    d. 5.5 degrees/sec

38. There are 177 control points in one complete rotation of the Varian RapidArc. After _____ degrees of rotation, the (angular) dose rate (expressed in MU/deg) changes.
    a. 90
    b. 40
    c. 10
    d. 2
    e. 1

39. _____ determines the MLC position in RaidArc delivery.
    a. Gantry speed
    b. Minimum dose rate per degree
    c. Maximum dose rate per degree
    d. Gantry position
    e. None of the above

40. Which of the following are true regarding VMAT?
    a. VMAT arc therapy can be delivered by a single arc or multiple arcs.
    b. Increasing the number of arcs gives additional flexibility in shaping of the dose distributions, but not all cases may require or benefit from multiple arcs.
    c. Multiple arcs have definite advantages in treating complex cases like the head and neck.
    d. VMAT treatment time is shorter than static IMRT method or tomotherapy.
    e. All of the above are true.

41. The _____ technique is used to image the patient just before treatment and compare position to the treatment plan.
    a. OBI
    b. IMRT
    c. IGRT
    d. IMAT
    e. none of the above

42. Which of the following statements regarding imaging in high-precision radiotherapy are true?
    a. On-board imaging (OBI) makes use of kV x-rays and flat panel image detectors.
    b. EPID is getting gradually replaced by OBI.
    c. A kV beam gives better image quality compared to an MV imager.
    d. Image guidance through the course of radiotherapy can eliminate inconvenient stereotactic or body frames.
    e. All of the above are true.

43. The objectives of radiosurgical treatment planning are _____.
    a. high conformality
    b. steep dose gradient surrounding the PTV
    c. sparing of the OARs bordering on the PTV
    d. high dose accuracy
    e. all of the above

44. SRS _____.
    a. delivers a large dose (>10 Gy) for a small stereotactically localized target
    b. is delivered in 1–5 fractions
    c. was developed for treating intracranial lesions and functional disorders
    d. involves high accuracy in targeting the tumor
    e. involves all the above

45. SRS treatments can be carried out with _____.
    a. gamma rays
    b. beta rays
    c. MV x-rays
    d. proton beams
    e. all the above radiation

46. Diagnostic imaging modalities used in SRS are _____.
    a. CT
    b. MRI
    c. digital subtraction angiography (DSA)
    d. US
    e. all the above

47. A _____ is generally the most economical means for SRS delivery.
    a. medical linac
    b. synchrotron
    c. Gamma Knife
    d. reactor
    e. neutron generator

48. The basic requirements for stereotactic radiosurgery include _____.
    a. accurate localization
    b. accurate planning
    c. accurate treatment delivery
    d. patient safety
    e. all of the above

49. The major reason for interest in heavy ions for SRS is due to _____.
    a. applications of Bragg peak
    b. improved radiation safety
    c. easy handling
    d. low cost
    e. higher cure rate compared to x-rays

50. The _____ imaging modality is not commonly used for SRS.
    a. MRI
    b. CT
    c. angiography
    d. computed radiography (CR)
    e. ultrasound

51. _____ is a non-malignant lesion commonly treated by SRS.
    a. Meningioma
    b. Erythema
    c. Melanoma
    d. Sarcoma
    e. Carcinoma

52. _____ is not treated by SRS.
    a. Acoustic neurinoma
    b. Pituitary adenoma
    c. Pinealoma
    d. Trigeminal neuralgia
    e. Ewing's sarcoma

53. Surgical resection of skull-based meningiomas is difficult because of _____.
    a. imaging difficulty
    b. proximity to critical structures
    c. tumor size
    d. tumor irregular shape
    e. tumor infiltration

54. The rationale for the treatment of benign tumors with fractionated SRT is _____.
    a. surgical removal is unsuccessful
    b. a single dose, however high, cannot destroy benign tumors
    c. to spare adjacent radiosensitive structures
    d. the tumor will turn malignant
    e. tumor responds better when treatment is fractionated

55. The most common functional disorder treated with SRS is/are _____.
    a. trigeminal neuralgia
    b. Parkinson's
    c. speech disorders
    d. epilepsy
    e. AVM

56. Trigeminal neuralgia typically involves the _____ cranial nerve.
    a. 2nd
    b. 4th
    c. 5th
    d. 6th

57. Common SRS treatment delivery systems in use in medical centers are _____.
    a. Novalis TX
    b. TomoTherapy
    c. CyberKnife
    d. Gamma Knife
    e. SPECT
    f. all of the above

58. The delivery techniques in present-day SRS treatment are _____.
    a. dynamic conformal arcs (DCA)
    b. static non-coplanar intensity-modulated radiotherapy (NCP-IMRT)
    c. volumetric-modulated arc therapy (RapidArc)
    d. Gamma Knife
    e. all of the above

59. SBRT _____.
    a. is a single-fraction treatment
    b. has a larger dose per fraction compared to conventional radiotherapy
    c. has fewer fractions compared to conventional radiotherapy
    d. is any treatment that uses image guidance
    e. is any treatment that uses a stereotactic frame

60. Linac-based radiosurgery _____.
    a. is a series of arc treatments with 5 to 40 mm collimators or micro-MLCs
    b. can be CRT-, IMRT-, or VMAT-based
    c. can be used to treat any site in the body
    d. has isocentric accuracy that is inferior to Gamma Knife
    e. features listed above are all true

61. Tomotherapy _____.
    a. involves fan beam intensity modulation
    b. technique is a "slit beam" that is modulated every 5–10 seconds as it rotates around the patient
    c. may be helical or sequential
    d. involves only two positions for the leaves, either open or closed
    e. is not very much in use these days
    f. features given above are all true

62. The Gamma Knife _____.
    a. is based on the use of ^{60}Co beams
    b. uses a fixed head frame
    c. can treat only cranium and upper C spine
    d. takes a long time to treat
    e. uses single isocenter
    f. features listed above are all true

63. Which of the following statements regarding CyberKnife treatment are true?
    a. The couch has five degrees of freedom.
    b. The robotic arm can deliver 100 to 300 static circular beams from 40–150 nodes.
    c. It can be used for both intracranial and extracranial targets.
    d. It provides continual image guidance throughout the treatment.
    e. It uses a stereotactic frame for intracranial treatments.
    f. All of the above are true.

64. The commonly treated body sites in CyberKnife treatments are _____.
    a. lung
    b. pancreas
    c. liver
    d. prostate
    e. all of the above

65. CyberKnife intracranial and spinal applications _____.
    a. provide sub-mm targeting accuracy
    b. provide unlimited spinal access and access to central regions of cranium
    c. provide continual image guidance to correct for positional errors
    d. corrects for respiratory motion
    e. stated above are all true

66. Total body irradiation (TBI) is commonly used for the treatment of _____.
    a. recurrent cases of brain tumor
    b. various types of leukemia
    c. malignant lymphomas
    d. aplastic anemia
    e. all of the above

67. Which of the following statements regarding total body irradiation (TBI) are true?
    a. TBI is used to deliver dose to the patient's whole body that is uniform to within ±10% of the prescribed dose.
    b. The treatment distance is the nominal SAD distance of the linac.
    c. Either ^{60}Co beam or high-energy x-ray beam from a linac is used for TBI.
    d. Some specific organs may be partially or fully shielded from the prescribed dose.
    e. All of the above are true.

68. Skin malignancies (skin lymphomas) can be treated in total skin electron therapy using _____.
    a. low-energy electron beams
    b. high-energy x-ray beams
    c. kV x-ray beams
    d. ^{60}Co beams
    e. all of the above

69. Which of the following considerations are important in TSET?
    a. The electron energy may have to be degraded to suit the depth of penetration required for treating the disease.
    b. The photon contamination accompanying the electron beam has much higher penetration and so must be within acceptable limits.
    c. Some organs like eyes, nails, etc., may require shielding.
    d. A beam spoiler may be necessary to prevent skin sparing.
    e. All of the above are true.

70. Match the following figures A, B and C to the concept of accuracy and precision.

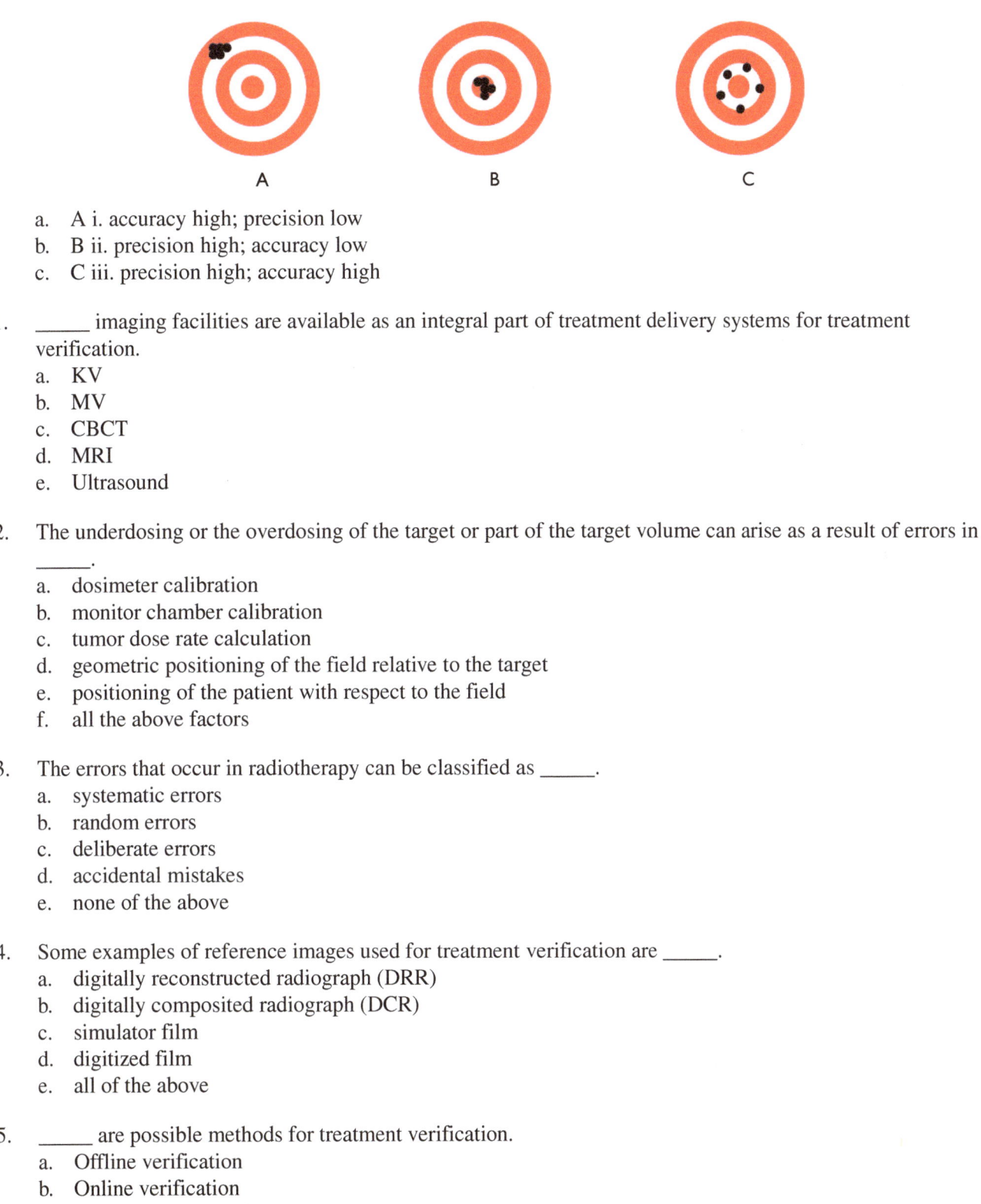

a. A i. accuracy high; precision low
b. B ii. precision high; accuracy low
c. C iii. precision high; accuracy high

71. _____ imaging facilities are available as an integral part of treatment delivery systems for treatment verification.
a. KV
b. MV
c. CBCT
d. MRI
e. Ultrasound

72. The underdosing or the overdosing of the target or part of the target volume can arise as a result of errors in _____.
a. dosimeter calibration
b. monitor chamber calibration
c. tumor dose rate calculation
d. geometric positioning of the field relative to the target
e. positioning of the patient with respect to the field
f. all the above factors

73. The errors that occur in radiotherapy can be classified as _____.
a. systematic errors
b. random errors
c. deliberate errors
d. accidental mistakes
e. none of the above

74. Some examples of reference images used for treatment verification are _____.
a. digitally reconstructed radiograph (DRR)
b. digitally composited radiograph (DCR)
c. simulator film
d. digitized film
e. all of the above

75. _____ are possible methods for treatment verification.
a. Offline verification
b. Online verification
c. Interfractional verification
d. Intrafractional verification
e. Real-time verification
f. All of the above

76. Pretreatment verification just prior to the start of first fraction will help detect _____.
    a. all systematic errors
    b. random errors
    c. accuracy of the whole treatment course
    d. gross errors
    e. all of the above

77. To correct for setup errors, the important factors to be identified on the reference and verification images are _____.
    a. all the structures in the image
    b. the isocenter
    c. the field borders
    d. the tumor surrogate (bone, implanted markers, etc.)
    e. all the above

78. The common sources of systematic errors are _____.
    a. target delineation
    b. target position (movement)
    c. data transfer to the treatment delivery system
    d. patient setup
    e. all of the above

79. Setup uncertainties arise due to _____.
    a. internal motion of the CTV
    b. day-to-day variations in patient positioning
    c. equipment defects like a sagging collimator, couch, etc.
    d. laser positioning error
    e. the transfer of treatment parameters from CT sim to the treatment unit
    f. all of the above

80. The setup margin can be reduced by _____.
    a. weekly beam calibration
    b. in-room imaging
    c. immobilization devices
    d. experienced technologists
    e. none of the above

81. Match the image to the imaging system.
    Image   *Imaging System*
    a. kV planar image   _____ i.   CT simulator
    b. DRR or DCR   _____ ii. MV portal imaging
    c. EPID   _____ iii. Simulator

III. Treatment Planning, Techniques, and Delivery ANSWERS

A. Treatment Planning Concepts

1. No

2. Yes

3. Yes

4. No

5. Yes

6. Yes

7. No   The two should be DICOM compatible.

8. Yes   Day-to-day images for the same treatment field can be compared.

9. Yes

10. No   It does. Reduction in normal tissue dose will help in escalating tumor dose.

11. Yes   See Khan (2010).

12. Yes

13. Yes

14. Yes

15. No

16. Yes

17. No   Simulation must simulate patient treatment in every respect.

18. Yes

19. No   The film processor is the main factor influencing the quality of a radiograph in any center, so the film processor, too, must be quality controlled by film sensitometry.

20. No   It is very useful for comparing with a port radiograph or DRR for treatment verification. Many centers keep a record of treatment fields using portal films. They also form part of the patient record and will come in handy in case of any litigation.

21. No

22. Yes

23. No   The patient does not hold his/her breath during treatment.

24. No   They are very much dependent on these factors.

25. Yes   The margins cannot be set too conservatively.

26. No   The margins must be optimally set.

27. No   Immobilization devices are often necessary.

28. No   It is absolutely necessary to avoid underdosing or overdosing near the field edges. The 1962 ICRU Clinical Dosimetry Report 10d recommended "Some method of immobilizing the patient and of checking the patient's position during treatment should be used." This assumes additional importance when we talk of conformal treatments, reduced CTV-PTV margins, and moving tumors and OARs in today's high-precision radiotherapy. Hence, the progress toward image guidance and automatic positional correction technology in the treatment delivery systems.

29. No   That is random error. For example, if you measure some length with a meter scale that is 1 cm short, any number of repeated measurements cannot eliminate this error. That is systematic error. It is systematic because it will be part of every measurement. Any planning error is a systematic error affecting treatment.

30. Yes   It is not exactly reproducible, and we can only talk of average error, standard deviation, etc., so it is a random error.

31. No   It is an accidental error (or mistake) and not systematic or random because it will not be repeated in daily fractions.

32. Yes   It will show which involves larger error. Many studies have shown larger errors in the transition from planning to patient treatment compared to day-to-day patient positioning errors, though it depends on several factors like accuracy of imaging, immobilization, the technologist's attention to detail, etc.

33. No   The patient must be resimulated.

34. No   The spread in the day-to-day variations in the treatment setup and systematic error transferred from planning can be estimated and corrected by taking several interfraction treatment verification images. Modern radiotherapy is moving toward the ideal situation of daily verification to avoid these errors, though imaging dose is not totally insignificant and needs to be given special attention.

35. Yes   Using online treatment verification protocols to quantify and correct for various types of errors can help in justifying tighter margins set for CTV-PTV. If we do not do treatment verification, obviously, the CTV-PTV margins must be set more liberally to avoid any geographical and dosimetric miss of the CTV by the treatment field.

36. Yes

37. No   They become more pronounced with increasing angle of obliquity.

38. No   Only tissue-like materials like wax, water, perspex, etc. are used as bolus.

39. Yes   Bolus is like extension of the patient tissue beyond the skin to derive clinical advantages while treating the patient. If it is kept away from the patient with significant air gap, lateral scattering will increase the penumbra and distribute the dose over a wider area. It will increase dose to normal tissues beyond the target.

40. Yes   For example, while treating the cheek, a shield is used inside the mouth, behind the cheek, to cut off dose to structures in the mouth.

41. Yes

42. No   They are for field sizes not large enough for scatter equilibrium. i.e., the side of the field $<R_p$.

43. No   Film dosimetry is the most convenient. In one exposure one can generate the dose distributions.

44. No   The tail is due to the photon contamination of the electron beams.

45. No

B. Photon Beam Isodose Curve Parameters

1. a   Answer b is wrong: The calibration depths are different (see the AAPM TG-51 protocol). Answer c is wrong: The buildup dose profile, including surface dose, very much depends on field size because of increasing electron contamination of photon beams. d_{max} shifts toward the surface for larger field sizes. d_{max} increases with increasing beam quality. Answer d is wrong: The range of Compton electrons increases with increasing photon energy.

2. a, b

3. a, c   PDD is normalized to d_{max} in SSD technique and is therefore 100% at d_{max}. PDD decreases with depth beyond buildup region, so b is wrong because of decreasing influence of the inverse square law (ISL) and beam attenuation.

4. b, c   The parameters that influence doses at d_{max} and at d also influence PDD(d).

5. d   PDD, being a ratio, is unaffected by the output of the machine.

6. a, c   Surface dose decreases with increase in beam quality, so b is wrong. Surface dose is much less for cleaner beams (i.e., beam devoid of electron contamination). A blocking tray considerably enhances surface dose. Using suitable filters below the blocking tray, the electron contamination and, hence, the surface dose can be reduced considerably. For further study, see Khan (2010). Low-energy x-rays from head scatter and patient backscatter are other sources of radiation contributing to surface and buildup dose.

    There is no charged particle equilibrium existing in the buildup region. Dose in such regions can be measured only with an extrapolation chamber or, alternately, with a thin entry window parallel plate chamber. In the latter case, suitable corrections for polarity and electron fluence perturbation must be applied, so d is wrong.

7. a

8. a, b, c, d

9. a, b, d

10. a, b, c

11. a, d

12. a, b, c, d

The shapes of isodose curves mainly depend on phantom scatter, beam quality change (because of beam flattening in linacs), the inverse square law effect on off-axis points, and the relative size of the field and penumbra. The shapes around field edges depend on the beam penumbra and side scatter. For further study, see Hendee and Ibbott (1996), pages 213–215.

13. b, c

14. a, b, c   Kilovoltage x-ray beams are effectively cut off by the diaphragm and deliver 5% to 10% dose outside the field due to side scatter. In the case of a ^{60}Co beam, the geometric penumbra significantly influences the dose beyond the field edges. Though the target size of the linac is only a few millimeters compared to the size of the ^{60}Co source, lateral transport of electrons beyond the borders influences the penumbra. Collimator transmission also contributes some dose beyond the borders.

15. a, b, c, d

16. a, b

17. b, d   Linacs are designed to give a flat isodose profile at a specified depth, usually 10 cm. This leads to underflattening at shallow depths and overflattening at deeper depths exhibiting "horns" (or excess dose) at shallow depths and rounded edges at deeper depths.

18. a, b, c   Electron transport increases physical penumbra and is not accounted for in geometric penumbra, as the name implies.

19. b   Actual dose is less since there is not adequate backscatter thickness. The difference, however, decreases for higher-energy photon beams due to decreasing backscatter.

20. a   Oblique incidence with respect to incident surface increases skin dose, so b is wrong. See Khan (2010). Decreasing tray-skin distance will increase electron contamination and hence skin dose, so d is wrong.

21. a, c

22. a, b, c

C. Photon Beam Isodose Distributions

1. b

2. b, d   Parallel opposed field treatment delivers fairly uniform dose to the whole volume (between the entry ports), the dose fall due to one field being compensated (though not exactly) by the opposing field. The dose uniformity, however, depends on beam quality and patient thickness. For more details on POP of fields treatment, see Bentel et al. (1991).

3. a, c

4. a

5. c

6. c

7. d

8.　a

9.　a, b, c　These are the general advantages of multiple field treatments.

10.　d

11.　a　Total PDD at target center is 65% + 25% + 25% = 115% = 200 cGy from which weighted dose to input ports—100%, 50%, and 50%—can be obtained.

12.　　i. a　115% is normalized to 100%, i.e., (115/115) × 100. PDDs at target center and at d_{max} also get scaled accordingly, i.e., (65/115) × 100, etc.
　　ii. b

13.　c　Since each field contributes equally to the tumor, (200/3) = 65%, 25%, and 25% for the three fields. So dose to input ports is given by (200/3) × 100/65, etc.

14.　b　In isocentric technique with ^{60}Co beam, beam output in air is defined by free space dose, D_{fs}, as defined in Khan (2010) or by equilibrium dose in air, $D_w (r_{eq})$, as defined in Rajan (1992). When given doses are equal, $\Sigma D_{fs} TAR(d_i)$ = tumor dose; D_{fs} = tumor dose / $\Sigma TAR(d_i)$.

15.　c　(200/3)/TAR of any field = given dose of that field.

D. Electron Beam Dose Distributions

1.　a, b, c, d
　　The dosimeter response to dose conversion involves several considerations.

2.　a

3.　a

4.　a

5.　a

6.　b

7.　d

8.　b

9.　a, b, c, d

10.　a, b, c, d

11.　c　Cerrobend density is about 20% less compared to lead.

12.　b　The density of lead is one order more compared to water. An electron beam loses about 2 MeV /cm in water.

13.　c　For side scatter equilibrium, an electron beam of energy E must have a minimum field size of about E cm × E cm, which is easy to remember.

14.　b

15. b   A 4 mm lead cutout will cut off 8 MeV electrons and transmit only the Bremsstrahlung radiation.

16. c   The beam width must be larger than the target width because of the constriction of the 90% isodose curve. For further study, see Levitt et al. (1999). The extent of constriction depends on field size and electron energy. Isodose charts must, therefore, be consulted before deciding on the adequacy of field size for target coverage.

17. a, b, c, d
            Electron dose perturbation caused by inhomogeneities is very complex. Some basic points have been discussed in Khan (2010).

18. a, b

19. a, b   The main purpose of this test is to ensure proper design of the linac to produce a beam of required flatness at clinical depths of interest. There are different specifications of flatness and depth given in the literature. See the figure below for one of the definitions. See Khan et al. (2010) for more details. The profile changes with depth, so c is wrong. Hence, it is defined at the depth of therapeutic interest. The peak dose in the flattened region should not exceed 103%.

20. a, b, c See the figure below.

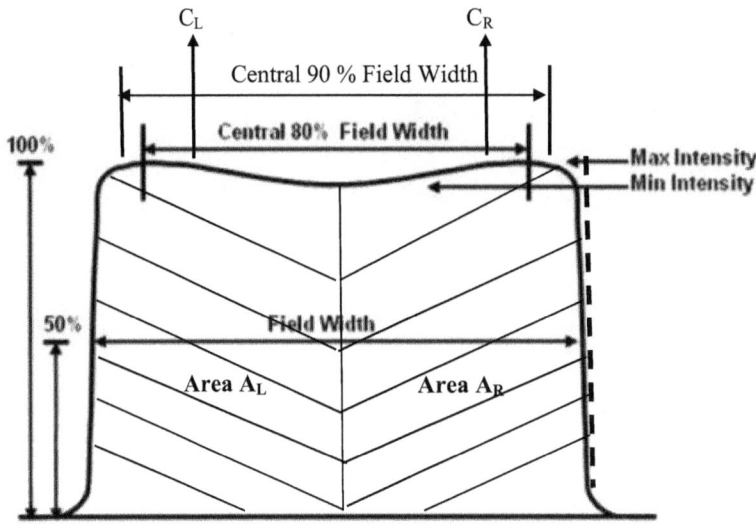

Cross-beam profile
(Figure from C.M. Able et al. at www.ncbi.nlm.nih.gov/pmc/articles/PMC3317858/)

Flatness Definition:

Field width: L_{50}; 90% and 80% field width: L_{90} and L_{50}

$L_{90} \times L_{90}$ along two perpendicular axes = A_{90}; $L_{50} \times L_{50}$ along two perpendicular axes = A_{50}.

Uniformity index = area of 90% isodose curve (A_{90}) / area of 50% isodose curve (A_{50}) > 0.7, i.e.,

$$L_{90} / L_{50} = \sqrt{A_{90} / A_{50}} > \sqrt{0.7} > 0.85$$

Depth: clinical depth of interest (near d_{max}, say, half of therapeutic depth)

% Symmetry = $[(C_R/C_L)] \times 100\%$ (symmetry points chosen within 90% field width)

21. a, b, c

22. a, b

23. b, c, d, e

24. c, d

25. a, b, c

26. a, b, c, d

27. d

28. e

29. d

30. b, c, d   The d_{max} decreases with increasing angle of obliquity.

31. e   The hot and cold spots (below bone) have only a small effect (<5%).

32. e   However, in this case, the hot and cold spot effects can be significant (about as much as 20%).

33. a, b, d, e
         Oblique incidence increases the surface dose.

34. a, b, c, d
         Field size must never be larger than necessary to cover the PTV.

35. a, b

36. f

37. b, e

38. c, d   Oblique incidence increases surface dose, dose at peak depth, and the practical range R_p.

39. e

40. a, c   The underdosing behind bone arises as a result of higher stopping power of bone compared to tissue. The tissue layers above bone actually receive higher dose (and not less) due to bone back scatter (which depends on Z). Dose to bone increases due to higher stopping power and absorption of Bremsstrahlung radiation.

41. b, c, d   Arc therapy actually decreases the surface dose, and a bolus is required to homogenize the dose. While planning a treatment for structures below the nose or ears, one must take the dose modifications into account. Otherwise the tumor will be overdosed or underdosed. Field abutment is possible with both photon and electron beams.

42. e $E_{p,0} = 3.5 \times d_{90} = 3.5 \times 4 = 14$ MeV. Adding a border on either side of PTV the required field size = 5 + 1 + 1 = 7 cm circular field or 7 x 7 square field.

E. Treatment Planning Systems

1.   e

2.   a, b, d, e

3.   a

4. a, b

5. a, b, c, d

6. a, b

7. b, c, d

8. c

9. a, c, d

10. a

11. a, b, d

12. a, b

13. a, b, c, d

14. a

15. a, b, c

16. c   TPS systems on the market are not identical, though they all serve the same purpose. They have different dose calculation algorithms, which are proprietary in nature, and other subtle differences.

17. a, b, c

18. e

19. a, b, c

20. b, c, d

21. a, b, c

22. c   The photon fluence Φ is given by dN/dA, where dN is the number of photons crossing a sphere of cross-section dA, as shown in the figure below. The definition ensures that the energy deposition is independent of the direction of travel of the photon.

23. b

24. c   Dose deposited at a point in a medium is given by $\Phi E [\mu_{en}(E)/\rho] = \psi [\mu_{en}(E)/\rho]$ under charged particle equilibrium (CPE) conditions. This concept is used to calculate dose in TPS systems. This quantity actually gives the collision kerma, K_c. K_c = Dose under CPE, i.e., the method is applicable beyond d_{max}, far from heterogeneities, etc.

$\psi [\mu_{tr}(E)/\rho]$ gives kerma, the energy of the interacting photons that is converted into kinetic energy of electrons released per unit mass. For energy absorption, we have to discount the part of the electron energy that appears as Bremsstrahlung and escapes from the point of interest. So we use $\mu_{tr}(E)(1-g) = \mu_{en}(E)$ giving collision kerma.

25. d   This term is often used in treatment planning algorithms. TERMA stands for **T**otal **E**nergy **R**eleased per unit **Ma**ss. It just includes the scatter photon energy also, along with electron energy released. It represents the total amount of energy available at a point for deposition. Since energy deposition at a point in a medium comes from both the primary photon and scatter photon, the term terma becomes useful for writing dose equations. Similarly the term "scerma" is also used to refer to "Scattered energy released per unit mass."

26. d   The energy deposition kernal gives the fraction of the released energy (from photons) that is deposited in another voxel in the medium. The whole of irradiated volume can be segmented into elementary voxels and fraction of all the energy released in these voxels (Terma) is deposited at the point in question by particle transport. The equation needs to be modified when heterogeneities are taken into account using the radiological pathlength concept, instead of simple pathlength (i.e., r must be replaced by ρ_r, r'). When this is done the method is called convolution superposition algorithm.

27. a, b, c, d
             For homogeneous phantoms, there is not much difference in accuracy between these dose calculation algorithms. For heterogeneous media, the accuracy depends on how well the kernel in an algorithm simulates the actual scatter conditions, whether electron transport is modeled, and the path length scaling in lateral direction is accounted for or not. Under charged particle equilibrium conditions, the dose can be computed from photon energy fluence. The algorithms that do not model electron transport can give good results (e.g., prostate, pelvis etc.). For regions where equilibrium may not exist (e.g., small fields, presence of heterogeneities), algorithms that model electron transport give better results (e.g., Helax–Nucletron (CCC), Varian Eclipse (AAA), etc.). These systems give better results for thoracic, head and neck regions, for IMRT treatments, etc.

28. a, b, c, d

29. a, c

30. e   The user has to provide both beam data and patient data and their accuracy will affect the accuracy of TPS dose calculations. While most of the beam data entered are relative dose factors, it is the machine output that directly affects the delivered dose and the TPS calculated dose.

31. a, b, c, d

F. Radiobiology and Clinical Oncology

1. No   The sensitivity of different phases differ in a cell cycle. For instance, cell injury during mitotic state is more detrimental compared to other phases.

2. No   Cell types having greater reproductive capacity or greater mitotic time are more radiosensitive

3. No   There is no difference in their inherent radiosensitivity. The cancer cell has only lost control on growth mechanism due to whatever reason.

4. Yes   But normal tissue cells recover faster compared to tumor tissue cells because of their cell regulatory mechanism, which is still intact.

5. Yes   Like the normal tissues. Fast growing tumors are more radiosensitive compared to slow growing tumors.

6. No   Because of the random nature of the interaction, only a fraction of cells gets killed for a given dose of radiation, the fraction increasing with increasing dose.

7. Yes

8. No   Not necessarily. Radioresponsive tumors will regress faster but whether it will lead to cure or not will depend entirely on the past history of the tumor. In fact, with radiosensitive tumors it is more likely that they would have metastasized early making cure more difficult.

9. No   Cell survival curves are exponential. Exposure to low-LET radiation (x-rays or electrons), however, exhibits an initial shoulder indicating cell recovery. (See the single dose curve in the figure that is part of answer #37 on page 182.)

10. No   Tumor in a patient is a different proposition altogether compared to *in vitro* study of a group of cells. Normal tissue complications often prevent delivery of required dose for better tumor control, while actually treating a patient.

11. No   As mentioned above, radiosensitive tumors are radioresponsive, but not all radioresponsive tumors are radiocurable.

12. Yes

13. Yes

14. No   Tolerance dose for organ increases if only a portion of the organ is irradiated.

15. No   Fractionation leads to better recovery of normal tissues compared to tumor recovery (but at the cost of higher dose to the tissues). This differential, the basis of radiation therapy can be explained in terms of (a) Repair of sublethal damage, (b) Redistribution of cells, (c) Repopulation, and (d) Reoxygenation.

16. a, b

17. a, b

18. d

19. b, c, d

20. d

21. b

22. a, b, c   See the figure below.

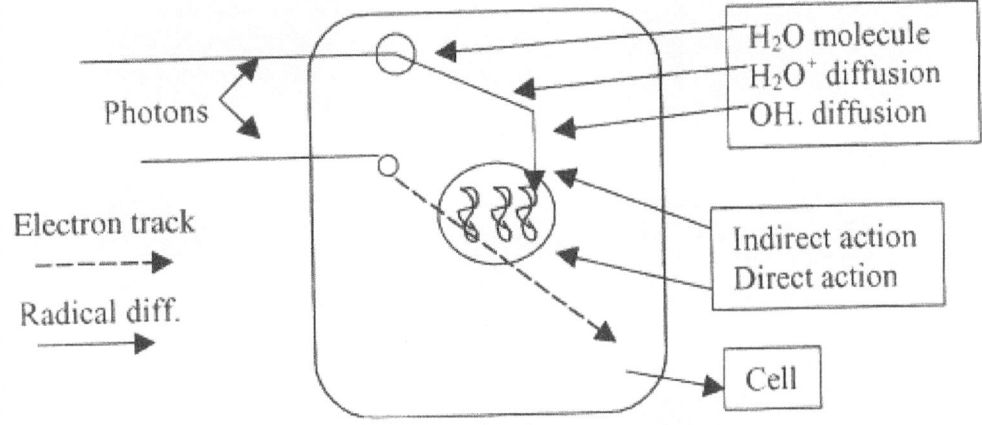

Direct and indirect action in a cell

23. b

24. c

25. c   A tumor of 1 cm^3 will contain about 10^9 cells. Let us say that 'n' doublings are required to reach a size of 1 cm^3. After n doublings, the number of cells = $2^n = 10^9$. 2^{10} is roughly 10^3 (1024). So, $2^{10} \times 2^{10} \times 2^{10} = 10^9$ or 'n' is 30. So, the time taken to reach 1 cm^3 size $\approx 30 \times 60$ days, approximately 5 years.

26. a, b, c

27. a, d

28. a

29. c

30. a

31. a, b

32. c

33. a, b

34. a, b, c, d

35. b, c

36. a, d

37. b, c

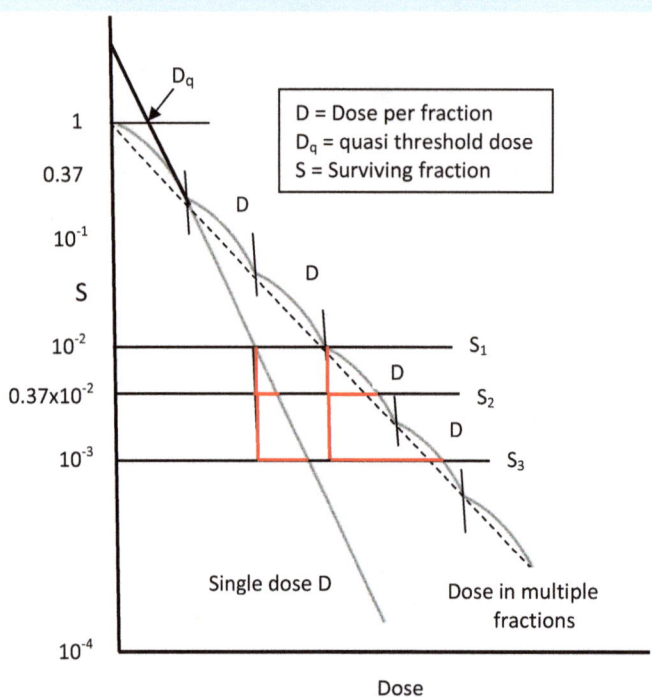

Survival curve for fractionated radiotherapy

38. b

39. b, c

40. b

41. b

42. a, b, c, d

43. d

44. c, d

45. a, b, c, d

46.   a. ii.   b. i.   c. iii.   d. iv. See the figure below.

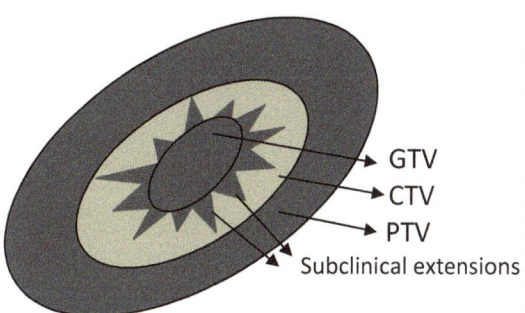

Definition of various volumes in radiation therapy as per ICRU Report No. 50 nomenclature (1993)

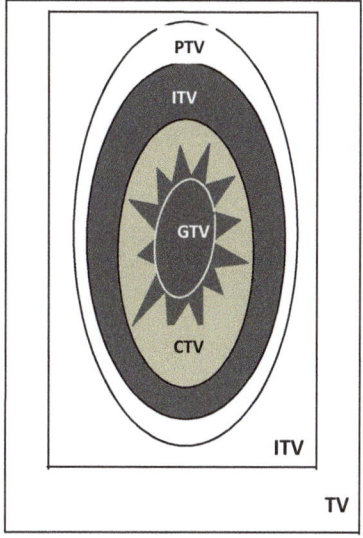

Definition of various volumes in radiation therapy as per ICRU Report No. 62 nomenclature (1999)

As per the revised target volume definitions (ICRU 62):
CTV + Internal margin (IM) = Internal Target Volume (ITV).
ITV + Setup margins (SM) = Planning target volume (PTV).
Treated volume (TV) = volume enclosed by a specified isodose surface, typically 95% isodose volume for curative radiotherapy.
Irradiated volume (IV) = volume receiving a dose that is significant compared to normal tissue tolerance dose.

47.   a

48.   a

49.   a

50.   d

51.   a, b, c, e

52.   a, b, c

53.   a, b, c, d

54.   b, d

55.   b, d, e   RBE obviously depends on the type of radiation since the biological damage depends on the density of ionization or LET. RBE depends on the **dose level** and the **number of fractions** because, in general, the shape of the dose-response relationship varies for radiations that differ substantially in their LET.

RBE can vary with the **dose rate** because the slope of the dose-response curve for low-LET radiations varies critically with a changing dose rate. For densely ionizing radiations, dose-rate is of little importance.

56.   c   RBE = 400 / 80 = 5.

57.   b

58.   e   The following figure illustrates the nature of ionization produced by low-LET and high-LET radiations.

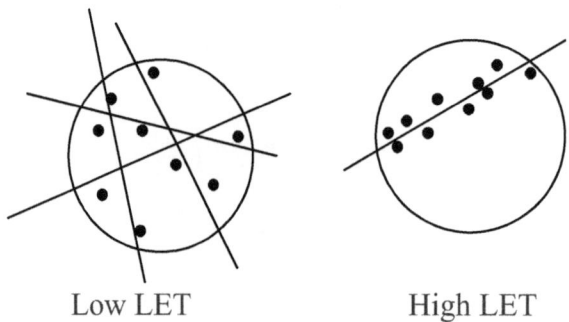

Low LET   High LET

Ion distribution for low-LET and high-LET radiations

59.   c   In general, the dose required to produce a certain biologic effect is reduced as the linear energy transfer (LET) of the radiation is increased.

60.   a, b, c

61.   a. ii   b. v.   c. i.   d. iii.   e. iv.

62.   a, b   They have very high relevance in dose fractionation.

63.   a, c, d

64.   a. iii.   b. i.   c. v.   d. ii.   e. iv.

65.   a. ii.   b. iii.   c. i.   d. v.   e. iv.
(From Emami et al. Tolerance of normal tissue to therapeutic irradiation. IJROBP 1991.)

66.   a, b, c   A simple model is used to explain the tolerance dose of normal tissues. The normal tissues are organized as functional sub units (FSU). Some tissues behave as though the FSUs are arranged in series and some tissues behave as though the FSUs are arranged in parallel. These organs are divided into serial organs and parallel organs.

67.   a, b, c

68.   d   Oxygen has no effect on radiosensitivity for high-LET radiation.

69. d OER occurs due to hypoxia.

70. c

71. a

72. c

73. c

74. b, c, d

75. e

76. d The extent and rate of reoxygenation in various types of tumors are, however, very variable and very difficult to predict.

77. d Survival fraction = 0.1 or 10%, so the percentage of cells killed is 90%.

78. c

79. a. iv. b. iii. c. i. d. ii.
All of these requirements cannot be satisfied at the same time. The parameters must be optimized depending on the situation.

80. b SF = 0.1. This corresponds to 0.9 fraction (i.e., 90%) cell killing.

81. a, b, c, d
See the black horizontal line in the figure. It shows that for the same SF, you need to give more dose if you fractionate. It is OK since the cells are being repaired in the interval. See the red vertical line in the figure. For the same dose, fractionation leads to higher cell survival. i.e., fractionated doses do less damage to normal tissues compared to a single large dose. This is important for reducing normal tissue complications.

The following figure illustrates the effect of fractionation on both the tumor and the normal tissue. One can see radiation effects more damage to the tumor compared to the normal tissues that also get irradiated in the treatment plan.

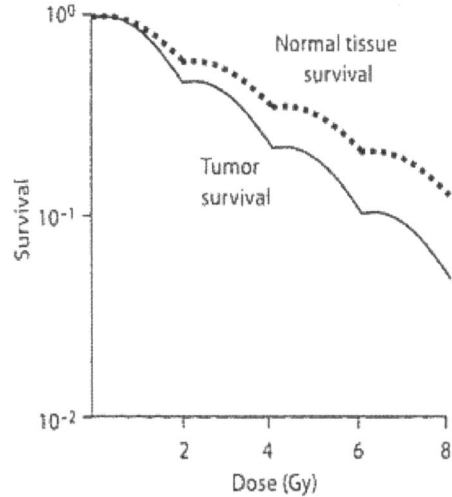

82. a, b, c

83.   e

84.   a, c, d   Some tumor cells are slow cycling.

85.   a, b, d, e

86.   b, c

87.   c   Two regimes are biologically equivalent if they lead to the same SF. The BED is derived from the L-Q model as BED = nd [1+ {d/(α/β)}]. For equal biological effect, $n_1 d_1$ [1+{d_1/(α/β)}] = $n_2 d_2$[1+ {d_2 /(α/β)}], i.e. 2×30 (1 + 2 /3) = $3 \times n_2$ (1 + 3/3) giving n_2 = 16.7. If the tissues are early-responding tissues, one has to only assume α/β = 10 (see the following problem).

88.   b

G. Treatment Techniques and Treatment Delivery

1.   b, d

2.   f

3.   e

4.   a, b, c   Sim-CT is preferred for IMRT planning.

5.   c

6.   a, e

7.   b, c

8.   c

9.   a, c, d

10.   a, b, d

11.   a, b, c, d

12.   e

13.   c, d

14.   a, b, d, e

15.   d

16.   a, b, c   There is no leaf motion in segmental delivery of IMRT.

17.   a. ii.   b. iii.   c. iv.   d. i.

18.   d   You have to just calculate how much dose would have been delivered in 0.05 sec., i.e., 600 MU/min $\times$ (1/60) $\times$ 0.05 = 0.5 MU or (0.5/2) $\times$ 100 = 25%.

19. a, b   Tongue and groove design refers to the "overlapping of adjacent leaves design" of commercial MLCs to reduce interleaf leakage. It has been clinically indicated that this design can cause an underdosing to the tune of 10% to 20% and not between the ends where the dose will, in fact, be higher due to thin edges.

20. a, c, d

21. b, c, d

22. a. ii.   b. i.   c. ii.   d. ii.   e. i.

23. c, d

24. d

25. a, c, d, e

26. a, b   In fact, the disadvantages of segmental IMRT are significant treatment time involved. Complex treatment problems may require a large number of segments per field.

27. e   The following figure (from the thesis work of Gordon Mark Mancuso, available at http://etd.lsu.edu) shows OARS bordering on PTV and the need for concave dose distribution for avoiding dose to bladder and rectum. Since the dose to OAR limits the dose to PTV, the PTV dose can now be escalated, or for the same dose as prescribed in 3D CRT, the NTC decreases in IMRT compared to 3D CRT.

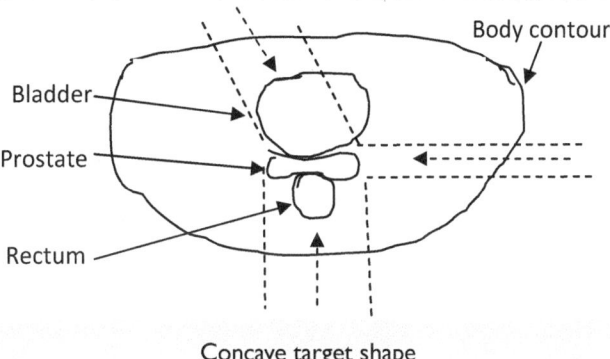

Concave target shape

28. a, b, c, e

29. f

30. a, b, c   Small positional errors can be detrimental in IMRT.

31. e

32. a, b, c

33. a, b   Though both dose rate and gantry speed can be varied, due to inertia of gantry, dose rate is preferentially modulated. Gantry speed is ideally kept at max to deliver dose quickly.

34. a, b, c

35. a, b

36. b   One rotation corresponds to 360° rotation. The angular speed is 5.5°/sec. If it takes x secs to complete one rotation, 5.5x = 360 or x = 360/5.5 = 65 secs.

37. b   Leaf speed is measured in cm/sec.

38. d   There are 177 control points in a 360 degree arc. The angle between control points = 360°/177 = 2°, i.e., after every 2° rotation of the gantry, the dose rate changes. The angular speed is generally kept the maximum.

39. d

40. e

41. c

42. e

43. e   Both geometric accuracy and dosimetric accuracy are equally important. Delivering the wrong dose to the right target or the right dose to the wrong target both constitute wrong treatment.

44. a, c, d   SRS involves single fraction.

45. a, c, d

46. a, b, c

47. a

48. e

49. a

50. e

51. a

52. e

53. b

54. c

55. a

56. c

57. a, b, c, d

58. e

59. b, c

60. e

61. a, b, c, d

62. a, b, c, d
                Gamma Knife can use multiple isocenters by moving the adjacent treatment spots to the focus, but the overlap of fields can cause hotspots.

63. a, b, c, d

CyberKnife performs frameless intracranial treatments, made possible by the continual image guidance.

64. e

65. a, c   CyberKnife provides unlimited access to both the spine and the cranium, where extreme peripheral lesions can also be accessed. With Gamma Knife, only a small region in the central part of the cranium can be accessed. Respiratory motion does not affect intracranial treatments.

66. b, c, d

67. a, c, d

68. a, c   kV x-ray beams were used in the past for TSET, but these days one uses low-energy linac electron beams. Since we have to treat the superficial layers of the skin, beams with higher penetration (^{60}Co or high-energy x-ray beams) are not used.

69. e

70. a. ii.   b. iii.   c. i.

71. a, b, c

72. f

73. a, b, d

74. e

75. f

76. d   Any gross error (say more than 10 mm) in patient field size, anatomical position, or orientation will defeat the whole purpose of treatment and can be detected and corrected by the first verification image obtained before the delivery of the first fraction.

77. b, c, d

78. e

79. b

80. b, c, d

81. a. iii.   b. i.   c. ii.

IV. Dose Calculation Methods

A. Applied Mathematics

Choose the right answer:

1. Similar triangles are triangles that have _____.
    a. equal sides
    b. equal angles
    c. always equal angles and equal sides
    d. equal vertex angles
    e. same base length

2. In the two similar triangles shown in the figure below, which of the following statements are true?
    a. a/b = a'/b'
    b. b/a = b'/a'
    c. a/a' = b/b'
    d. ab' = a'b
    e. all the above

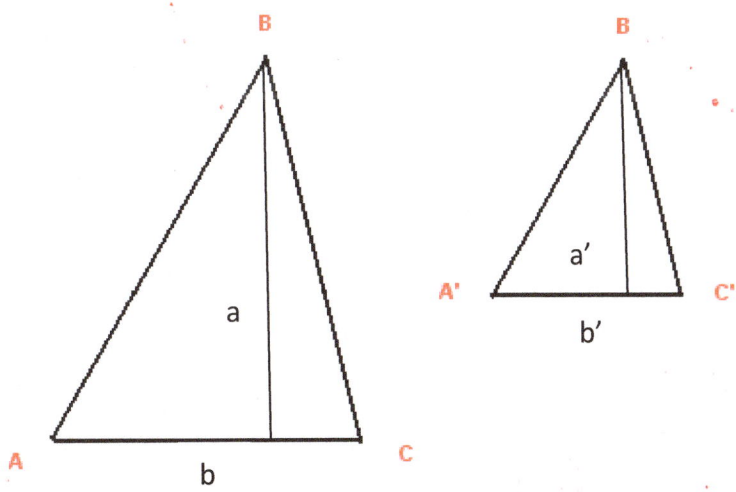

Properties of similar triangles

3. An 8 x 12 field is prescribed to treat a patient at 120 cm SSD. _____ are the required jaw settings. (The machine field size is defined 1 m.)
    a.   8, 12
    b.   10.4, 14.3
    c.   8, 14
    d.   6.6, 10
    e.   4, 6

4. An x-ray machine projects a field size of 15 x 15 cm^2 at an SAD of 100 cm. _____ are the X and Y diaphragm openings (in cm). Given: the X and Y diaphragms are at 46.5 and 34.5 cm from the focal spot, respectively.
    a.   12, 12
    b.   10, 12.4
    c.   7, 5.2
    d.   6.6, 10
    e.   5, 10.5

5. A patient is set up in the simulator at an SSD of 115 cm. The field size on the skin was recorded as 15 x 24. A collimator setting of _____ should be used on the 100 cm SAD linac to produce the treatment field. (The collimator settings define the field size at SAD distance.)
    a.   14 x 20
    b.   13 x 20
    c.   10 x 32
    d.   8 x 18
    e.   7.5 x 12

6. The field size set on the collimator is 10 x 10. For a patient setup with the isocenter on the midline, _____ are the entrance and exit field sizes, respectively. (The patient thickness is 24 cm.)
    a.   10 x 10, 20 x 20
    b.   9 x 9, 12.5 x 12.5
    c.   8.8 x 8.8, 11.2 x 11.2
    d.   6.6 x 6.6, 10.5 x 10.5
    e.   5 x 5, 15 x 15

7. _____ are the geometric penumbra of an x-ray beam on the patient surface and at a depth of 10 cm, respectively (see the figure below). The following parameters are given (in cm):
   SSD = 100; SDD = 42; s = 0.2.
   a. 1.0, 1.0
   b. 0.78, 0.93
   c. 0.276, 0.324
   d. 0.124, 0. 237
   e. 0.05, 0.08

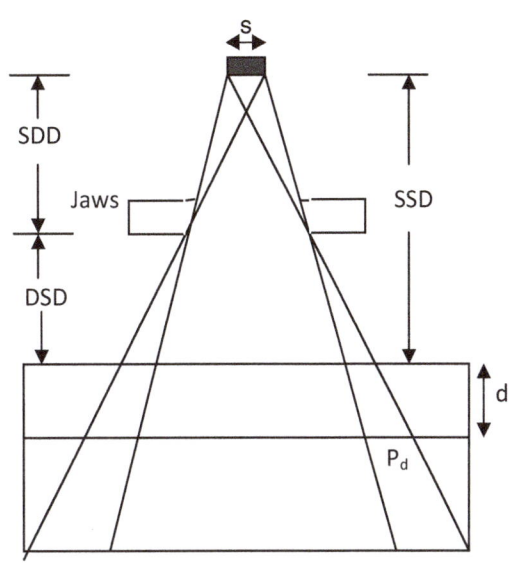

Penumbra width of a therapy beam

SSD: source-to-surface distance
SDD: source-to-diaphragm distance
DSD: diaphragm-to-surface distance
P_d: penumbra at depth d
s = source diameter

8. The field size at 100 cm from the source is 10 x 10. _____ is the field size at 110 cm.
   a. 11 x 11
   b. 10 x 11
   c. 11 x 10
   d. 15 x 15
   e. 20 x 20

9. Assuming a point source of radiation, how does the air kerma rate vary with respect to the distance from the source in air (assuming negligible attenuation)? It varies as _____.
   a. d
   b. d^2
   c. 1/d
   d. $1/d^2$
   e. d^3

10. Assuming a point source of collimated radiation, how does the "air kerma rate x field Area (A)" vary with respect to the distance from the source in air (assuming negligible attenuation)? It _____.
    a. varies as d
    b. varies as d^2
    c. varies as A/d
    d. varies as A^2/d^2
    e. does not vary

11. If the distance from the source is doubled, the air kerma rate _____.
    a. remains the same
    b. is doubled
    c. becomes 4 times
    d. becomes 1/4
    e. depends on the source spectrum

12. The calibration dose rate for a 6 MV linear accelerator is typically 1.00 cGy/MU at a depth of d_{max} (1.5 cm) for an SSD of 100 cm. The dose rate at d_{max} at 110 cm SSD changes the calibration dose rate by a factor of _____.
    a. $110^2 / 100^2$
    b. $100^2 / 110^2$
    c. $111.5^2 / 101.5^2$
    d. $101.5^2 / 111.5^2$
    e. $111.5^2 \times 101.5^2$

13. A 15 MV linear accelerator is calibrated in SSD setup., i.e., D_{cal} (3 cm, 10 x 10, 100 SSD) = 1 cGy/MU. The calibration dose rate for an SAD setup, i.e., D_{cal} (3 cm, 10 x 10, 100 SAD) can be obtained by multiplying 1 cGy/MU by a factor of _____.
    a. $103^2 / 100^2$
    b. $100^2 / 103^2$
    c. $103^2 \times 100^2$
    d. $100^2 / 103^2$
    e. $103 / 100$

14. A lead wire of 1.5 cm length is kept on the patient (see the figure below). _____ will be the length of its image (in cm) on a simulator film. The following data are given:
    source-to-wire distance = 100 cm; source-to-film distance = 140 cm.
    a. 3.4
    b. 2.9
    c. 2.1
    d. 1.7
    e. 1.0

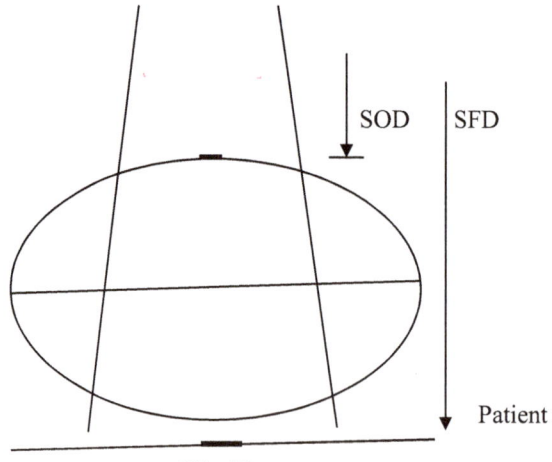

SOD: Source to Object Distance
SFD : Source to Film Distance

Object magnification during imaging

15. _____ is the film magnification factor in the above problem.
    a. 2.9
    b. 2.3
    c 1.9
    d. 1.4
    e. 0.5

16. A beaded tray projects a 1 cm grid at 100 cm machine isocenter. The tray slides in the simulator head at source-to-tray distance of 40 cm. _____ is the distance between the beads (in cm) on the tray.
    a. 2.5
    b. 1.3
    c. 0.4
    d. 0.2
    e. 0.05

17. A lead block of 3 cm diameter base is kept on the shielding tray which is at a distance of 40 cm from the source. _____ will be the size of the shadow on the film (in cm) at a source-to-film distance of 140 cm.
    a. 4.2
    b. 4.7
    c. 5.3
    d. 10.5
    e. 12.6    L×W

18. Two adjacent fields of 28 x 15 are used to treat a longer length at 5 cm depth as shown in the figure below (not to scale). A _____ cm field gap on the skin, shown for the larger field dimension in the figure, will ensure matching of the two fields at the desired depth.
    a. 12
    b. 10.4
    c. 3.2
    d. 2.8
    e. 1.4

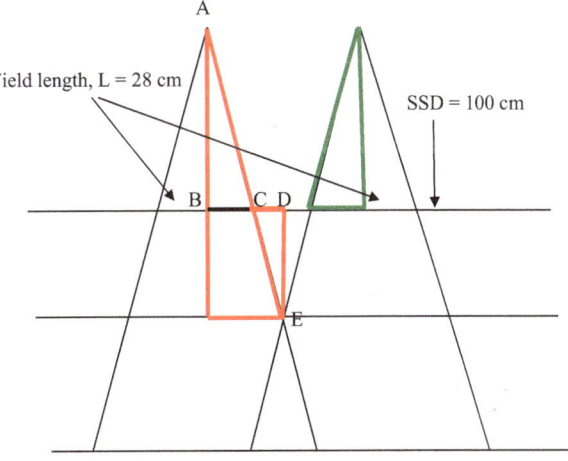

Similar triangle properties for field gap evaluation

19. The TMR (10) values for 6 MV beam for the 10 x 10 and 12 x 12 fields are respectively 0.778 and 0.788. By linear interpolation, the TMR(10) value for an 11x 11 field is _____.
    a. 1.566
    b 0.788
    c. 0.783
    d. 0.778
    e. 0.010

20. The TMR (12) values for 6 MV beam for the 10 x 10 and 15 x 15 fields are respectively 0.720 and 0.748. The TMR(10) value for a 13 x 13 field is _____.
    a. 0.748
    b 0.745
    c. 0.737
    d. 0.726
    e. 0.023

21. Part of the TMR table for the 6 MV beam is given below.

Depth (cm)	Field Size (cm x cm)		
	8 x 8	10 x 10	12 x 12
8	0.826	X1	0.843
9	Y1	Z	Y2
10	0.765	X2	0.788

The TMR (9, 10 x 10) value is _____.
    a. 0.844
    b. 0.806
    c. 0.794
    d. 0.746
    e. 0.695

22. The output of a linac beam is 300 cGy/min. If the dose rate is doubled the time to deliver a dose will _____.
    a. remain the same
    b. also become doubled
    c. become half
    d. depend on the original dose rate
    e. depend the linac model and the beam quality

23. The output dose rate of a linac is 300 cGy/min. The treatment of a patient requires 2.6 minutes on the machine. _____ will be the treatment time (in minutes) for the same treatment conditions if only the dose rate is set to 400 cGy/min.
    a. 3.0
    b. 1.95
    c. 1.4
    d. 1.0
    e. 0.5

24. The tumor dose prescribed by the physician is 6000 cGy. The total number of treatment days is 30. _____ is the dose (in cGy) per fraction.
    a. 200
    b. 130
    c. 125
    d. 100
    e. 30

25. The tumor dose at the tumor center is 200 cGy. The tumor dose rate is 300 cGy/min. The treatment time (in minutes) is _____.
    a. 3.4
    b. 2.6
    c. 1.4
    d. 0.67
    e. 0.44

B. Beam Calibrations (Photon and Electron Beams)

Choose the right answer(s) (more than one may be correct):

1. Which of the statements are true regarding the kV dosimetry using the AAPM TG-61 code of practice?
    a. Basic measurements can be made in air or in a water phantom depending on beam quality.
    b. Cylindrical chambers or PP chambers can be used depending on beam quality.
    c. N_x or N_K calibrated chambers can be used.
    d. The reference depth of measurement is the same as that of ^{60}Co.
    e. All of the above are true.

2. _____ is the dose (in cGy) on the surface of a phantom from an in-air measurement using a PP chamber in the kV x-ray machine used for superficial therapy (see the figure below). Use the data given below:
   Chamber used: PTW PP chamber
   Beam Quality: 30 kV (0.648 mm AL HVL); $N_K = 69.34$ mGy/nC at 22 °C and 760 mm Hg
   $B_W = 1.081$; $[(_m\mu_{en})_{w,air}]_{air} = 1.025$. $P_{stem,air} = 1.001$
   Chamber reading corrected for (T,P), polarity and saturation, $M_c = 13.26$ nC
    a. 134.6
    b. 102.0
    c. 87.6
    d. 45.2
    e. 33.5

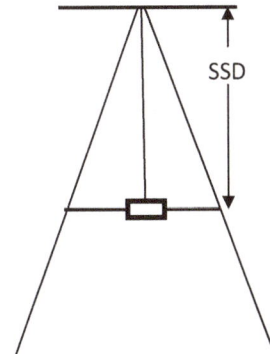
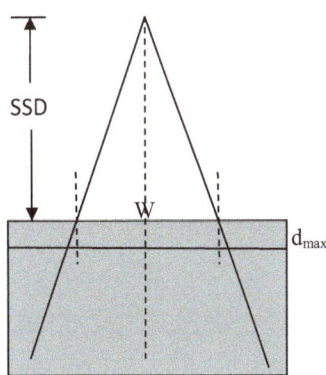

In-air calibration to determine dose on the phantom (AAPM TG-61 COP)

3.   _____ is the dose (in cGy) on the surface of a phantom from an in-air measurement using a cylindrical therapy level chamber (NE 2571) in the kV x-ray machine used for superficial therapy (see the figure in the previous problem). Use the data given below:

Chamber used: NE 2571 cylindrical chamber
Beam quality: 120 kV (0.143 mm Cu HVL); N_K = 69.34 mGy/nC at 22 °C and 760 mm Hg
B_W = 1.321; $P_{stem,air}$ = 1.001; $[(_m\mu_{en})_{w,air}]_{air}$ = 1.023
Chamber reading corrected for (T,P), polarity, and saturation M_c = 11.21 nC

a.   67.8
b.   77.6
c.   86.3
d.   105.1
e.   184.2

4.   The above problem assumes an open-ended applicator with an SSD of 20 cm. Assuming the applicator was close ended and that the above measurement reading was obtained when the chamber was placed touching the applicator face (see the figure below), _____ is the correction that needs to be applied in the above problem to derive the dose on the phantom surface. (The outer radius of the chamber = 4.3 mm.)

a.   2.4
b.   1.68
c.   1.02
d.   0.94
e.   0.67

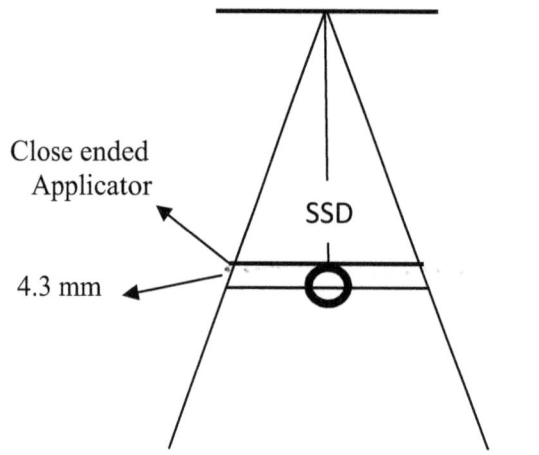

Measurement for close ended applicator

5.   _____ is the correction in the above problem if SSD = 15 cm and the chamber outer diameter = 1 cm.

a.   3.03
b.   2.4
c.   1.88
d.   1.07
e.   0.64

6. _____ is the dose in cGy at 2 cm in a water phantom from an in-phantom measurement using a cylindrical therapy level chamber (NE 2571) in the kV x-ray machine used for orthovoltage therapy (see the figure below). Use the data given below:

Chamber used: NE 2571 cylindrical chamber

Beam quality: 200 kV (1.036 mm Cu HVL); $N_K = 41.2$ mGy/nC at 22 °C and 760 mm Hg

$P_{sheeeth} = 0.999$; $[(_m\mu_{en})_{w,air}]_w = 1.061$; $P_{Q,ch} = 1.023$

Chamber reading corrected for (T,P), polarity, and saturation $M_c = 20.49$ nC

   a. 100.6
   b. 91.5
   c. 88.0
   d. 78.8
   e. 50.3

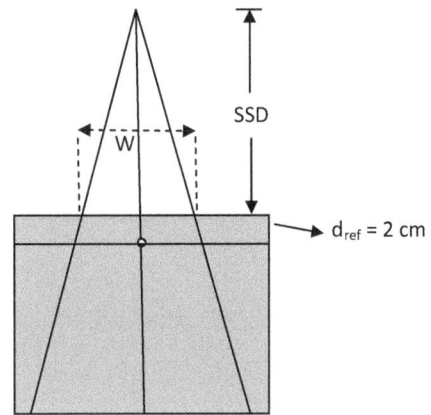

In-phantom dose calibration (AAPM TG-61 COP)

7. In the above problem, if PDD (2 cm) is 90.2%, _____ is the dose in cGy at d_{max} (i.e., at the surface).
   a. 110.2
   b. 108.3
   c. 105
   d. 103.2
   e. 101.5

8. _____ is the dose (in cGy) at the reference depth of 5 g/cm^2 in a water phantom using the NE 2571 chamber for a ^{60}Co beam given the following data:

$N_{Dw,Co} = 45.1$ mGy/nC at 22 °C and 760 mm Hg

Chamber reading corrected for (T,P), polarity, and saturation = 24.82 nC

The reference field conditions are SSD = 100 cm and field size 10 x 10 at the given SSD
   a. 132
   b. 112.6
   c. 111.9
   d. 65.3
   e. 1.8

9. A Roos PP chamber was calibrated against a cylindrical reference chamber (NE 2571) at the reference depth in water in a ^{60}Co beam (see the figure below). _____ is the absorbed dose to water calibration factor, $[N_{Dw,Co}]_{Roos,pp}$ (in mGy/nC) of the Roos PP chamber using the following data:

    $[N_{Dw,Co}]_{NE2571}$ = 45.1 mGy/nC at 22 °C and 760 mm Hg

    Chamber readings—corrected for (T,P), polarity, and saturation—of NE 2571 and Roos PP chamber are 22.23 nC and 12.12 nC, respectively

    a. 123.4
    b. 105
    c. 82.7
    d. 73.9
    e. 43.4

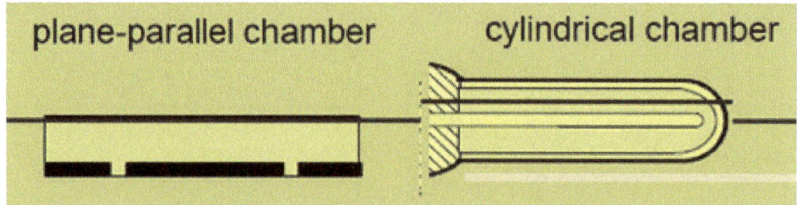

In-air calibration of PP chamber in ^{60}Co beam (AAPM TG-51 COP)

10. What is the fundamental difference between TG-51 and the earlier protocol (TG-21) of the AAPM?
    a. New chamber factors have been included in TG-51.
    b. TG-21 applies for x-ray beams <10 MV.
    c. The dosimetry data used in TG-51 are more accurate, making significant differences in the dose derived compared to the earlier protocol.
    d. TG-51 is based on $N_{Dw,Co}$ calibration, while TG-21 is based on N_X calibration for the chamber.
    e. The dosimetry formalisms are quite different.

11. The ionization chambers recommended for photon dosimetry are _____.
    a. 0.6 cm^3 Farmer type ion chambers
    b. 0.125 cm^3 scanning chambers
    c. 0.02 cm^2 micro chambers
    d. Roos type PP chambers
    e. extrapolation chambers

12. The phantom recommended by TG-51 for x-ray beam calibration is of _____.
    a. water
    b. acrylic
    c. tissue-equivalent material
    d. graphite
    e. aluminum

13. In order to use a therapy level Farmer-type ion chamber for the dosimetry of x-ray beams as per TG-51 protocol, one should have the _____ for the chamber of interest.
    a. $N_{Dw,Co}$ calibration factor provided by an ADCL or a PSDL laboratory
    b. exposure calibration factor
    c. air kerma calibration factor
    d. air kerma strength calibration factor
    e. k_Q factor for the user beam quality

14. _____ are the corrections to be applied to the "raw" chamber reading.
   a. Temperature/pressure correction
   b. Saturation correction
   c. Polarization correction
   d. Electrometer charge/current calibration
   e. All of the above

15. Apart from PDD$(10)_X$, _____ is the other parameter very much used for beam quality specification.
   a. PSF
   b. NPSF
   c. TAR
   d. TPR$_{20,10}$
   e. PDD $(20)_X$

16. Which of the following statements are true regarding the PDD in TG-51 protocol for the dosimetry of linac beams >10 MV?
   a. The effective center of the therapy chamber must be used for PDD measurement.
   b. The PDD $(10)_X$ must be determined at the nominal SSD of the teletherapy unit.
   c. The 10 cm depth has been chosen to eliminate the influence of electron contamination on the beam quality specifier.
   d. PDD $(10)_X$ must be derived by measuring PDD$(10)_{Pb}$ with an a 1 mm acrylic sheet in the beam and using the equation given in TG-51.
   e. Electron contamination will significantly influence k_Q values.

17. Match the dosimetric parameters to their values in x-ray beams.
   Dosimetric Parameters   *Values*
   a. calibration depth   _____ i. $0.6 \, r_{cav}$
   b. beam quality specifier   _____ ii.   100 cm
   c. beam quality measurement SSD   _____ iii. PDD $(10)_X$
   d. shift to effective center from geometric center _____ iv. 10 cm
   e. calibration field size   _____ v.   10 x 10

18. When making measurements using TG-51 and a cylindrical chamber for the dosimetry of linac beams, the

   _____.
   a. geometric center of the chamber must be placed at a defined d_{ref} depth for electron beams and at fixed depth of 10 cm for photon beams
   b. effective center of the chamber must be placed at d_{ref} for electron beams and at fixed depth of 10 cm in photon beams
   c. geometric center of the chamber must be placed at d_{max} for electron beams and at fixed depth of 10 cm in photon beams
   d. effective center of the chamber must be placed at d_{max} for both electron and photon beams
   e. the geometric center of the chamber can be kept any depth beyond d_{max}

19. The TG-51 recommends keeping the geometric center of the cylindrical chamber at the reference depth for x-ray beam output measurements because the shift to the effective center is _____.
   a. not necessary since both coincide while measuring in a phantom
   b. accounted for in the chamber calibration factor
   c. canceled by calibrating the chamber with respect to the geometric center in the calibration lab
   d. separately accounted in the k_Q factor
   e. going to make negligible error in beam calibration

20. The ion chamber readings for –300 volts and +300 volts were -4.073×10^{-8} and $+4.061 \times 10^{-8}$ C, respectively. _____ is the polarity correction for the chamber.
    a. 1.000
    b. 0.999
    c. 0.950
    d. 0.924
    e. 0.898

21. The ion chamber readings for –300 volts and –150 volts were -4.073×10^{-8} and -4.053×10^{-8} C, respectively. _____ is the saturation (ion recombination) correction for the chamber.
    a. 1.005
    b. 0.995
    c. 0.937
    d. 90
    e. 0.85

22. The ionization chamber used in the hospital has a calibration factor referring to 760 mm Hg and 22 °C. If the measurement conditions in the hospital are 20.7 °C and 750.1 mm Hg, _____ is the temp/pressure correction, P_{TP}, that must be applied to the chamber reading.
    a. 1.322
    b. 1.009
    c. 1.000
    d. 0.983
    e. 0.9543

23. The hospital electrometer was calibrated at an ADCL. Calculate the electrometer calibration using the following data: input charge to the electrometer = 1.000 nC and the electrometer reading = 0.997 nC. The electrometer calibration is given by _____.
    a. 1.010 nC
    b. 1.003
    c. 1.000 nC/C
    d. 0.997
    e. 0.9925

24. Calculate the dose at 10 cm depth (in cGy) for the 23 MV clinical linac beam in an SAD setup using the TG-51 protocol and the following data: Chamber used: NE 2571 : $N_{Dw,Co} = 4.533 \times 10^9$ cGy/C
    Chamber reading at 10 cm depth, with chamber g.c. kept at 10 cm depth = -4.073×10^{-8} C

P_{ion}	P_{pol}	P_{elect}	P_{TP}	k_Q
1.004	0.999	1.000	1.008	0.971

    a. 200.54
    b. 195.64
    c. 181.25
    d. 160.44
    e. 150.35

25. In the above example, _____ is the dose per MU (in cGy/MU) at the dose maximum depth (at SAD) using the following data: Monitor unit set on the console to deliver the above dose = 200 : TMR (10, 10 x 10, 100 SAD) for the set field = 0.901
    a. 1.783
    b. 1.546
    c. 1.14
    d. 1.006
    e. 1.000

26. _____ is the dose at 10 cm depth (in cGy) for the 6 MV, 10 x 10 field clinical linac beam in an SSD setup using the TG-51 protocol and the following data:
Chamber used: PTW N30006 : $N_{Dw,Co} = 5.347 \times 10^9$ cGy/C at 22 °C and 760 mm Hg;
Chamber reading at 10 cm depth with chamber g.c. kept at 10 cm depth = -12.78×10^{-9} C

P_{ion}	P_{pol}	P_{elect}	P_{TP}	k_Q
1.004	1.000	1.000	0.997	0.990

    a. 67.72
    b. 56.2
    c. 52
    d 43.0
    e. 41.7

27. In the above example, _____ is the dose per MU (in cGy/MU) at the dose maximum depth using the following data:
Monitor unit set on the console to deliver the above dose = 100: PDD (10, 10 x 10, 100 SSD) = 0.676
    a. 100.117
    b. 45.43
    c. 12.64
    d. 1.002
    e. 0.994

28. _____ is the dose at 10 cm depth (in cGy) for the 18 MV, 10 x 10 field clinical linac beam in an SSD setup using the TG-51 protocol and the following data:
Chamber used: PTW N30006: $N_{Dw,Co} = 5.347 \times 10^9$ cGy/C
Chamber reading at 10 cm depth with chamber g.c. kept at 10 cm depth = -15.21×10^{-9} C

P_{ion}	P_{pol}	P_{elect}	P_{TP}	k_Q
1.003	1.000	1.001	0.998	0.975

    a. 95.32
    b. 84.3
    c. 79.46
    d, 68.2
    e. 23.0

29. In the above example, _____ is the dose per MU (in cGy/MU) at the dose maximum depth using the following data:
Monitor unit set on the console in the above problem = 100: PDD (10, 10 x 10, 100 SSD) = 0.801
    a. 99.92
    b. 67.23
    c. 15.4
    d. 1.006
    e. 0.992

30. Electron dosimetry is more complicated than photon dosimetry because _____.
    a. measurements are more difficult
    b. the ion chambers are more complicated in design
    c. no reference electron beam is available for electron calibration at ADCLs making the user derive an additional calibration from reference photon to reference electron beam.
    d. electron theory is more complex
    e. all the above are true

31. When measuring the PDD with a cylindrical chamber for electron beams, the shift to effective depth from the geometric center depth in terms of cavity radius, r, is _____.
    a. 0.6 r
    b. 0.6 r up to d_{max} and 0.7 r beyond d_{max}
    c. 0.5 r at all depths
    d. 0.6 r at all depths
    e. depends on the depth

32. The effective point of a PP chamber is at the _____.
    a. geometric center of the chamber cavity
    b. inner surface of the chamber wall on the entry side
    c. outer surface of the chamber wall on the entry side
    d. inner surface of the chamber wall on the exit side
    e. outer surface of the chamber wall on the exit side

33. Which of the following statements are true regarding the ion chambers used in photon and electron dosimetry protocols?
    a. Farmer chambers are not suitable for the dosimetry of low-energy (<6 MeV) electron beams.
    b. PP chambers are not suitable for the dosimetry of photon beams.
    c. For Farmer type chambers in photon dosimetry, wall corrections are negligible for thin walls (≤0.5 mm).
    d. For PP chambers in electron dosimetry, wall corrections are negligible for thin walls (≤0.5 mm).
    e. All of the above are true.

34. Match the dosimetric parameters to the corresponding quantities for electron beams.

Dosimetric Parameter		Corresponding Quantity
a. calibration depth	_____	i. $0.5\, r_{cav}$
b. beam quality specifier	_____	ii. 100 cm
c. beam quality measurement SSD	_____	iii. R_{50}
d. shift to effective center from geometric center	_____	iv. d_{ref}
e. calibration field size	_____	v. ≥10 x 10

35. The electron energies are represented by their in-water penetration depth parameter R_{50} (in cm) in the TG-51 protocol, while the earlier protocols represented the incident electron energies by $\bar{E}_0$ the mean energy of the electron beam, incident on the patient. _____ equation/s give a good estimate of the electron beam energy $\bar{E}_0$ from R_{50}.
    a. $\bar{E}_0 = R_{50}$
    b. $\bar{E}_0 = 2\, R_{50}$
    c. $\bar{E}_0 = 2.4\, R_{50}$
    d. $\bar{E}_0 = 3\, R_{50}$
    e. $\bar{E}_0 = R_{50} / 2.33$

36. _____ is the nominal energy (in MeV) of a clinical electron beam having an R_{50} of 4.3 cm. (Use the formula: nominal energy $E_n = 2.4\, R_{50}$.)
    a. 1
    b. 2.4
    c. 4.3
    d. 10
    e. 12.6

37. TG-51 recommends that for electron beam calibrations or beam quality determinations, a minimum field size of 10 x 10 must be used for R_{50} up to 8.5 cm. This corresponds to an electron energy of _____.
    a. 20
    b. 10
    c. 8.5
    d. 4.3
    e. 2

38. _____ is the R_{50} (in cm) range of an $\overline{E}_0 = 18$ MeV clinical electron beam.
    a. 20
    b. 10
    c. 7.5
    d. 5.7
    e. 4

    $E_0/2.4 \approx R_{50}$

39. _____ is the practical range of an electron beam having an R_{50} of 2.6 cm.
    a. 4.5
    b. 3.1
    c. 2.6
    d. 1.3

40. _____ is the practical range (in cm) of a 10 MeV clinical electron beam.
    a. 20
    b. 15
    c. 10
    d. 5
    e. 3.5

41. _____ is the maximum electron energy (in MeV) that can be used if an organ beyond 6 cm depth must be protected from radiation.
    a. 18
    b. 12
    c. 6
    d. 4
    e. 1.8

42. _____ is the minimum field size that must be used for determining the beam quality or for beam calibration of the clinical electron beams.
    a. any field size
    b. 10 cm x 10 cm
    c. field size adequate to establish lateral scatter equilibrium
    d. field size >2 R_p
    e. minimum available field size

43. _____ is the therapeutic depth (in cm) of a 6 MeV clinical electron beam.
    a. 6
    b. 4
    c. 3
    d. 1.5
    e. 0.6

44. If the tumor requires a therapeutic depth of 3 cm, _____ is the energy (in MeV) of the electron beam you would choose.
    a. 2
    b. 4
    c. 8
    d. 12
    e. 24

45. _____ is the approximate mean energy (in MeV) of a 10 MeV electron beam at a depth of 4 cm.
    a. 0.5
    b. 2.0
    c. 3.0
    d. 4.0
    e. 8.0

46. _____ is the approximate range (in cm) of a 12 MeV electron beam in a water phantom.
    a. 12
    b. 8
    c. 6
    d. 4
    e. 2

47. _____ is the energy of a 12 MeV electron beam at a depth of 4 cm in water using Harder's relation (given $R_p = 6.1$ cm).
    a. 12
    b. 7.8
    c. 5.9
    d. 4.1
    e. 0.6

48. _____ is the MU required to deliver a dose of 200 cGy at $d_{max} = 2.9$ cm for a 12 MeV, 10 x 10 field electron beam at 100 cm SSD. D_{cal} (d_{max}, 10 x 10, 100 SSD) = 1.005 cGy/MU).
    a. 200
    b. 199
    c. 185
    d. 144.9
    e. 100

49. Which of the following statements regarding the reference conditions for electron beam calibration are true according to the TG-51 protocol?
    a. Calibration depth = d_{max}.
    b. Reference field size = minimum field size required for scatter equilibrium.
    c. SSD = nominal SSD used in the clinic.
    d. Calibration depth = $d_{ref} = 0.6 R_{50} - 0.1$.
    e. All of the above are true.

50. Given Q = x-ray beam quality and Q_{ecal} = reference electron beam quality, k_{ecal} is defined as _____.
    a. $N_{Dw,Co}$
    b. $N_{Dw,Q}$
    c. $[N_{Dw,Q} / N_{Dw,Co}]$
    d. $[N_{Dw,Q} / N_{Dw,Qecal}]$
    e. $[N_{Dw,Qecal} / N_{Dw,Co}]$

51. The factor k_{ecal} is _____
    a. photon to electron calibration conversion factor (or photon to electron beam quality conversion factor)
    b. unique for each chamber
    c. is defined only for PP chamber
    d. is available in the protocol for the chambers that are in common use in dosimetry
    e. always required for electron beam dosimetry

52. k_{ecal} for a Roos (PP) chamber is 0.901 determined for a reference electron beam quality R_{50} = 7.5 cm. $N_{Dw,Co}$ for the Roos chamber is by 83.17 mGy/nC at 22 °C and 760 mm Hg pressure. The chamber reads 33.7 nC at the reference depth for the reference electron beam quality. Other corrections are assumed to be 1. _____ is the dose (in cGy) at the reference depth.
    a. 312
    b. 284.6
    c. 252.5
    d. 212.9
    e. 134.8

53. k_{ecal} for the NE 2571 cylindrical therapy chamber is 0.903 determined for a reference electron beam quality R_{50} = 7.5 cm. $N_{Dw,Co}$ for the NE chamber is by 45.1 mGy/nC at 22 °C and 760 mm Hg pressure. The chamber reads 67.9 nC at the reference depth for the reference electron beam quality. The room temperature was 24 °C and the pressure 774 mm Hg. _____ is the dose (in cGy) at the reference depth.
    a. 154.2
    b. 173.6
    c. 189.9
    d. 273.4
    e. 293.2

54. In the TG-51 electron dosimetry protocol, the P_{gr}^Q correction factor _____.
    a. applies explicitly in the dosimetry equation
    b. corrects for the dose gradient in water at the chamber position in the absence of the chamber
    c. depends on the cavity size
    d. must be determined by the user
    e. is mentioned but no procedure for its measurement is mentioned

55. An NACP PP chamber is cross calibrated against a PTW cylindrical chamber in a 20 MeV electron beam at a reference depth in a water phantom. The following data are given:
    $[N_{Dw,Co}]_{PTW}$ = 51.87 mGy/nC; k_{ecal} = 0.897; $(k_{R50})'$= 0.906; $P_{gr}(20 \text{ MeV})$ = 0.993 chamber reading, M_{PTW} = 59.3 nC; M_{NACP} = 17.12 nC. Assume chamber correction factors as unity. The $[N_{Dw,Co} \, k_{ecal}]$ factor (in mGy/nC) for the PP chamber is _____.
    a. 180.3
    b. 145.0
    c. 124.9
    d. 111.7
    e. 95.8

56. A Roos chamber (PP) was calibrated in a calibration laboratory (in mGy/nC) against NE 2571 reference chamber at 5g/cm² in a water phantom for a ^{60}Co beam (see the figure below). The entrance window of the PP was kept at the calibration depth. using the following data:
$[N_{Dw,Co}]_{NE}$ = 45.28 mGy/nC at 22 °C and 760 mm Hg pressure.
M_{NE} = 27.6 nC; M_{Roos} = 15.1 nC; Room temp. = 22 °C; P = 760 mm Hg
The $[N_{Dw,Co}]$ factor for the Roos PP chamber is _____.
   a. 123.6
   b. 108.5
   c. 82.8
   d. 45.7
   e. 23.6

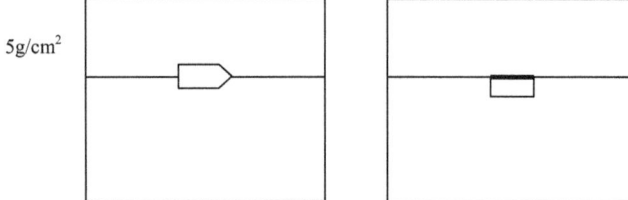

5g/cm²

PP chamber calibration against NE reference chamber in ^{60}Co beam

57. _____ is the dose at the reference depth of 3.9 cm for a 16 MeV clinical electron beam (in cGy) using the following data for the NE chamber: $[N_{Dw,Co}]$ = 45.28 mGy/nC at 22 °C and 760 mm Hg pressure
M = 78.2 mGy/nC; T = 20.7°; P = 750.1; k_{ecal} = 0.903; $(k_{R50})'$ = 0.911; k_{sat} = 1.012; k_{pol} = 1.003;
P_{gr} = 0.98
   a. 378.6
   b. 365.5
   c. 323.9
   d. 310
   e. 292.3

58. According to the TG-51 protocol, _____.
   a. PP chambers are preferable over cylindrical chambers
   b. cylindrical chambers should not be used for electron beam dosimetry
   c. for R_{50} ≤2.6 cm, PP chambers are required
   d. the beam calibration depth is d_{max}
   e. the effective center of the cylindrical chamber is kept at d_{ref}

59. The electron beam calibration conditions for MU calculations during treatment in the SSD setup are _____.
   a. SSD = clinical SSD used
   b. depth of measurement 10 cm
   c. d_{max} depth along the beam central axis
   d. d_{ref} depth along the beam central axis
   e. measuring medium acrylic or perspex

C. Basic External Beam Calculations (Photon Beams)

Choose the right answer (Yes or No):

1. (Yes / No)   In SSD techniques, the beams are weighted at d_{max} point.

2. (Yes / No)   In SAD techniques, the beams are weighted at d_{max} point.

3. (Yes / No) In SSD techniques, the beams are weighted at target center.

4. (Yes / No) In SAD techniques, the beams are weighted at target center.

5. (Yes / No) Planning of multiple field treatment is much easier compared to parallel opposed (pair of) fields treatment.

6. (Yes / No) To measure the isodose distribution for a 10 cm × 10 cm field size, at least a 40 cm x 40 cm x 40 cm water phantom must be made use of.

7. (Yes / No) Isodose or PDD data can be used directly for a patient, without any modification.

8. (Yes / No) Doses prescribed for palliative treatments are much larger compared to curative treatments.

9. (Yes / No) TAR concept is usually not used with accelerator photon beams because the dose in free air is not well defined due to conceptual difficulties.

10. (Yes / No) TAR can be measured directly or derived from the PDD data.

11. (Yes / No) TAR at d_{max} and PSF are not the same.

12. (Yes / No) In general TAR increases with increase in depth.

13. (Yes / No) At shallow depths, TAR is greater for low beam qualities (kV x-ray region) because of increased side scatter.

14. (Yes / No) At larger depths, TAR is smaller for high-energy photons compared to low beam qualities.

15. (Yes / No) At low energies (< about 0.5 mm Cu), BSF increases with increase in energy.

16. (Yes / No) At orthovoltage beam qualities (> about 0.5 mm Cu HVL) BSF increases with an increase in energy.

17. (Yes / No) TAR decreases with increase in field size.

18. (Yes / No) TAR increases with increasing field asymmetry.

19. (Yes / No) The output of a kV or a ^{60}Co unit can be measured in terms of air kerma rate.

20. (Yes / No) For clinical dosimetry, the accelerator photon beam calibration is carried out in a water phantom at a reference depth of 10 cm, for a 10 cm x 10 cm field, defined at 100 cm SSD or SCD (source to chamber distance).

21. (Yes / No) Exit dose calculated using the standard PDD tables will give the correct exit dose.

22. (Yes / No) The off-axis ratio is the dose at any point in a plane perpendicular to the central axis to that of the dose on the central axis in that plane.

23. (Yes / No) Off-axis ratios can be used to determine dose at off-axis points.

24. (Yes / No) At a phantom depth of 10 cm, on the central axis, scatter contribution to the dose is more for an accelerator photon beam compared to a ^{60}Co beam.

25. (Yes / No) Isodose curves give the dose distribution in a two-dimensional plane.

26. (Yes / No) A flattening filter is used in a ^{60}Co unit to obtain uniform energy fluence across the field size.

27. (Yes / No) At any given depth, the scatter contribution to total dose decreases with increase in incident photon energy.

28. (Yes / No) Cross-beam profiles can give information regarding field size, physical penumbra, beam flatness, and symmetry.

29. (Yes / No) A universal wedge is a wedge of large angle that produces any lesser wedge angle by combining wedged and open fields during treatment.

30. (Yes / No) A wedge in motion is known as a dynamic wedge.

31. (Yes / No) The energy fluence of the photons falls according to the inverse square law of distance with the source or target position as the origin.

32. (Yes / No) Photon beams have a finite range in a patient.

33. (Yes / No) In AAPM TG-51 protocol, beam quality for photons is specified by $PDD(10)_X$.

34. (Yes / No) The field size dependence of the PDD of an accelerator photon beam is very much influenced by the electron contamination of the beam.

35. (Yes / No) Surface or buildup region doses must be measured using an extrapolation chamber.

36. (Yes / No) Beam divergence and beam attenuation in phantom decrease PDD.

37. (Yes / No) Scatter generally decreases PDD.

38. (Yes / No) Half blocked tangential breast treatments can be easily accomplished with independent jaw movements.

39. (Yes / No) One of the advantages of a dynamic wedge over a physical wedge for treatment is that there is no change in beam quality across the field size.

40. (Yes / No) TMR is a special case of TPR where $d_{ref} = d_{max}$.

41. (Yes / No) When measuring Pion with the half voltage technique for pulsed beams, if the ratio of M_H / M_L —equation 12 in the protocol—is less than 1.02, $P_{ion} = M_H / M_L$ to within 0.1%.

42. (Yes / No) P_{ion} and P_{pol} corrections must be performed weekly by the user for accurate dosimetry.

43. (Yes / No) The hospital x-ray beams are calibrated at d_{max} while the protocol recommends measuring the dose at a reference depth of 10 cm. The dose at d_{max} is derived by using $PDD(10)_X$ measured for beam quality specification

44. (Yes / No) TG-51 recommends against the use of non waterproof chambers in photon dosimetry.

45. (Yes / No) P_{ion} is uniquely related to the beam quality.

46. (Yes / No) TG-51 protocol is applicable to all photon beams used in radiotherapy.

47. (Yes / No) For x-ray energies <10 MV, $PDD(10)_X$ can be directly measured without adopting the "lead insert method" recommended by TG-51 for deriving $PDD(10)_X$.

48. (Yes / No) Wedged beams and open beams have the same S_c factors for various field sizes.

49. (Yes / No)   For a wedged beam, the majority of the head scatter arises from both the flattening filter and the wedge.

50. (Yes / No) d_{max} increases with increasing beam energy.

51. (Yes / No)   Exit dose decreases with increasing beam energies.

52. (Yes / No)   For POP of fields, the advantage of using higher MV x-ray beam is lower skin dose and greater dose uniformity across the volume.

53. (Yes / No)   The disadvantage of using higher MV x-ray beams is lower dose in the buildup region and problem in treating superficial nodes.

54. (Yes / No)   A 5 x 20 field has a greater TMR compared to a 10 x 10 field.

55. (Yes / No)   Increasing SSD can decrease the skin dose.

56. (Yes / No).   Peak scatter factor is depth dependent.

57. (Yes / No)   Peak scatter factor is independent of SSD.

58. (Yes / No)   PSF is dependent on beam quality.

59. (Yes / No)   The S_c factors are machine independent.

60. (Yes / No)   Tray factor exhibits significant field size dependence.

61. (Yes / No)   The attenuating filter correction factors always appear in the denominator of the dose equation.

62. (Yes / No)   Beam hardening is more for dynamic wedge fields compared to physical wedged fields.

63. (Yes / No)   A dynamic wedge does not alter the open field PDD.

64. (Yes / No)   The open field and wedge field of a large wedge angle can be combined to produce a wedged field of any desired intermediate angle.

Choose the right answer(s) (more than one may be correct):

(Field sizes are specified in cm unless otherwise specified.)

65. A superficial x-ray unit has a cone size of 10 cm diameter and an SSD of 18 cm. To treat a patient with a field size of 14 cm diameter, the required SSD (in cm) is _____.
  a. 36
  b. 30
  c. 25.2
  d. 18
  e. 10

66. A skin lesion is treated using 80 kV, 2 mm Al HVL x-ray beam using a 2 cm diameter cutout in a 3 cm diameter collimator. _____ is the treatment duration (in minutes and seconds) to deliver 450 cGy to the superficial lesion. The following data are given:
the beam calibration is given by D_{cal} (0 cm, 3 cm cone, 15 cm SSD) = 280.4 cGy/min
BSF (3 cm) = 1.14; BSF (2 cm) = 1.11
   a. 10' 5"
   b. 6' 20"
   c. 3' 15"
   d. 1' 38"
   e. 30"

67. A larger-size skin lesion was treated by the same x-ray beam as in the last problem with an 8 cm cutout in a 10 cm diameter collimator. _____ is the treatment duration (in minutes) to deliver 425 cGy to the superficial lesion. The following data are given:
The beam calibration is given by D_{cal} (0 cm, 10 cm cone, 15 cm SSD) = 246.5 cGy/min
BSF (10 cm) = 1.24; BSF (8 cm) = 1.22
   a. 1.75
   b. 2.8
   c. 4.9
   d. 8.3
   e. 10

68. Which of the following is a true statement?
   a. TAR increases with increasing SSD.
   b. TAR decreases with increasing SSD.
   c. TAR increases with decreasing field size.
   d. TAR decreases with increasing beam energy.
   e. None of the above are true.

69. In the figure below, TMR is defined as _____.
   a. dose at A / dose at B for the same field size
   b. dose at A / dose at B for two different field sizes
   c. dose at B / dose at A for same field size
   d. dose at B / dose at A for different field sizes
   e. dose at A / dose at A without the phantom

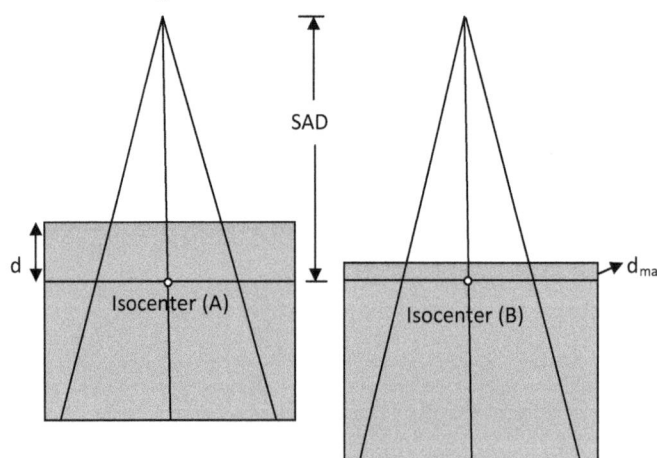

Concept of tissue maximum ratio (TMR)

70. In the figure below, PDD is defined as _____.
    a. dose at A / dose at B with and without phantom
    b. dose at A / dose at B for the same field size
    c. dose at A / dose at B for two different field sizes
    d. dose at B / dose at A for same field size W, and SSD
    e. dose at B / dose at A for different field sizes

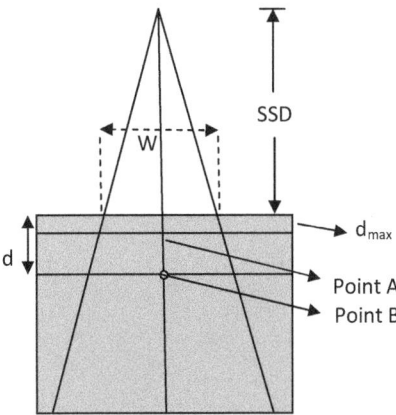

Concept of percentage depth dose

71. In an SSD setup of treatment, the dosimetric quantity to be used for MU calculation is _____.
    a. PSF
    b. TMR
    c. TPR
    d. SMR
    e. PDD

72. The dose at depths d and d_{max} measured in a water phantom at the same point and for the same field size represents the quantity _____.
    a. TAR
    b. TPR
    c. TMR
    d. PDD
    e. SMR

73. TMR varies with _____.
    a. field size
    b. SSD
    c. depth
    d. beam quality
    e. none of the above parameters

74. For a clinical x-ray beam, as the field size increases, _____.
    a. the output remains constant
    b. the output increases
    c. PDD increases
    d. scatter contribution decreases
    e. the x-ray energy decreases

75. The equivalent square of a 10 x 22 cm^2 is _____.
    a. 22 x 22
    b. 13.8 x 13.8
    c. 10 x 10
    d. 12 x 20
    e. 8.5 x 8.5

76. The equivalent square field for a circularly shaped field of 8 cm diameter is _____.
    a. 8 x 8
    b. 12 x 4
    c. 7.6 x 7.6
    d. 7.1 x 7.1
    e. 4 x 4

77. _____ is the side of the equivalent square field that is equivalent to a circle whose diameter is 5 cm.
    a. 5
    b. 4.4
    c. 4
    d. 2.5
    e. 1.25

78. _____ is the side of the equivalent square of the blocked field (bf) in the figure below.
    a. 16
    b. 13.3
    c. 10.8
    d. 10
    e. 8.7

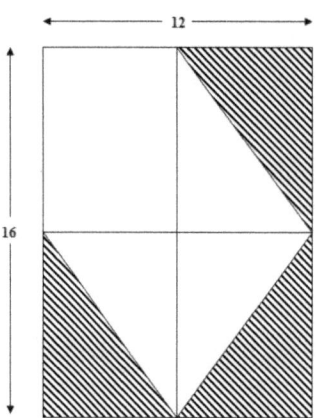

Equivalent square of a blocked field

79. For the same MU setting, the dose at a given depth (along the central axis) reduces with blocking of the field, keeping the collimator setting the same. This is because _____.
    a. the scatter seen by the point of interest reduces and hence the dose
    b. the primary radiation reaching the point reduces
    c. both primary and scatter reduce and, hence, the dose
    d. primary dose increases, but the scatter dose decrease is much more pronounced, causing effective dose reduction

80.   _____ is the side of the equivalent square of the blocked field in the figure below.
   a.   15
   b.   13
   c.   5.4
   d.   3.3
   e.   2.0

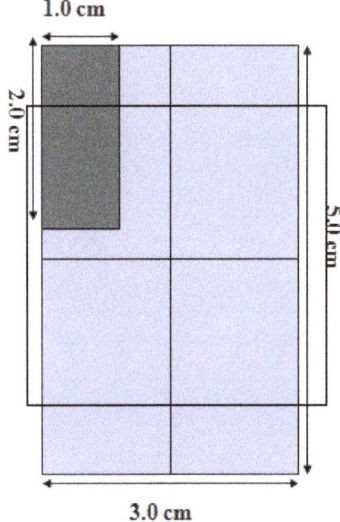

Equivalent square of a blocked field

81.   _____ is the side of the equivalent square of the blocked field in the figure below.
   a.   13
   b.   12.7
   c.   11.5
   d.   8
   e.   6.3

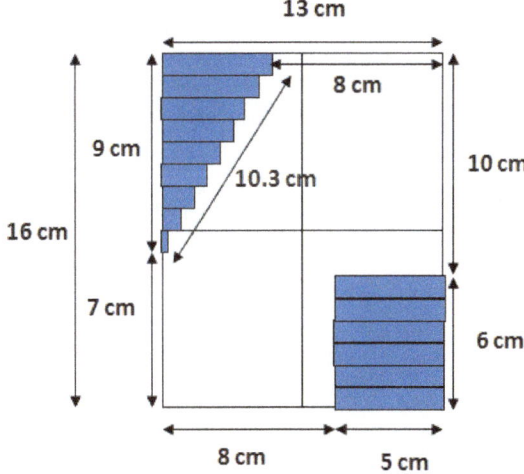

Equivalent square of a blocked field

82. _____ is the output factor for the blocked field OF_{bf} given in the above problem using the required data from below:

OF$_{of}$ in air = 1.024; NPSF$_{bf}$ = 1.005; TMR (12, 15 x 15) = 0.748
a. 1.03
b. 1.02
c. 1.0
d. 0.76
e. 0.2

83. _____ is the MU required to deliver 200 cGy at 12 cm depth for the blocked field given in the above problem in an SAD setup (see the figure below). The beam calibration is also done for the SAD setup, namely D_{cal} (1.5, 10 x 10, 100 SAD) = 1.01 cGy/MU. The following data are given: OF_{bf} = 1.03; TMR (12, bf) = 0.728.
a. 283
b. 264.1
c. 145.6
d. 141.4
e. 123.9

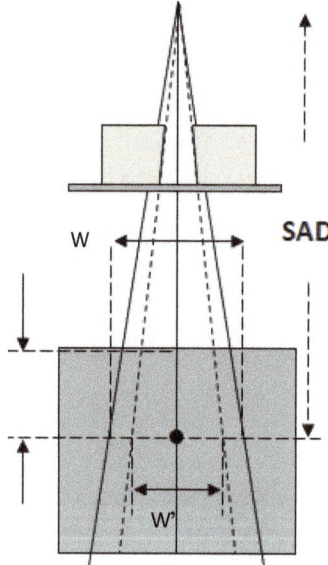

MU calculation for a blocked field in SAD setup

84. _____ is the side of the equivalent square of the blocked field shown in the following figure.
    a. 13
    b. 12.5
    c. 10.3
    d. 9.7
    e. 8

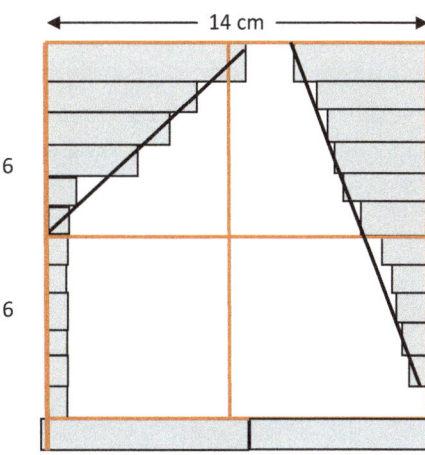

Determining the equivalent field

85. The desired method for measuring output factors is _____.
    a. measure at d_{max}
    b. measure at $d < d_{max}$
    c. measure beyond transient equilibrium depth and derive the factor for d_{max} using appropriate factor
    d. measure close to the exit point to avoid contamination electrons
    e. all the above methods are equally applicable

86. A 6 MV linac x-ray beam at 100 cm from the source delivers a dose rate of 316 cGy/min at d_{max} depth (see the figure below.) At _____ distance (in cm) from the source, the dose rate be 100 cGy/min.
    a. 300
    b. 213.5
    c. 177.8
    d. 164
    e. 100

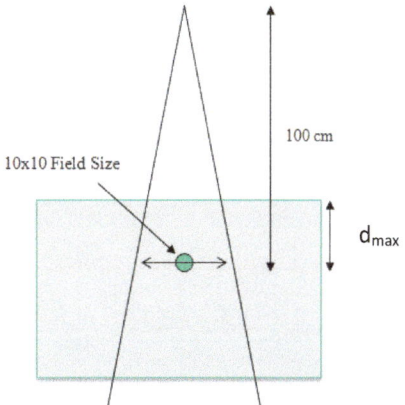

d_{max} dose follows the inverse square law

87. If the air kerma rate in air at 100 SSD for a ^{60}Co beam is 180 cGy /min., _____ is the exposure rate (in cGy/min) if the SSD is changed to 115 cm.
    a. 75.6
    b. 80.2
    c. 112.6
    d. 125
    e. 136.1

88. A 15 MV x-ray beam of a linac was calibrated for SSD treatment. The calibration dose rate was measured to be D_w (d_{max}, 10 x 10, 100) = 1.01 cGy/MU. If it is required to treat the patient in SAD setup (see the figure below), _____ is the calibration dose rate (in cGy / MU) for the SAD treatment.
    a. 1.03
    b. 1.07
    c. 1.10
    d. 1.20
    e. 1.32

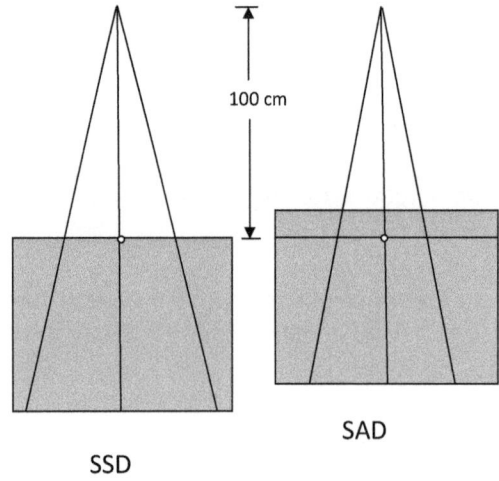

100 cm

SAD

SSD

Concept of SSD and SAD calibrations

89. Which of the following statements is false?
    a. Percent depth dose increases with HVL or beam energy.
    b. Percent depth dose increases with increase in depth of lesion.
    c. Percent depth dose increases with increasing field area.
    d. Percent depth dose increases with SSD/

90. _____ are the important components of the dose received at a depth "d" in a water phantom for a linac photon beam (see the figure below).
   a. Primary contribution
   b. Collimator or head scatter
   c. Phantom scatter
   d. Room scatter
   e. Head leakage

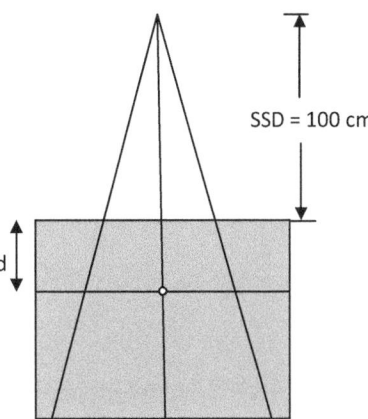

Factors contributing to dose, D(d) in a water phantom, on central axis

91. Collimator scatter factor is determined using a therapy level ion chamber along the beam central axis, irradiated _____.
   a. in air
   b. in a phantom at d_{max}
   c. in a phantom at 10 cm depth
   d. in a mini phantom
   e. with a buildup cap of sufficient wall thickness

92. Collimator scatter factor is also known as _____.
   a. head scatter factor
   b. output factor in air
   c. relative output factor in air
   d. total output scatter
   e. phantom scatter factor

93. The S_c factor for a 10 x 10 field is _____.
   a. 0
   b. 1.0
   c. >1.0
   d. <1.0
   e. dependent on the beam quality

94. The output factor (OF) for a 10 x 10 field is _____.
   a. 0
   b. <0
   c. >1.0
   d. =1.0
   e. dependent on the beam quality

95. The scatter factors are normalized to _____ with respect to the field size (in cm x cm).
    a. 5 x 5
    b. 5 x 10
    c. 10 x 10
    d. 25 x 25
    e. 40 x 40

96. Which of the following are true with respect to S_c (the collimator scatter factor). S_c is independent of _____.
    a. SSD
    b. any mini phantom depth
    c. mini phantom depth $d > d_{te}$ (transient equilibrium thickness)
    d. collimator setting
    e. moderate field blocking
    f. field symmetry/asymmetry

97. Which of the following are true about the mini phantom used for S_c measurements? It must be _____.
    a. preferably of low Z to mimic tissue
    b. narrow enough to use with smaller field sizes
    c. have transient equilibrium thickness for all beam qualities with respect to the beam entrance face
    d. square shaped in cross-section

98. The diameter of the mini phantom is 3 cm. The chamber center is at 10 cm depth placed at SAD = 100 cm. _____ is approximately the minimum field size (in cm) for which S_c can be measured.
    a. 3 x 3
    b. 3.4 x 3.4
    c. 10 x 10
    d. 20 x 20
    e. >20 x 20

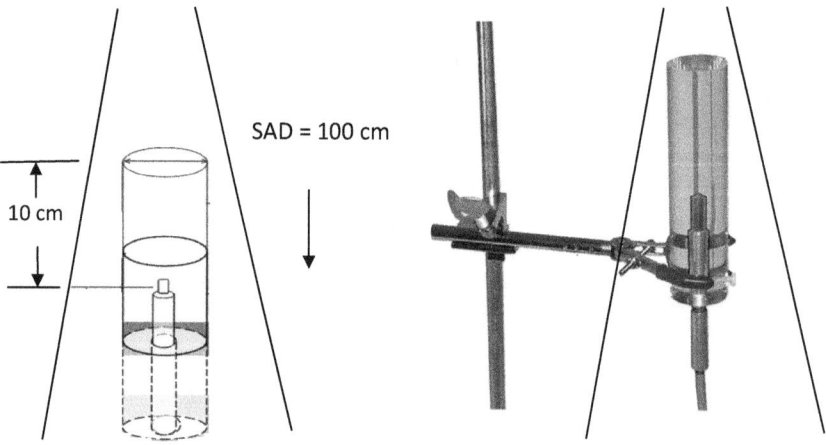

"Mini phantom" for collimator scatter measurements

99. _____ corrections need to be applied for MU calculations for typical clinical treatment conditions that differ from beam calibration conditions.
    a. Field size
    b. Depth
    c. Beam modifier influence
    d. Off axis correction
    e. Non-standard distance
    f. All of the above are true

100. For open beams, the maximum contribution to S_c arises from the _____.
    a. flattening filter
    b. collimator
    c. vacuum window
    d. transmission chambers
    e. light optics

101. The relative contribution of various components of the linac head to head scatter, S_c, _____.
    a. cannot be individually determined
    b. can be measured by separate experiments
    c. can be measured if we can have access to the various components
    d. can be theoretically determined by Monte Carlo computations
    e. can be obtained from the machine manufacturer

102. Separation of output factor into S_c and S_p is _____.
    a. required to simplify calculations
    b. not very essential for any type of calculations
    c. required for blocked field dosimetry
    d. required for the MU calculations of all fields

103. S_c for open beams, with respect to field size changes, can vary _____.
    a. only less than 1%
    b. by 2% to 3%
    c. by 5% to 8%
    d. by 10% to 12%
    e. by 20% to 30%

104. The MV beam calibration conditions for MU determination in the SSD setup are _____.
    a. SSD = 100 cm
    b. field size at the SSD = 10 cm x 10 cm
    c. depth of measurement 10 cm
    d. point of measurement at d_{max} on the central axis
    e. in air measurement at SSD + d_{max} distance from the source

105. Given: 6 MV beam (100 SAD machine), 12 cm x 15 cm field size, separation is 20 cm, tray factor is 1. _____ are the MU settings required to give 200 cGy to midline through AP/PA ports weighted 3:2 if an SSD of 100 cm is to be used for both. PDD (1.5, 13.5 x 13.5, 100 SSD) = 68.3%; OF (13.5 x 13.5) = 1.022; D_{cal} (1.5, 10 x 10, 100 SAD) = 1.01 cGy/MU.
    a. AP = 162; PA = 108
    b. AP = 192; PA = 128
    c. AP = 182; PA = 121
    d. AP = 170; PA = 114
    e. AP = 160; PA = 160

106. Given: 6 MV beam (100 SAD machine), 15 cm x 15 cm field size, depth to calculation point is 7 cm, and a treatment distance of 100 cm SSD. _____ is the MU setting required to give 300 cGy through a single port. PDD (7, 15 x 15, 100 SSD) = 79.9%; OF (15 x 15) = 1.031; D_{cal} (1.5, 10 x 10, 100 SAD) = 1.01 cGy/MU.
    a. 497.0
    b. 486.5
    c. 394.3
    d. 392.6
    e. 360.6

107. The output factor and the NPSFs for a 6 MV beam are given below:
OF (6 MV, 30 x 30) = 1.079; NPSF (6 MV, 30 x 30) = 1.033.
The collimator component of the output factor (S_c) is given by _____.
   a. 1.079
   b. 1.079 × 1.033
   c. 1.079 / 1.033
   d. 1.033 / 1.079
   e. 1.079 − 1.033

108. The output factor and the NPSFs for a 6 MV beam are given below:
OF (6 MV, 20 x 20) = 1.050; NPSF (6 MV, 20 x 20) = 1.023.
The collimator component of the output factor (S_c) is given by _____.
   a. 1.05
   b. 1.026
   c. 1.023
   d. 1.0
   e. 0. 974

109. A field (in cm) of 20 x 20 was created by blocking a 25 x 25 field set on the collimator (See the figure below).
_____ is the OF of the blocked field. The following data are given:
OF (25 x 25) = 1.065
NPSF (25 x 25) = 1.028 NPSF (20 x 20) = 1.023
   a. (1.065 / 1.028) × 1.023
   b. 1. 065 × 1.028 × 1.023
   c. 1.028 × 1.023 / 1.065
   d. 1.065 / 1.023
   e. 1.028 × 1.023

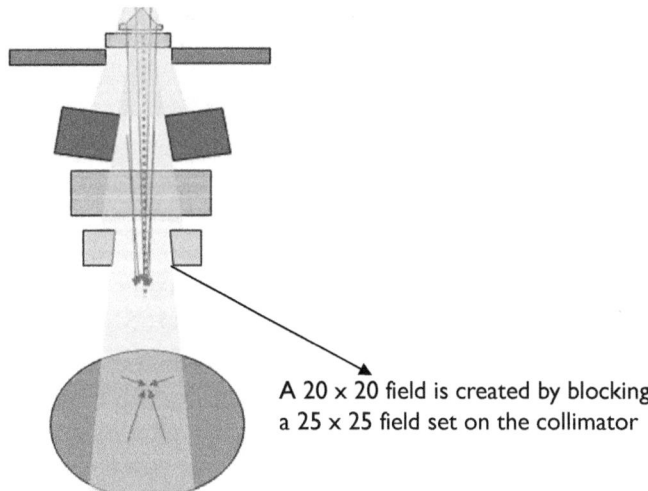

A 20 x 20 field is created by blocking
a 25 x 25 field set on the collimator

Field blocking to modify the
collimator field size

110. _____ is the MU required to deliver 130 cGy at d_{max} for a 10 x 20 field produced by 6 MV. The following data are given:
Table to find the side of an equivalent square for a given field:
The OF table for square field sizes for 6 MV
Beam calibration D_{cal} = 1.01 cGy/MU at d_{max} for a 10 x 10 field

		Short Axis (cm)			
		1.0	**5.0**	**10.0**	**20.0**
	1.0	1.0			
Long Axis (cm)	**5.0**	1.8	5.0		
	10.0	2.2	6.7	10.0	
	20.0	2.3	7.9	13,0	20.0
	30.0	2.4	8.2	13.9	23.3

Equivalent Square Table for Rectangular Fields

	Side of Square Field								
	0	4	6	8	10	15	20	30	40
Output Factor	.863	.931	.961	.983	1.000	1.031	1.053	1.084	1.104

Field Size Dependence of OFs

a. 130
b. 126.3
c. 120
d. 100
e. 98.4

111. Using the OF table given in the above problem, _____ is the MU required to deliver 200 cGy at 10 cm depth for a 20 x 20 field produced by a 6 MV beam in an SSD setup. PDD (10, 20 x 20) = 70.1%. Beam calibration D_{cal} = 1 cGy/MU at d_{max} for a 10 x 10 field.
a. 300.7
b. 285
c. 271
d. 210
e. 196.7

112. In the above problem, _____ is the MU required to deliver the same dose at the same depth for a 5 x 10 field. Use the Day formula for equivalent square determination. PDD (10) for the given field = 64.7%.
a. 354.4
b. 321
c. 270
d. 200
e. 176.9

113. The dose at 5 cm depth for a 6 MV 15 x 15 field in an SSD setup is 180 cGy. _____ is the dose (in cGy) at 10 cm depth, given the following data:
PDD (5, 15 x 15, 100 SSD) = 87.6%
PDD (10, 15 x 15, 100 SSD) = 69.0%
    a. 174
    b. 141.8
    c. 109.2
    d. 90.0
    e. 54.8

114. In a palliative treatment the physician describes 200 cGy to the spine at 6 cm depth for a 6 MV, 10 x 10 field. D_{cal} (d_{max}, 10 x 10, 100 SSD) = 1.01 cGy/MU and PDD (6, 10 x 10, 100 SSD) = 82.4%.
The MU to be set on the machine to deliver the required dose is _____.
    a. 400
    b. 342.8
    c. 240.3
    d. 200
    e. 128.6

115. In the above problem, _____ is the dose (in cGy) at 10 cm depth, given PDD (10, 10 x 10, 100 SSD) = 66.9%.
    a. 267.8
    b. 200
    c. 185.8
    d. 162.4
    e. 120.6

116. For a 10 x 10 field, both TMR and FDD (PDD/100) = 1, but at 12 cm depth TMR > FDD because _____.
    a. TMR does not account for attenuation
    b. TMR is affected by ISL fall to a lesser extent
    c. FDD is not affected by ISL
    d. FDD is not influenced by attenuation
    e. FDD accounts for both attenuation and ISL fall in dose with depth while TMR is influenced only by attenuation

117. A patient's spine was treated with a 6 MV beam at 6 cm depth using an extended distance of 135 cm for a larger field coverage. _____ is **the percentage change in** PDD (6 cm, 135 cm SSD) compared to PDD (6 cm, 100 cm SSD).
    a. an increase of about 20
    b. an increase of about 2
    c. a decrease of about 10
    d. an increase of about 5
    e. zero

118. In the above problem, _____ is the factor the MU increases compared to the standard SSD to deliver the prescribed dose.
    a. 10
    b. 5
    c. 1.8
    d. 1
    e. 0.5

119. _____ is the PDD value for a 6 x 5 field at a depth of 7.5 cm depth using the following table:
    a. 85.5
    b. 85
    c. 84.4
    d. 83.7
    e. 82

	Field Size (cm^2)		
	4 x 4	6 x 6	10 x 10
Depth (cm)			
7.,0	86.4		87.6
7.5			
8.0	82.4		83.7

120. The PDD tables for 6 MV beam show that PDD (10, 20 x 20) for the open field is 70.1% while PDD (10, 20 x 20) for a 45 degree wedge field is 71.4%. Regarding the influence of the wedge, you can conclude _____.
    a. it not only attenuates the beam but also changes the PDD of the open field
    b. it increases the depth of penetration of the beam incident on the patient
    c. wedge hardens the beam
    d. it increases the average energy of the linac beam
    e. it decreases the average energy of the linac beam

121. Often one requires to treat a patient at non-standard SSDS. The typical beam calibration of a clinical 6 MV photon beam is D_{cal} = 1.01 cGy/MU at d_{max} = 1.5 cm for a 10 x 10 field at 100 cm SSD. _____ is the reference dose rate (in cGy/MU) at d_{max} in a 10 x 10 field if the patient has to be treated at 110 cm SSD (see the figure below).
    a. 1.01
    b. 1.00
    c. 0.837
    d. 0.50
    e. 0.439

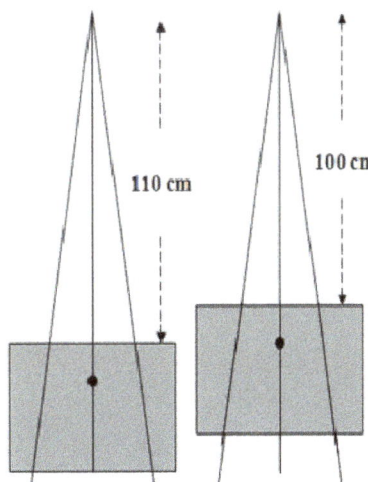

Beam output for non-standard SSD

122. _____ is the PDD for the new SSD in the above problem assuming the field size on the phantom remains the same for both SSDs (see the figure below). PDD (6 MV, 8, 10 x 10, 100) = 74.3%.
    a. 75.1
    b. 74.3
    c. 72.6
    d. 68.3
    e. 61.0

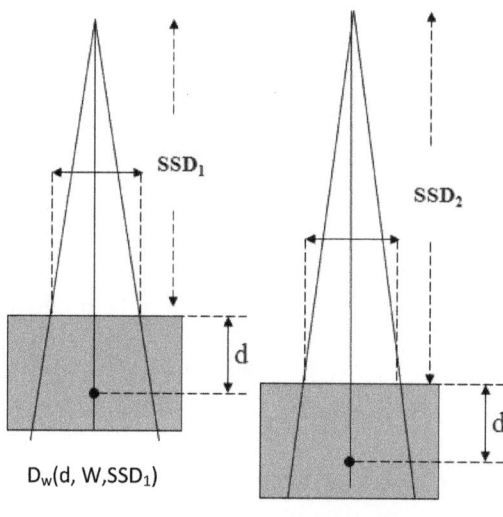

PDD for the non-standard SSD

123. Often the hospital machines are calibrated in SSD setup but are used in SAD setup for most of the cases. A hospital machine is calibrated in the SSD setup for a 15 MV beam, and the beam calibration = D_{cal} = 1.005 cGy/MU at d_{max} = 3.0 cm for a 10 x 10 field at 100 cm SSD. If the physicist now wishes to use an isocentric, or SAD technique (see the figure below), _____ is the new beam calibration for the SAD setup.
    a. 1.066
    b. 1.035
    c. 1.025
    d. 1.0
    e. 0.98

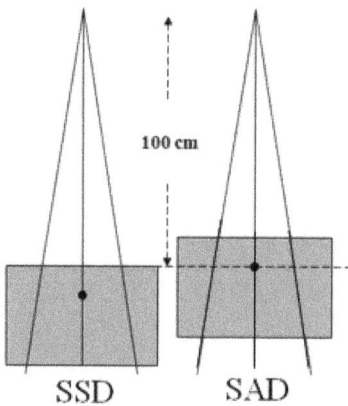

Use of SSD calibration for SAD setup

124. _____ are the MU for delivering 200 cGy at a depth of 12 cm for a 6 MV 15 x 15 field in the SAD setup. The following data are given:
    OF (15 x 15) = 1.030 TMR (12, 15 x 15) = 0.748
    D_{cal} (1.5, 10 x 10, 100 SSD) = 0.99 cGy/MU
    a. 259.6
    b, 254.3
    c, 200
    d. 178.7
    e. 100.0

125. _____ will be the dose rate in (cGy/MU) at d_{max} at 110 cm SSD for a 20 x 20 field size in the above problem.
    OF (20 x 20) = 1.051.
    a. 1.05
    b. 1.0
    c. 0.99
    d. 0.86
    e. 0.75

126. The tray attenuation (K_{tray}) factor is about _____.
    a. 1%
    b. 2% to 3%
    c. 5% to 10%
    d. >20%
    e. depends on the field size chosen

127. A patient is treated using a 15 MV beam at 120 cm SSD. The field size on the patient was 15 x 15 (see the figure below). _____ MU are required to deliver a dose of 200 cGy at the tumor center at 11 cm depth. The following data are given:
    PDD (11, 15 x 15, 120 SSD) = 75.8%; S_c (12.5 x 12.5) = 1.011; S_p (15 x 15) = 1.013;
    D_{cal} (3, 10 x 10, 100 cm SSD) = 1.005 cGy/MU
    a. 400
    b. 385.7
    c. 365.6
    d. 300
    e. 180.7

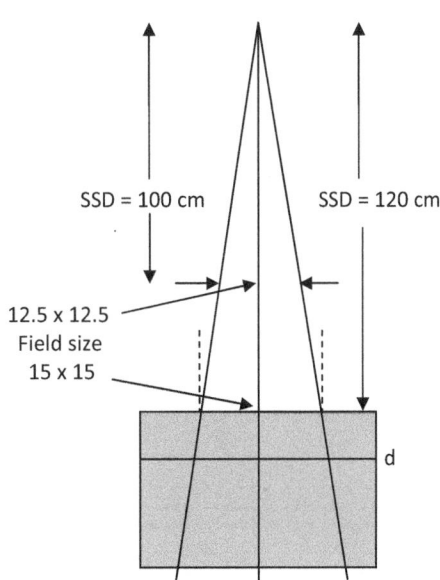

SSD = 100 cm   SSD = 120 cm

12.5 x 12.5
Field size
15 x 15

d

Treatment calculations for non-standard SSD

128. A 10 MV linac is calibrated to give 0.99 cGy /MU at d_{max} = 2.5 cm for SSD = 100 cm and field size
10 cm x 10 cm. _____ is the MU to be set to deliver a dose of 200 cGy to a patient using a treatment field size
of 20 cm x 20 cm, SSD of 100 cm, and a target depth of 15 cm. S_c (20 x 20) = 1.033, S_p(20 x 20) = 1.020,
PDD (15, 20 x 20, 100) = 60.4%.
    a.   400
    b.   365.4
    c.   317.4
    d.   200
    e.   165

129. The patient in the above problem has to be treated at a larger distance of 120 cm for the same treatment dose,
tumor depth, and field size conditions. _____ are the MUs to be set on the machine.
PDD (15, 20 x 20, 120) = 61.9%. S_c(16 x 16) = 1.023.
    a.   623
    b.   500
    c.   486
    d.   447
    e.   325

130. The beam output of a 10 MV x-ray beam at d_{max} depth (2.5 cm) on the beam central axis, at SAD = 100 cm and
collimator field size 10 cm x 10 cm, is 1.01 cGy/mu. _____ are the monitor units required to deliver a dose of
200 cGy to the target center at a depth of 10 cm in the patient using a 20 cm x 20 cm field size.
TMR(10, 20 x 20) = 0.858
S_c (20 x 20) = 1.032; S_p(20 x 20) = 1.02
    a.   314
    b.   219
    c.   184
    d.   123
    e.   87

131. _____ are the monitor units for the above problem if the unit had been calibrated at constant SSD instead of
isocentric setup. (Refer to the figure under question 123 on p. 226.)
    a.   209
    b.   184
    c.   167
    d.   147
    e.   108

132. A 6 MV treatment was planned in a 100 cm SSD setup, but due to an error in the ODI the patient was treated at
98.5 cm. _____ is the dose escalation (in %) at the dose maximum depth.
    a.   0.99
    b.   1.6
    c.   3.1
    d.   5
    e.   6.2

133. The reduction in scatter due to the blocking of an open field is the greatest for _____.
    a.   18 MV beam at d_{max}
    b.   18 MV beam at 10 cm depth
    c.   10 MV beam at d_{max}
    d.   10 MV beam at 6 cm depth
    e.   4 MV beam at 10 cm depth

134. Scatter increases with decreasing beam quality and increasing depth. The dose under 1.6 cm width cord block, at 5 cm depth, in a water phantom, for a 6 MV beam is about _____% of the open field dose. (The block is 5 HVL thick.)
    a. 100
    b. 50
    c. 15
    d. 10
    e. 3

135. The tray on a simulator is a distance of 65 cm from the focal spot. The shadow of a block on the tray marked on the film measures 4 cm across. The film magnification is 1.4. The size of the block on the simulator and its projection on the patient skin, respectively, are _____.
    a. 1.9, 2.9
    b. 4, 4
    c. 4, 2.9
    d. 2, 4
    e. 1,4

136. _____ are the MU required to deliver a dose of 150 cGy at d_{max} using the following data.
    Beam used: 15 MV, 30 degree, 10 x 10 wedge field. SSD = 100 cm. WF = 0.617.
    The beam calibration = D_{cal} = 1.015 cGy/MU at d_{max} = 3.0 cm for a 10 x 10 field at 100 cm SSD.
    a. 300.75
    b. 253.75
    c. 239.5
    d. 154.3
    e. 150

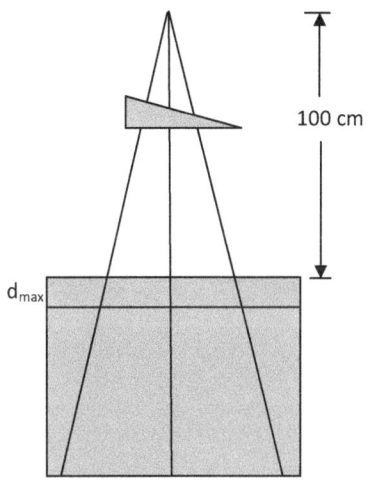

Treatment calculation for a wedged beam

137. If the wedge factor is 0.5, the MU required to deliver the prescription dose, compared to the open field, will

    _____.
    a. increase by a factor of 2
    b. decrease by a factor of 2
    c. increase by a factor of 2^2
    d. remain the same
    e. possibly increase or decrease depending on the beam quality

138. _____ are the MU required to deliver a tumor dose of 200 cGy at a depth of 12 cm in an isocentric treatment using the following data:

    Beam used: 15 MV, 30 degree, 15 x 15 wedge field. SAD = 100 cm. WF (30°, 15 x 15) = 0.759

    The beam calibration = D_{cal} = 0.99 cGy/MU at d_{max} = 3.0 cm for a 10 x 10 field at 100 cm SAD

    TMR (15 MV, 12 cm) = 0.839; OF (15 MV, 15 x 15) = 1.033; D_{cal} = 0.99 cGy/MU

    a. 307
    b. 250.7
    c. 200
    d. 168
    e. 135.7

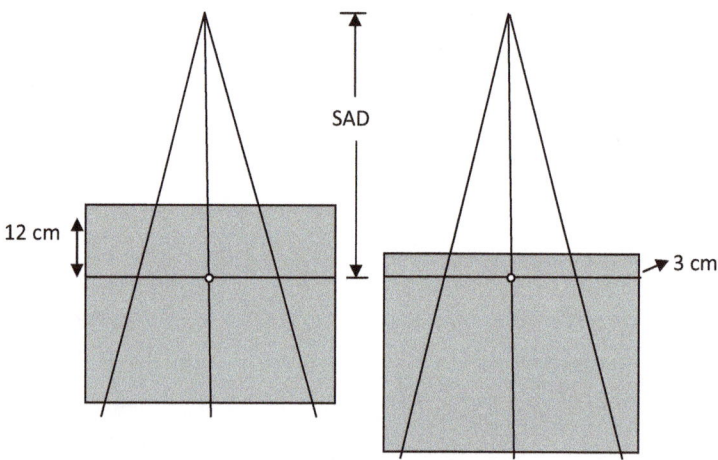

Relating treatment geometry to calibration geometry

139. Combining the open field and a wedged field of a large angle, the wedge profile of any intermediate angle can be created (see the figure below). A linac has an internal 60 degree wedge. To produce θ degree wedge isodose profile, _____ percentage of the dose must be delivered using the wedge.

    a. (tan θ / tan 60) × 100
    b. tan 45
    c. tan 60 × tan θ
    d. 0,30
    e. tan 60 / tan θ

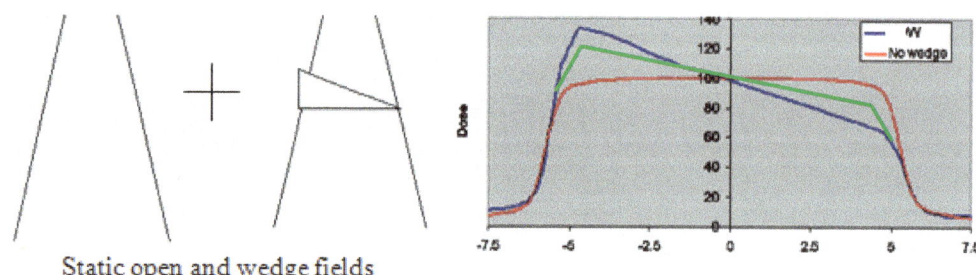

Static open and wedge fields

Intermediate wedge angle profile generation

140. To produce a 40° wedge profile using a 60° internal wedge, the fraction of the dose to be delivered using the wedge is given by _____.

    a. 1
    b. 0.5
    c. 0.48
    d. 0.4
    e. 0.05

141. A dose of 200 cGy is to be delivered to the target center using a 40° wedge field. The linac has an internal 60° wedge. _____ dose (in cGy) must be delivered by the wedge and open field, respectively.
    a. 96 and 104
    b. 104, 96
    c. 200, 0
    d. 100, 100
    e. 90, 110

142. A dose of 200 cGy is to be delivered to the target center at 12 cm depth using a 30°, 15 x 15 wedge field. The linac has an internal 60° wedge. _____ MU are to be set on the machine for the wedge and open fields, respectively, for delivering the required wedge field dose. The following data are given:
    PDD $(12, 15 \times 15)_{wf}$ = 64.1; OF (15 x 15) = 1.030; PDD $(12, 15 \times 15)_{of}$ = 62.4
    WF = 0.410; D_{cal} (1.5, 10 x 10, 100 SSD) = 0.995 cGy/MU
    a. 96, 104
    b. 200, 100
    c. 120, 80
    d. 245, 210
    e. 225, 289

143. _____ angle (WA), in degrees, of wedge must be used in the figure below to give uniform dose in the region of overlap of the two field. The hinge angle (HA) = 90°.
    a. 180
    b. 90
    c. 45
    d. 30
    e. 15

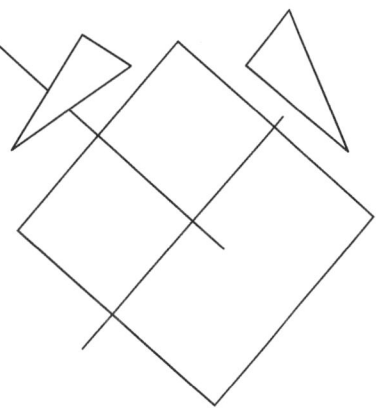

90° hinge angle for the two adjacent wedged fields irradiation

144. _____ angle (in degrees) of wedge must be used to give uniform dose in the region of overlap of the two fields. The hinge angle = 60°.
    a. 175
    b. 120
    c. 90
    d. 60
    e. 15

145. _____ angle of wedge must be used in the figure below to give uniform dose in the region of overlap of the two fields. The hinge angle = 60°.
   a. 180
   b. 120
   c. 90
   d. 60
   e. 30

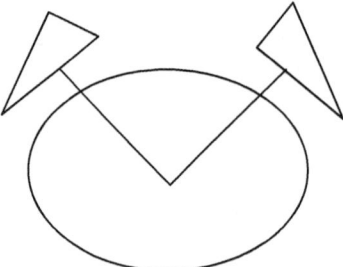

Relating hinge and wedge angles for dose uniformity in the region of overlap

146. An open field treatment dose of 200 cGy has been prescribed for the tumor depth in a single field treatment. A wedge needed to be used for the treatment. The wedge factor is 0.75. The open field MU must be increased _____ times to deliver the prescribed dose.
   a. 2
   b. 3
   c. 1
   d. 0.75
   e. 1.33

147. A dose of 200 cGy is to be delivered at 5 cm depth in a 6 MV, 12 x 12 POP of fields in an isocentric setup (see the figure below). _____ MU are to be set up on the machine for the A/P and P/A fields, respectively, using the following data. The fields are equally weighted at the isocenter.
   TMR (5, 12 x 12) = 0.929; TMR (10, 12 x 12) = 0.788; OF (12 x 12) = 1.017;
   D_{cal} (1.5, 10 x 10, 100 SAD) = 0.99 cGy/MU
   a. 130, 123.6
   b. 124.6, 124.5
   c. 106.9, 126.0
   d. 100, 100
   e. 90.6, 12.5

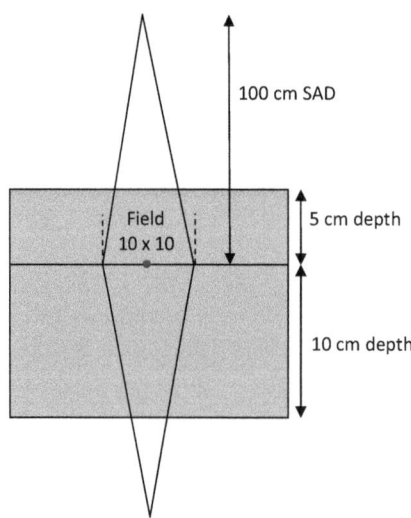

100 cm SAD

Field
10 x 10

5 cm depth

10 cm depth

Treatment calculation for equally weighted POP of fields

148. _____ will be the MUs to be set for the A/P and P/A fields, respectively, in the above problem if the fields are weighted 2:1 at the isocenter.
    a. 234.6, 123.3
    b. 220, 117
    c. 184, 184
    d. 170.5, 97.4
    e. 142.5, 84

149. A whole brain treatment of 3600 cGy by 6 MV, POP of fields, in SAD setup was planned for a separation of 17 cm while the actual head thickness turned out to be 15 cm. The escalated dose to the midline in cGy is _____.
    a. 126.9
    b. 108
    c. 96
    d. 84.7
    e. 69

150. For an 18 MV 20 x 20 field POP of fields, the minimum depth of the 95% of the midplane dose occurs at a depth of _____ cm from the surface. (The d_{max} for the beam occurs at 3.5 cm.)
    a. 6
    b. 5.4
    c. 4
    d. 3
    e. 0.5

151. For a treatment of the whole brain by a 6 MV, 18 x 18 equivalent square, POP of fields, in an SAD setup, the physician prescribes a dose of 200 cGy at the midline. The width of the patient's head is 16 cm. _____ are the MU to be set on the machine **per field**, given TMR (8, 18 x 18) = 0.861 and OF (18 x 18) = 1.042. D_{cal} (1.05, 10 x 10, 100 SAD) = 0.995 cGy/MU.
    a. 200
    b. 112
    c. 89.5
    d. 64
    e. 26

152. In the above problem, _____ is the dose at d_{max} (in cGy) for one of the brain fields (see the figure below).
    a. 189.6
    b. 167
    c. 144.7
    d. 132.6
    e. 87.0

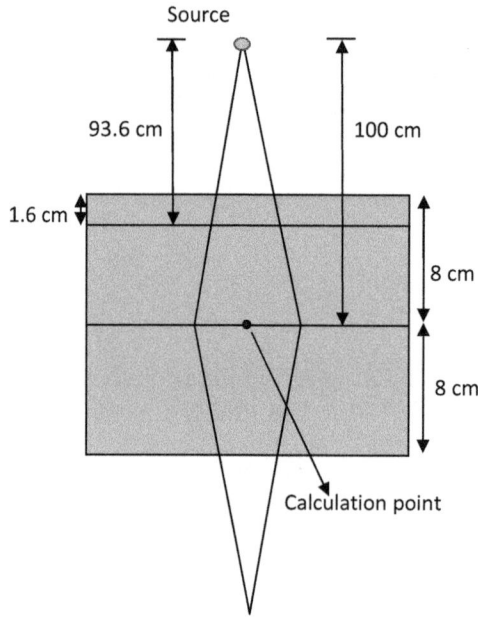

d_{max} dose calculation for POP of fields in an SAD setup

153. In an isocentric treatment, a dose of 60 cGy is prescribed to the tumor center. The parameters of the treatment field are: WF = 0.58; D_{cal} (d_{max}) = 0.90 cGy/MU; TMR = 0.782.
    The MU to be set on the machine is _____.
    a. 195
    b. 147
    c. 110
    d. 89
    e. 64

154. The surface dose for 6 MV photon beam at 100 cm SSD is about _____.
    a. 2% to 3%
    b. 5% to 10%
    c. always 25% to 26%
    d. 15% to 40% depending on field size
    e. 70% to 90%

155. A POP of fields treatment was set up for a patient at 100 cm SAD in isocentric mode. The AP thickness of the patient is 24 cm and the field size at midplane is 15 x 15. The field size on the skin is _____.
    a. 13.2
    b. 11.7
    c. 10.6
    d. 8.9
    e. 6.1

156. A pregnant woman is being treated with AP/PA 6 MV mantle fields to a total dose of 4000 cGy. The fetus is at a distance of 15 cm from the edge of the field without any additional shielding on the patient. The maximum fetus dose of the open field dose is around _____.
    a. 20%
    b. 15%
    c. 10%
    d. 5%
    e. 2%

157. The breast is treated with POP of 6 MV fields. The treatment plan shows that 10% of the PTV receives less than 90% dose. The possible reason for this is _____.
    a. the field arrangement is not proper
    b. 6 MV is too high an energy to give uniform dose to the PTV
    c. the field size needs to be increased
    d. planning error is the likely cause
    e. the PTV bordering on the skin lies in the buildup region

158. _____ beam quality gives the highest skin dose for 10 x 10 100 cm SSD.
    a. ^{60}Co
    b. 6 MV
    c. 18 MV
    d. 6 MeV
    e. 20 MeV

159. In an SSD treatment, the dose at 20 cm depth is 200 cGy. PDD values at 20 and 5 cm depths are 0.415 and 0.876 respectively. _____ is the dose at 5 cm depth.
    a. 368.8
    b. 123.0
    c. 94.7
    d. 67.7
    e. 54.6

160. A 6 MV POP of fields treatment in SSD setup was used to deliver 200 cGy to the patient midline (see the figure below). Calculate the total dose (in cGy) to d_{max} and to the cord, respectively. Use the following data: Field size 15 x 15 blocked to 8 x 8. $d_{max} = 1.5$ cm : cord depth = 3 cm; field separation = 20 cm.
    PDD (1.5, 8 x 8, 100) = 100; PDD (3, 8 x 8, 100) = 95
    PDD (10, 8 x 8, 100) = 66.7; PDD (17, 8 x 8, 100) = 45.2; PDD (18.57, 8 x 8, 100) = 41.6
    a. 200, 200
    b. 212, 210
    c. 215,100
    d. 198.5, 168.7
    e. 280, 126.3

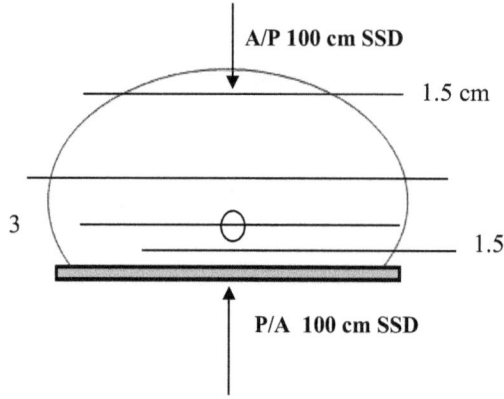

Dose calculation for POP of fields treatment in an SSD setup

161. A 6 MV, 15 x 15, POP of AP/PA fields is used to treat at an AP depth of 10 cm in an isocentric setup (see the figure below). The prescription dose is 200 cGy. _____ is the MU to be delivered through AP and PA fields, respectively. The following data are given:
    PA depth = 12 cm. Fields weightage 1 : 1
    TMR (10,15 x 15) = 0.802; TMR (12, 15 x 15) = 0.748; OF (15 x 15) = 1.030
    D_{cal} (1.5, 10 x 10, 100 cm SSD) = 1.01 cGy/MU
    a. 120, 80
    b. 132, 121.6
    c. 119.8, 128.5
    d. 108.9, 122.4
    e. 102.9, 137.7

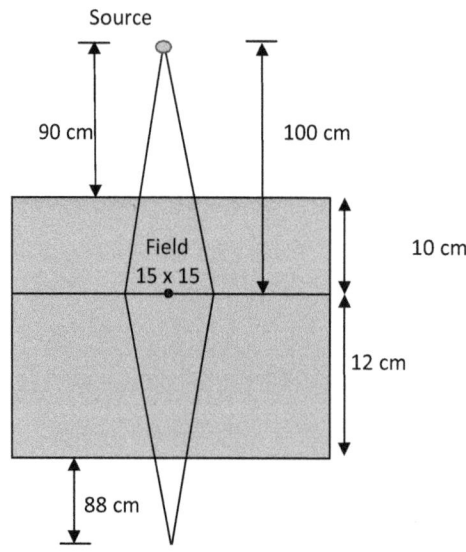

Non-midline dose calculation for POP of fields treatment in an SAD setup

162. A 6 MV, 15 x 15, four fields technique is used to treat a pelvis tumor in the SAD setup (see the figure below). The fields weightage is given as AP/PA : RL/LL fields = 2 : 1. In addition, the AP and PA fields are equally weighted. If the tumor prescription is 250 cGy per treatment, _____ is the dose (in cGy) delivered to the tumor by AP, PA, LL, and RL fields, respectively.
    a. 83.3, 83.3, 41.7, 41.7
    b. 75, 75, 50, 50
    c. 166.6, 166.6, 83.3, 83.3
    d. 150, 100, 125, 125
    e. 250, 200, 100, 50

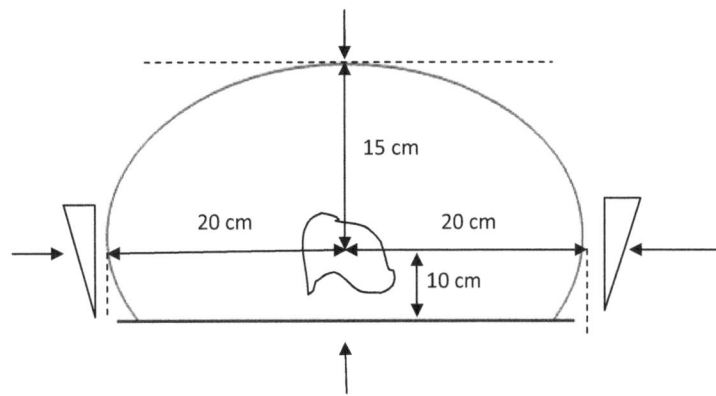

Treatment calculation for a four-field SAD technique (opposed pairs of wedge and open fields)

163. In the above problem, the LL and RL are 15° wedge fields. _____ are the doses delivered through each of the fields (AP, PA, LL, and RL, respectively), assuming the treatment was carried out in an SAD setup. The following data are given:
    AP depth = 15 cm; PA depth = 10 cm; LL depth = RL depth = 20 cm;
    TMR (15,15 x 15) = 0.672; TMR (10, 15 x 15) = 0.802; TMR (20, 15 x 15) = 0.554;
    OF (15 x 15) = 1.030; WF(15°) = 0.713; D_{cal} (1.5, 10 x 10, 100 cm SSD) = 1.01 cGy/MU
    a. 123.6, 122.5, 98.4, 99.6
    b. 110, 98.4, 99.6, 112.1
    c. 119.2, 99.8, 101.4, 101.4
    d. 123, 97, 106, 97.6
    e. 98.4, 99.6, 116.5, 124.3

164. A 6 MV, 7 x 7, three fields SAD technique was used to deliver a boost therapy dose of 180 cGy to the lung (see the figure below). The AP and RL were wedged fields with 15° and 45° wedges, respectively. The field weightage was AP : PA : RL = 3 : 2 : 1. _____ are the MU to be set on the machine for AP, PA, and RL fields, respectively. The following data are given:
TMR (8, 7 x 7, 100) = 0.819; TMR(15, 7 x 7, 100) = 0.614; TMR(5, 7 x 7, 100) = 0.918
WF (15°) = 0.705; WF(45°) = 0.490. D_{cal} (1.5, 10 x 10, 100 cm SSD) = 1.008 cGy/MU
OF (7 x 7) = 0.969.
   a. 90, 60, 30
   b. 115.5, 74.8, 42.9
   c. 123.4, 96.4, 73.3
   d. 142.4, 100, 76.5
   e. 152.8, 102, 73.4

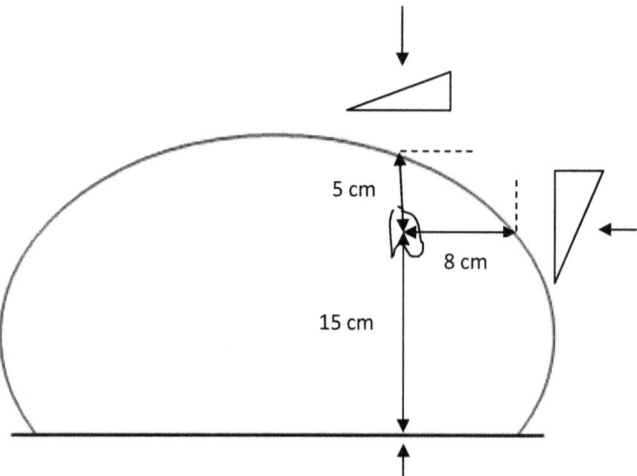

Treatment calculation for a three-field SAD technique (two wedged fields and one open field)

165. A patient is being treated using a POP of RL and LL fields delivering a dose of 200 cGy to the midplane of the patient (see the figure below). Patient thickness along the beam central axis (CA) is 14 cm. The SSD setup is used for the treatment, and the fields are equally weighted. _____ are the doses (in cGy) at the mid plane for the RL and LL fields, respectively.
   a. 200, 200
   b. 150, 250
   c. 100, 100
   d. 50, 150
   e. 20, 180

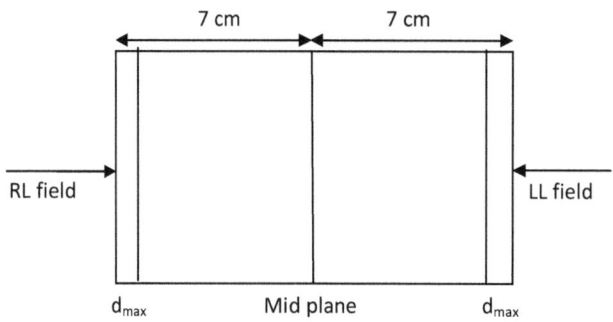

Midplane dose for POP of equally weighted lateral fields in SSD setup

166. In the above problem, _____ are the total MU delivered. Given:
     PDD (7, 12 x 12, 100) = 79.2%; OF (12 x 12) = 1.012; D_{cal} (1.6, 10 x 10) = 1.01 cGy/MU.
     a. 310.6
     b. 285.7
     c. 262.3
     d. 247

167. In the above problem, _____ are the MU delivered for the RL and LL fields, respectively, if the fields RL:LL
     are weighted as 2:1.
     a. 200,100
     b. 164.7, 82.3
     c. 180, 97.5
     d. 143.6, 102.2
     e. 138, 76.4

168. In the above problem, the $D(d_{max})$ in cGy for the RL and LL fields, respectively, are _____ if the treatment
     plane is at 5 cm depth for the RL field. Given PDD (5, 12 x 12, 100) = 87.1% and
     PDD (9, 12 x 12, 100) = 71.1%.
     a. 163, 81.5
     b. 150, 89.4
     c. 143.2, 98.6
     d. 130, 120
     e. 98.4, 167.7

169. In the above problem, _____ are the tumor dose (TD) contributions for the RL and LL fields, respectively, at
     the treatment plane.
     a. 150, 49.6
     b. 148.6, 63.3
     c. 145, 62.8
     d. 142, 57.9
     e. 128.4, 56.2

170. In problem 168 above, _____ are the entry and exit dose (ED & ExD) in cGy for the patient from the RL field
     (see the figure below). Given d_{max} = 1.6 cm and PDD (12.4, 12 x 12, 100) = 59.2%.
     a. 200, 100.5
     b. 187.6, 103.9
     c. 163, 96.5
     d. 143.2, 107.6
     e. 123.8, 84.6

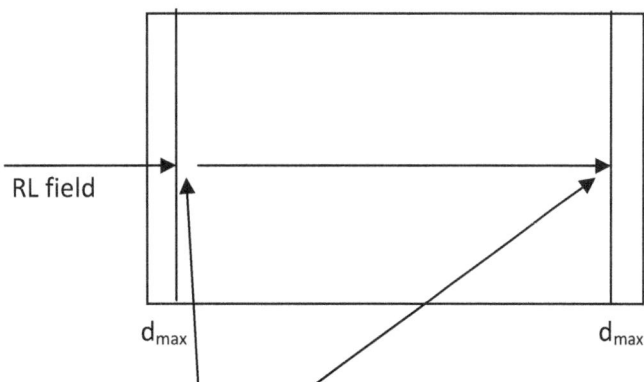

Entrance and exit dose for the RL field in POP of fields treatment

171. In problem 168 above, _____ are the entry and exit dose (ED and ExD) in cGy for the patient from the LL field (see the figure below). Given $d_{max} = 1.6$ cm and PDD (12.4, 12 x 12, 100) = 59.2%.
    a. 94.6, 54.7
    b. 81.5, 48.2
    c. 76.3, 52.9
    d. 50, 94
    e. 34.2, 134.7

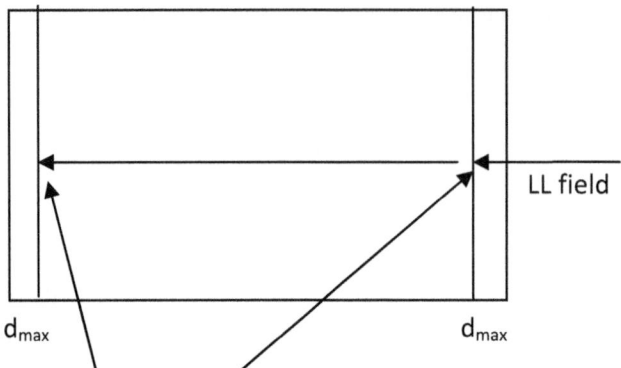

Entrance and exit dose for the LL field in POP of fields treatment

172. In problem 168 above, _____ is the total dose in cGy – ED + ExD on the right side of the patient (see the figure below).
    a. 274.6
    b. 268.4
    c. 242
    d. 237.7
    e. 211.2

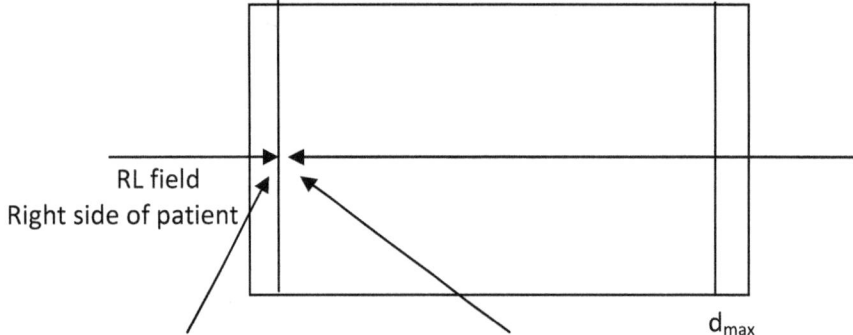

Total dose at d_{max} point on the RHS of patient (in the POP of lateral fields treatment)

173. In problem 168 above, _____ is the total dose in cGy – ED + ExD on the left side of the patient (see the figure below).
    a. 240.6
    b. 203.2
    c. 178
    d. 156.7
    e. 123.3

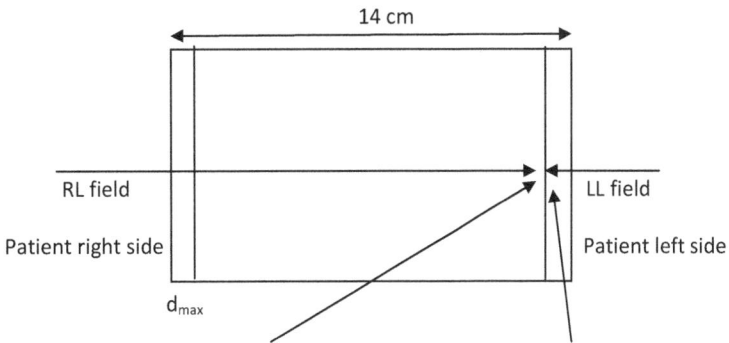

Total dose at d_{max} point on the LHS of patient (in the POP of lateral fields treatment)

174. In problem 168 above, _____ are the MU for the RL and LL fields, respectively, to deliver 200 cGy at the treatment plane. Given OF (12 x 12) = 1.012. D_{cal} (1.6, 10 x 10) = 1.01 cGy/MU.
    a. 159, 79.7
    b. 148.5, 98.2
    c. 134.3, 96
    d. 128.7, 95.2
    e. 123.3, 86.3

175. In a four-field pelvis treatment, the AP/PA fields are weighted twice as much as the RL/LL fields, keeping the total $D(d_{max})$ for the four fields the same (400%). _____ are the weighted $D(d_{max})$ doses in percentage for the AP, PA, RL, and LL fields, respectively.
    a. 100, 100, 100, 100
    b. 125, 75, 125, 75
    c. 133.3, 133.3, 66.7, 66.7
    d. 125, 125, 75, 75
    e. 120, 140, 80, 80

176. In the above problem, _____ are the doses to the d_{max} point of the AP, PA, RL, and LL fields, respectively, if 200 cGy is delivered to 210% isodose line that encloses the treatment target.
    a. 127, 127, 63.5, 63.5
    b. 150, 150, 100, 100
    c. 120, 80, 120, 80
    d. 124, 124, 73.5, 73.5
    e. 100, 100, 150, 150

177. In the above problem, _____ are the MUs delivered by the AP, PA, RL, and LL fields, respectively, that would result in delivering 200 cGy to the 210 isodose line. Given:
    treatment beam: 6 MV, 15 x 15 field; OF (15 x 15) = 1.03; D_{cal}(1.6, 10 x 10, 100) = 1.01 cGy/MU.
    a. 133, 132.5, 67, 69
    b. 150.5, 148, 87.6, 88.3
    c. 130, 132, 76, 67
    d. 126, 128, 72, 76
    e. 122, 122, 61, 61

178. A patient is being treated using a three field technique, a 12.5 x 12.5 open anterior field and 9 x 9 RL and LL fields using 30° wedges. The fields are weighted as AP:RL:LL = 200%:100%:100%.
_____ are the doses (in cGy) to d_{max} points of the AP, RL, and LL fields, respectively, if a dose of 200 cGy is prescribed to the 230° isodose line that encloses the treatment target.
   a. 245, 245, 245
   b. 230, 200, 180
   c. 200, 189.6, 178.3
   d. 173.9, 87, 87
   e. 150, 85,85

179. In the above problem, _____ are the MUs to be delivered through each field to deliver the prescribed dose.
Given: Treatment beam: 6 MV; Field size: AP field 12.5 x 12.5; RL and LL fields 9 x 9
OF (12.5 x 12.5) = 1.015; OF (9 x 9) = 0.991; WF = 0.547
D_{cal} (1.6, 10 x 10, 100) = 0.99 cGy/MU
   a. 212, 198, 187
   b. 173, 162, 162
   c. 185, 176, 176
   d. 162.5, 162.5, 156
   e. 200, 100, 100

180. A patient is being treated using a POP of RL and LL fields delivering a dose of 200 cGy to the midplane of the patient (see the figure below). Patient thickness along the beam central axis (CA) is 14 cm. The SAD setup is used for the treatment, and the fields are equally weighted.
_____ is the dose (in cGy) at the mid plane for the RL and the LL fields, respectively.
   a. 200, 200
   b. 150, 250
   c. 120, 100
   d. 100, 100
   e. 20, 180

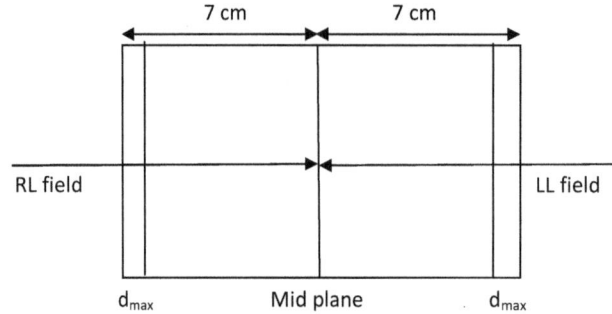

POP of RL & LL fields delivering dose to mid plane in an SAD setup

181. In the above problem, _____ are the MU delivered through each field. Given:
TMR (7, 12 x 12, 100) = 0.871; OF (12 x 12) = 1.017; D_{cal} (1.6, 10 x 10, 100 SAD) = 1.01 cGy/MU.
   a. 200
   b. 178.4
   c. 154.6
   d. 111.8
   e. 100

182. In the above problem, _____ are the MU delivered through the RL and the LL fields, respectively, if the fields RL and LL are weighted as 2:1 at the isocenter.
    a. 149, 74.5
    b. 136, 84
    c. 120, 80
    d. 100, 60
    e. 84, 148

183. In the following problem, the same treatment (as in problem #180) is delivered at 5 cm depth instead of the midplane. The weighting and the field size remain the same. Given:
    TMR (5, 12 x 12, 100) = 0.929; TMR (9, 12 x 12, 100) = 0.815
    OF (12 x 12) = 1.017; D_{cal} (1.6, 10 x 10, 100 SAD) = 1.01 cGy/MU
    _____ are the MUs delivered by the RL and the LL fields, respectively.
    a. 234.7, 92.7
    b. 100, 185.4
    c. 139.7, 79.6
    d. 200, 100
    e. 87.6, 74.8

184. For problem #180 above, _____ are the entrance and exit doses from the POP of fields.
    Given TMR (17.4, 12 x 12) = 0.592.
    a. 300, 100
    b. 238, 245.6
    c. 208.3, 164.2
    d. 184.5, 178.6
    e. 100, 100

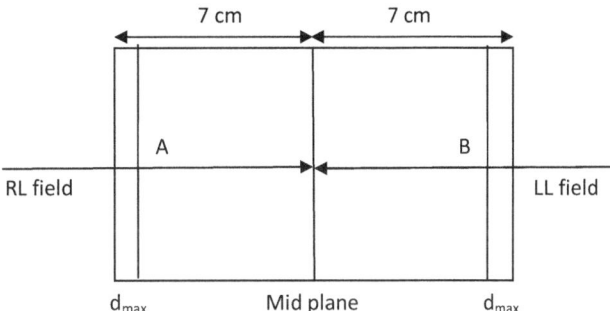

Midplane dose in SAD setup

185. A patient is treated in a four-field technique using a 6 MV beam. The field size used is 12.5 x 12.5. The fields are equally weighted at the isocenter and total dose is normalized to 100%. (Each field contributes 25%.) A dose of 200 cGy is prescribed to the 95% isodose line as shown in the figure below.

_____ are the MUs that must be delivered from the AP/PA fields and RL/LL fields, respectively. The following data are given:

AP depth = 9 cm; Lateral depth = 14 cm; TMR (9, 12.5 x 12.5) = 0.819;

TMR (14,12.5 x 12.5) = 0.683; OF (12.5 x 12.5) = 1.019; D_{cal} (1.6, 10 x 10, 100 SAD) = 0.99 cGy/MU

  a.   120/120, 100/100

  b.   100.6/100.6, 87.3/87.3

  c.   95/95, 95/95

  d.   91.3/91.3, 82.6/82.6

  e.   63.7/63.7, 76.3/76.3

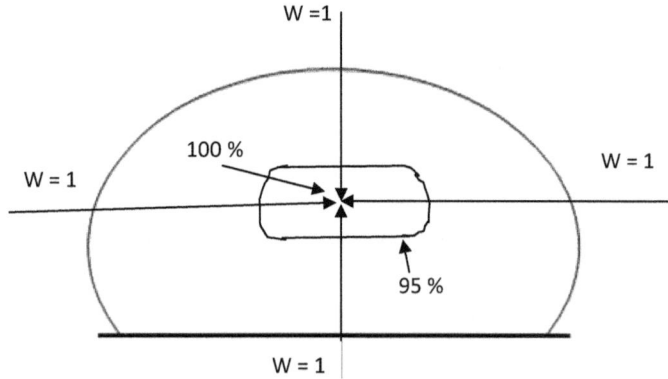

Four-field isocentric setup

186. _____ are the MU for the AP/PA and RL/LL fields for the above problem assuming the AP/PA fields and RL/LL fields are weighted 2:1. The total dose at the isocenter from all the fields is normalized to 100% (see the figure below).

  a.   200/200, 100/100

  b.   84/84, 51/51

  c.   189/96, 189/96

  d.   87/78, 78/87

  e.   84/96, 61/63

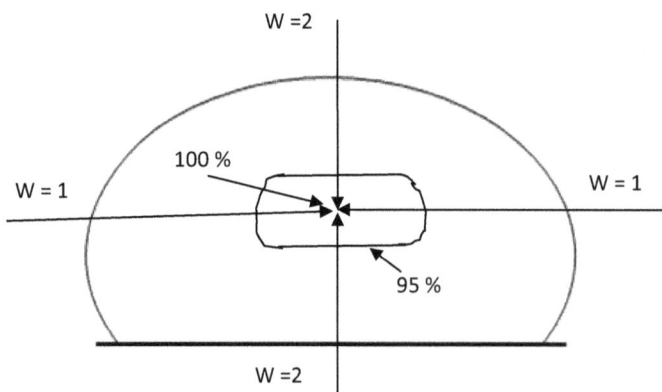

Four-field isocentric setup

187. Given $S_c(6 \times 6)$ and $S_p(6 \times 6)$, the output factor (OF) is given by _____.
    a. $S_c + S_p$
    b. $S_c - S_p$
    c. $S_c \times S_p$
    d. S_c / S_p
    e. S_P / S_c

188. Given $S_c(6 \times 6) = 0.970$ and $S_p(6 \times 6) = 0.990$, the output factor (OF) is given by _____.
    a. 0.960
    b. 1.5
    c. 1.0
    d. 0.6
    e. 0.5

189. TAR (12×12) – TAR (0×0) gives _____.
    a. TMR (12×12)
    b. scatter dose for the 12 x 12 field
    c. primary dose for 12 x 12 field
    d. SAR
    e. SMR

190. TMR (15×15) – TMR (0×0) gives _____.
    a. TAR (15×15)
    b. primary dose for the 15 x 15 field
    c. scatter dose for 15 x 15 field
    d. SMR
    e. SAR

191. Given the TAR = 0.721 and TAR_0 = 0.538 for a given field size of a ^{60}Co beam, SAR is equal to _____.
    a. 0.721
    b. 0.538
    c. 1. 26
    d. 0.183
    e. 0

192. Separating the dose into the primary and secondary or scatter component in radiotherapy _____.
    a. is not useful
    b. helps in the dosimetry of conventional x-ray beams with square fields
    c. helps in the dosimetry of irregular fields
    d. helps in determining the dose under the blocked regions of a field
    e. is not always possible

193. The TAR $(10, 10 \times 10)$ and TAR $(10, 0 \times 0)$ values for a ^{60}Co beam are 0.704 and 0.536, respectively. The scatter component of the dose at 10 cm depth along the beam central axis is _____.
    a. 1.24
    b. 0.704
    c. 0.168
    d. 0.094
    e. 0

194. The SAR (11, 12 x 12) and TAR (11, 0 x 0) values for a ^{60}Co beam are 0.186 and 0.501, respectively. The TAR value at 10 cm depth for a 12 x 12 field along the beam central axis is _____.
    a. 0.186
    b. 0.034
    c. 0.022
    d. 0.687
    e. 0

195. The off-axis-ratio (OAR) for the point A is defined as _____ (see the figure below).
    a. D(A) / D(C)
    b. D(C) / D(A)
    c. D(C) x D(A)
    d. D(A) / D(C)
    e. D(C) / D(A)

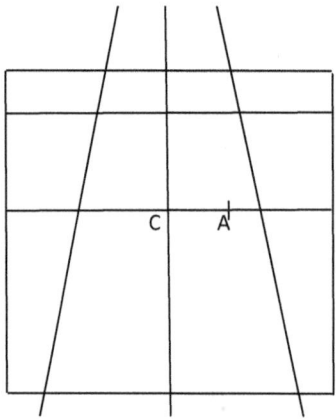

Off axis ratio concept

196. In the above problem, PDD(A) = 73.8% and PDD(C) = 70.7%. The OAR (A) is given by _____.
    a. 1.223
    b. 1.044
    c. 1
    d. 0.98
    e. 0.74

197. A patient is being treated at 120 cm SSD. The field size at 120 cm SSD is 13.5 x 13.5. The treatment depth is 9 cm. The prescribed tumor dose (TD) for this field is 100 cGy. The following data are given:
    S_C (11.3 x 11.3) = 1.007; S_p (13.5 x 13.5) = 1.009; PDD (9,13.5 x 13.5, 120) = 73.7%
    D_{cal}(1.6, 10 x 10, 100 SSD) = 0.99 cGy/MU
    The MU required to deliver this dose is given by _____.
    a. 223.4
    b. 210.3
    c. 200
    d. 194.8
    e. 84.6

198. Calculate the MUs for the above problem if the treatment had been carried out by SAD technique. The following data are given: D_{cal} (1.6, 10 x 10, 100 SSD) = 0.99 cGy/MU; S_C (11.3 x 11.3) = 1.007; S_p (14.5 x 14.5) = 1.012; TMR (9,14.5 x 14.5, 130 SAD) = 0.824.
    a. 232.6
    b. 212.6
    c. 206
    d. 200.2
    e. 172.3

199. A 6 MV blocked field shown in the figure below is used for a patient treatment in the SSD setup. _____ is the dose in cGy at the point P below the block at 10 cm depth, assuming a dose of 100 cGy is delivered at d_{max} for a 10 x 10 open field. The data are given: OF (12 x 12) = 1.012; OF (12 x 4) = 0.959; Block transmission = 3%; PDD (10, 12 x 12, 100) = 67.9%; PDD (10, 12 x 4, 100) = 64.3%.
    a. 50
    b. 24.8
    c. 16.5
    d. 8.9
    e. 0

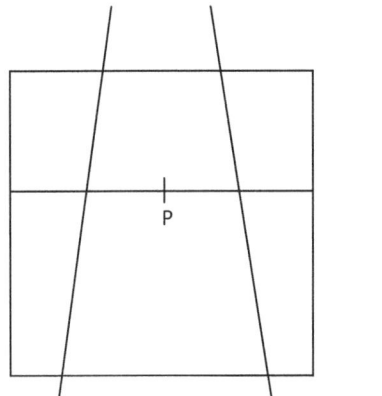
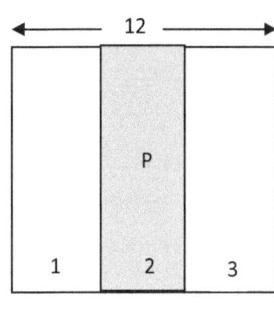

Dose determination under a block

200. The dose under the block in the above problem comes out to be about 14%, while the block transmission is only 3% of the open field dose. This is because _____.
    a. the calculations are not accurate
    b. the scatter has not been estimated with high accuracy
    c. of the scatter from the open field areas reaching the point P
    d. the primary dose reaching point P has been neglected
    e. the calculations are correct only for lower x-ray energies

201. The beam output (cGy/min) measured in air for a ^{60}Co unit increases with field size mainly due to _____.
    a. collimator scatter
    b. room scatter
    c. couch backscatter
    d. chamber wall scatter

202. A beam flattening filter is used with _____.
    a. a photon beam
    b. an electron beam
    c. both beams
    d. none of the above

203. Wedged beams are used _____.
    a. to correct for sloping patient surface
    b. in case of irregularly shaped tumor volumes
    c. while combining two angled fields for obtaining uniform target dose
    d. in breast treatment

204. The independent movement of jaws is necessary for _____.
    a. creation of blocked fields
    b. avoiding beam divergence at one edge when necessary
    c. creating dynamic wedge fields
    d. none of the above

205. Disadvantages of using a physical wedge include _____.
    a. hardening of the beam, which also depends on wedge thickness
    b. difficulty in positioning the wedge reproducibly in the beam path
    c. attenuation of the beam and an increase in treatment time
    d. inconvenience in handling heavy wedges

206. Single field treatments _____.
    a. give unacceptable dose nonuniformity across the target volume
    b. give too high a dose to normal tissues
    c. are not used in curative therapy
    d. give good tumor control

207. A dose of 300 cGy is prescribed to the center of the target volume in a POP field treatment. The beam output (i.e., at d_{max} depth) is 1.02 cGy/MU. If the total PDD at the target center is 150%, the monitor units to be set for each of the two POP fields are given by _____.
    a. 100 / 1.02
    b. 150 / 1.02
    c. 200 / 1.02
    d. 300 / 1.02

208. In the case of isocentric treatment, compared to constant SSD method, _____.
    a. setup time decreases
    b. setup error decreases
    c. treatment accuracy increases
    d. overall time to treat a patient increases

209. The TAR _____.
    a. concept is usually not preferred with accelerator photon beams because the dose in free air is not well defined for these beams
    b. can be derived from PDD data
    c. at d_{max} and PSF are practically the same quantity
    d. always increases with depth

210. The side of an equivalent square of a rectangular field (a cm × b cm) is approximately given by _____.
    a. 2ab / (a+b)
    b. ab
    c. (a+b) / 2
    d. none of the above

211. TMR depends on _____.
    a. depth
    b. field size
    c. beam quality
    d. SSD

212. TMR data are generally obtained from _____.
    a. measurements
    b. Monte Carlo computations
    c. TAR data
    d. PDD data

213. Collimator scatter factor, Sc(r), provides the _____.
    a. output factor in air
    b. influence of head scatter on beam output, with increasing collimator field size
    c. influence of phantom scatter on beam output, with increasing collimator field size
    d. none of the above

214. Collimator-phantom scatter factor, Sc,p(r), gives the _____.
    a. output factor measured in air
    b. output factor measured in phantom
    c. influence of head and phantom scatter on beam output measured in phantom, with increasing collimator field size
    d. none of the above

215. When utilizing an SSD technique, the _____.
    a. patient is set up with respect to the patient skin at constant SSD
    b. patient is set up with respect to the target center at constant SAD
    c. concept of PDD is used to determine the dose at the input port
    d. concept of TAR is used to determine the dose at the input port

216. The isocentric treatment technique assures _____.
    a. more clearance to the external accessories attached on the collimator
    b. patient comfort during treatment
    c. reproducibility of treatment setup
    d. none of the above

217. The depth of dose maximum in a patient for kV x-rays is _____.
    a. 0 mm or skin surface
    b. 5 mm
    c. 5 cm
    d. none of the above

218. The depth of dose maximum in a patient for a relatively clean ^{60}Co beam is _____.
    a. 0 mm or skin surface
    b. 5 mm
    c. 5 cm
    d. none of the above

219. _____ is the recommended depth for accelerator photon beam calibration.
    a. 0 mm or phantom surface
    b. d_{max}
    c. 5 cm
    d. 10 cm

220. The AAPM TG-51 protocol is based on calibration in ^{60}Co beam in terms of _____.
    a. absorbed dose to water
    b. exposure
    c. air kerma
    d. dose equivalent

221. The reference phantom for beam calibration (AAPM TG-51) is _____.
    a. Perspex™
    b. polystyrene
    c. solid water
    d. water

222. The point of measurement for the Farmer-type chamber for beam calibration (AAPM TG-51) is _____.
    a. the effective point of measurement for the chamber
    b. the geometric center of the chamber
    c. the outer surface of the chamber
    d. none of the above

223. During photon beam calibration, 1 mm Pb foil is placed in the beam path because it _____.
    a. cuts off all contamination electrons from reaching the patient
    b. cuts of all unwanted photons
    c. introduces known electron contamination, facilitating the determination of PDD only due to photons, PDD $(10)_X$, specifier of beam quality
    d. increases the PDD

224. Reference conditions for photon beam calibration in a water phantom for the constant SSD geometry are given by _____.
    a. reference depth: 10 g/cm2
    b. reference depth: d_{max}
    c. field size: 10 cm x 10 cm at the nominal treatment SSD
    d. field size: 10 cm x 10 cm at SAD = SCD (source to chamber distance)

225. Match the beam quality to the depth of dose maximum, d_{max}, in cm, for a 10 cm x 10 cm field.
Beam Quality	d_{max} (cm)
a. kilovoltage	_____ i. 0.5
b. ^{60}Co	_____ ii. 0
c. 10 MV	_____ iii. 5
d. 25 MV	_____ iv. 2.5

226. Match the beam quality to the percentage depth dose (10 cm × 10 cm field, 10 cm depth; SSD = 80 cm for ^{60}Co and 100 cm for accelerator photon beams).
Beam Quality	Percentage Depth Dose
a. ^{60}Co beam	_____ i. 64.8
b. 4 MV	_____ ii. 55.6
c. 10 MV	_____ iii. 73

227. Match the beam quality to the PSF for a 15 cm x 15 cm field and SSD as in question #88.
Beam Quality	PSF
a. ^{60}Co beam	_____ i. 1.015
b. 6 MV	_____ ii. 1.05
c. 18 MV	_____ iii. 1.008

228. Physical penumbra depends on _____.
    a. geometric penumbra
    b. collimator transmission
    c. lateral photon scatter (in the patient)
    d. lateral electron transport (in the patient)

229. A patient is treated with a set of POP fields by an isocentric technique with the isocenter placed at the mid plane. If the patient thickness along the CA is 24 cm and SAD = 100 cm, _____ is the SSD of the patient.
    a. 76 cm
    b. 88 cm
    c. 100 cm
    d. 124 cm

230. The air kerma rate of a ^{60}Co beam at 1 m from the source for a 10 cm x 10 cm field is 100 cGy/min. The output factor for a 6 cm x 6 cm field is 0.92. _____ is the output for the 6 cm x 6 cm field in cGy/min.
    a. 108
    b. 100
    c. 92
    d. none of the above

231. The output factor (OF) of a ^{60}Co unit for a 6 cm x 6 cm field is 0.92. The OF of the unit, when the source is replaced by a new source of twice the activity, is given by _____.
    a. <0.92
    b. >0.92
    c. 0.92
    d. none of the above

232. The output of a ^{60}Co unit at 1 m from the source was measured to be 108 cGy/min for a 10 cm x 10 cm field. After changing the source, the output for the 10 cm x 10 cm field was measured to be twice the old value. The new output for a 15 cm x 15 cm field will be _____.
    a. the same as the old value
    b. twice the old value
    c. more or less than the old value
    d. none of the above

233. According to Sterling's formula, two fields are considered equivalent if they have _____.
    a. the same area
    b. the same (area/perimeter) value
    c. the same perimeter value
    d. none of the above

234. The output of a clinical photon beam was determined to be 1.01 cGy/MU at the dose maximum point for a 20 cm x 12 cm field for an SSD of 1 m. The side (in cm) of an equivalent square field that would have the same output is given by _____.
    a. 20
    b. 16
    c. 15
    d. 12

235. A circular field and a square field are considered equivalent if they have _____.
    a. the same area
    b. the same (area/perimeter) value
    c. the same perimeter value
    d. none of the above

236. The side of an equivalent square field of a circular field of diameter D is approximately given by _____.
    a. 0.9 D
    b. D
    c. 1.5 D
    d. none of the above

237. The radius (in cm) of the equivalent circular field of a 10 cm x 10 cm field is approximately given by _____.
    a. 5.6
    b. 10
    c. 31.8
    d. none of the above

238. PDD for use in SSD type of treatments is normalized to _____.
    a. skin dose
    b. dose at d_{max}
    c. dose at d_5 cm
    d. dose at d_{10} cm

239. For high-energy ($>$ ^{60}Co energy) clinical photon beams, skin dose is _____.
    a. less than the peak dose
    b. more than the peak dose
    c. equal to the peak dose
    d. none of the above

240. Field weighting is carried out in planning radiation therapy treatment to _____.
    a. homogenize the dose in the target volume
    b. limit doses to the critical structures in the beam path
    c. reduce the treatment time
    d. none of the above

241. In a constant SSD three-field technique with the three fields equally weighted, the dose at the input ports due to the three fields is _____.
    a. 1 : 1 : 1
    b. 100% : 100% : 100%
    c. 1 : 0.5 : 0.5
    d. 1 : 1 : 0.5

242. Dose distribution in multiple levels may be necessary when the _____ is treated.
    a. pelvis of a thin patient
    b. breast of a thin patient
    c. breast of a large patient
    d. brain of an obese patient

243. To minimize the gap between two adjacent fields abutting at a depth, _____.
    a. a posterior field alone can be used
    b. an anterior field alone can be used
    c. a half beam block can be used

244. When adjacent areas must be treated, it is best to _____.
    a. calculate the gap necessary between the fields to prevent an overlap
    b. set a 2 cm gap between all fields
    c. rely on tattoos to prevent an overlap

245. The approximate depth of dose maximum in water for a 10 MV clinical photon beam is _____.
    a. 0
    b. 5 mm
    c. 25 mm
    d. none of the above

246. With increase in field size, the (effective) primary radiation incident on a patient _____.
    a. increases
    b. decreases
    c. remains the same

247. The output of a ^{60}Co unit measured in a phantom compared to measurement in air at the same point will be _____.
    a. more
    b. less
    c. the same

248. A collimator _____.
    a. flattens the beam at a specified depth in tissue
    b. defines the beam
    c. determines the dose rate
    d. stops the scatter radiation

249. The collimator scatter for accelerator photon beams _____.
    a. cannot be measured in air
    b. can be measured in air with the Farmer chamber with ^{60}Co buildup cap
    c. can be measured in air with the Farmer chamber with a cap of appropriate buildup thickness
    d. must be measured in a water phantom at calibration depth

250. The entrance dose is defined as _____.
    a. dose at d_{max} for the incident field
    b. skin dose for the incident field
    c. dose at the calibration depth for the incident field
    d. none of the above

251. The exit dose is defined as the _____.
    a. dose at d_{max} for the exiting field
    b. skin dose for the exiting field
    c. dose at calibration depth for the exiting field
    d. none of the above

252. A patient is being treated with POP fields (A/P and P/A). The total dose at the dose maximum point on the anterior side for this treatment is given by _____.
    a. entrance dose (due to A/P field)
    b. exit dose (due to P/A field)
    c. entrance dose (A/P field) + exit dose (P/A field)
    d. none of the above

253. A patient is being treated with POP fields (A/P and P/A) through an SSD technique. The midline dose to be delivered is 200 cGy. The patient separation along the central axis is 20 cm. The treatment beam PDD (10 cm) is 68% and PDD (18.5 cm at exit point) is 42%.

     _____ is the total dose at the dose maximum point on the anterior side for this treatment.
     a. $[200 / 68] \times 100$
     b. $[100 / 68] \times 100$
     c. $[200 / 42] \times 100$
     d. $[100 / 68] \times 100 + [100 / 68] \times 0.42$

254. Clarkson's Method is used for the dose calculation of _____.
     a. a rectangular field
     b. an irregular field
     c. a circular field
     d. a square field

255. Tangential parallel-opposed fields are mostly used to treat _____.
     a. cancer of the cervix
     b. head and neck tumors
     c. breast cancer
     d. extremities

256. Wedges are commonly used in _____.
     a. orthogonal head and neck fields
     b. four fields in cancer of the cervix
     c. three oblique fields in cancer of the esophagus
     d. all the above

257. The clearance between the patient skin surface and the collimator accessories of the teletherapy machine is required _____.
     a. to reduce the electron contamination
     b. for accurate immobilization
     c. to increase the setup reproducibility
     d. none of the above

258. A four-field box technique is traditionally preferred in the _____.
     a. head and neck region
     b. pelvic region
     c. thoracic region
     d. extremities

259. Extended SSDs are used for _____.
     a. total body irradiation
     b. large mantle fields
     c. treating head and neck cases
     d. none of the above

260. A patient is to be treated with a 6 MV photon beam for a field size of 12 cm x 12 cm on patient surface at an extended SSD of 125 cm. The depth of target center is 9 cm. In this case (compared to standard SSD treatment), _____.
     a. the reference dose rate (or dose/MU) at the input port (d_{max} position) decreases
     b. the PDD (9, 12 x 12, 125) increases due to increased SSD
     c. the PDD correction is approximately given by Mayneord's F factor
     d. there is no change in phantom scatter

261. A patient is to be treated with a 6 MV photon beam at an extended SSD of 125 cm (see the figure below). The calibration dose rate, D_{cal} (1.5, 10 x 10, 100) is given by 0.993 cGy/MU. _____ is the reference dose rate (in cGy/MU) for the new SSD, D_{ref} (1.5, 10 x 10, 125).
    a. 0.993
    b. $0.993 \times (126.5 / 101.5)$
    c. $0.993 \times (101.5 / 126.5)^2$
    d. none of the above

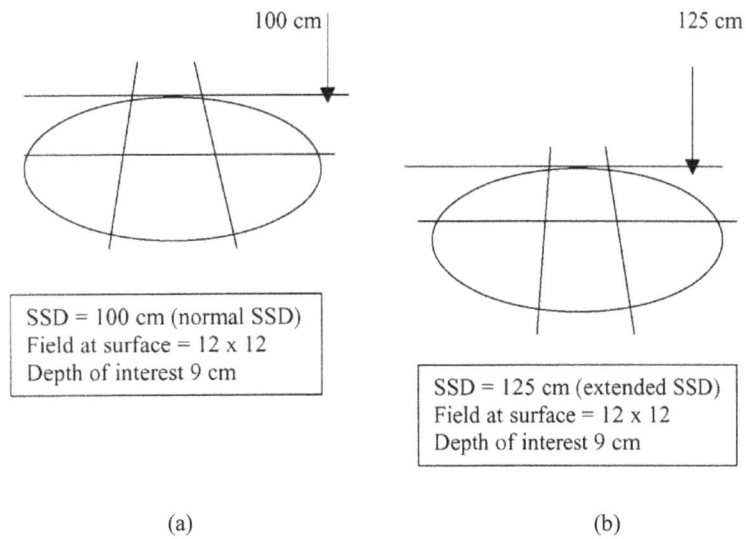

(a)   (b)

Nominal SSD and extended SSD geometry for treatment
(same field size on surface of patient)

262. A patient is to be treated with a 6 MV photon beam at an extended SSD of 125 cm, the collimator setting being 10 cm x 10 cm. The reference dose rate (in cGy/MU) for the new SSD, D_{ref} (1.5, 12.5 x 12.5, 125), for the same collimator setting (see figure a below) is to be calculated. D_{cal} (1.5, 10 x 10, 100) = 0.993 cGy/MU. OF (15, 12.5 x 12.5, 100) = 1.018. The reference dose rate for the new SSD is _____.
    a. 0.993
    b. $0.993 \times (101.5 / 126.5)2$
    c. $0.993 \times 1.018 \times (101.5 / 126.5)2$
    d. none of the above

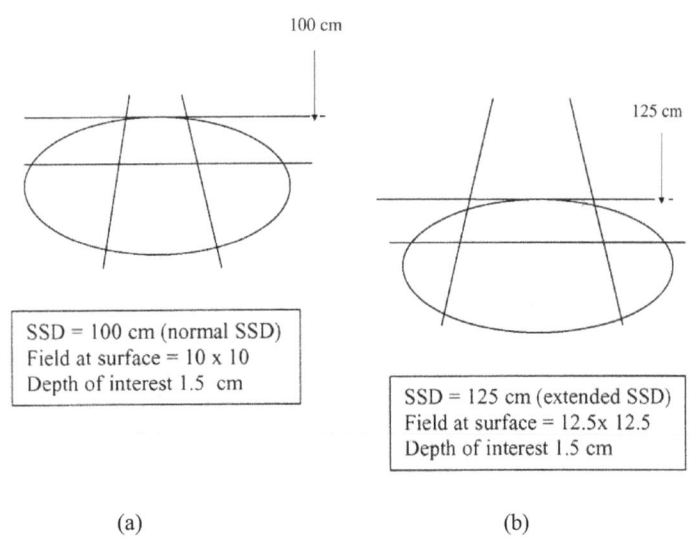

(a)   (b)

Nominal SSD and extended SSD geometry for treatment
(same collimator opening)

263. A patient is to be treated using a blocked field. The open field without blocking is 12 cm x 12 cm. After blocking, the equivalent field of the blocked field is 9 cm x 9 cm. The dose at 10 cm depth must be calculated for the anterior field. Which statements given below are true?
   a. There will no change in the effective primary reaching the patient.
   b. The collimator scatter will decrease due to blocking.
   c. The phantom scatter will decrease due to blocking.
   d. The output for 12 cm x 12 cm would give the output of the treatment field.

264. A patient is to be treated using a 10 MV x-ray beam at 100 cm SAD. A dose of 100 cGy has to be delivered to the target center depth of 10 cm using a 15 cm x 15 cm anterior field. The beam output (0.996 cGy/MU) is known at d_{max} for a 10 cm x 10 cm field in a water phantom, for SAD = 100 cm. Other factors required for determining the MU to be set for treatment are _____.
   a. TMR(10, 15 x 15, 100)
   b. OF(15 cm x 15 cm)
   c. PSF(15 cm x 15 cm)
   d. treatment time

265. A patient is to be treated using two opposing (A/P-P/A) fields. The target center must receive a total dose of 300 cGy. The two fields are weighted A/P: P/A = 2:1 at the target center. The target dose delivered through the _____.
   a. A/P port is 100 cGy
   b. A/P port is 200 cGy
   c. A/P port is 300 cGy
   d. P/A port is 100 cGy

266. Field weighting _____.
   a. gives the relative contribution of beams to dose at target center or d_{max} point
   b. changes the dose rates at input ports
   c. improves dose uniformity across the target
   d. reduces dose to normal tissues or critical structures

267. Field weighting is used _____.
   a. when tumors are centrally located
   b. when contribution from any of the fields needs to be reduced or increased with respect to other fields
   c. in rotation therapy
   d. in single field treatments

D. Basic External Beam Calculations (Electron Beams)

Circle the right answer (Yes or No):

1. (Yes / No) In the AAPM TG-51 protocol, the electron beam quality is specified by E_0, the mean energy of the electrons incident on the patient.

2. (Yes / No) Buildup in electron beams is due to electron scattering and the finite range of delta rays produced.

3. (Yes / No) The tail of the electron beam depth dose distribution is due to some long-range electrons in the beam.

4. (Yes / No) The energy fluence of an accelerator electron beam falls according to the inverse square law of distance with the target position as the origin.

5. (Yes / No) Electron beams have a finite range in the medium.

6. (Yes / No) The range of electrons of energy E (MeV) in water is roughly about E/2 cm.

7. (Yes / No) Jaw positions do not influence the electron beam output of a therapy accelerator.

8. (Yes / No) One can use lead cutouts at the end of the electron applicators to treat tumors of irregular cross section.

9. (Yes / No) Low-energy electrons are more easily scattered compared to high-energy electrons.

10. (Yes / No) P_{gr}^Q can be determined by making measurements at $(d_{ref} + r_{cav})$ and d_{ref} or from the PDD curve of the user electron beam from the dose values at these depths.

11. (Yes / No) While using diode for electron beam PDD measurement, the depth ionization curve actually represents the depth dose curve.

12. (Yes / No) Depth ionization curve measured using a Farmer-type chamber can be taken to represent depth dose curve of the beam.

13. (Yes / No) The PDD of electron beam does not depend on the angle of incidence of the beam on the patient.

14. (Yes / No) P_{gr} for electron beams is always <1.

15. (Yes / No) According to TG-51, a cylindrical chamber can be used for all electron and x-ray beams in clinical practice.

16. (Yes / No) The tail of electron PDD curve is caused by the x-ray contamination of the beam produced outside and from within the phantom.

17. (Yes / No) The choice of field size in electron beam therapy should be based on the isodose coverage of the target volume.

18. (Yes / No) In the TG-51 protocol, k_Q factors are determined with respect to a reference photon beam quality of ^{60}Co beam. For electron beams **a reference electron beam quality** was not chosen which would have simplified electron beam calibration and dosimetry. This is because the calibrators at calibration laboratories are equipped with a ^{60}Co beam for therapy chamber calibrations, but not an electron beam (i.e., a linac).

19. (Yes / No) The x-ray contamination is defined as dose at R_p + 2 cm.

20. (Yes / No) Both photon beam and the electron beam spread the same way with respect to the x-ray target.

21. (Yes / No) The concept of "virtual source position" is not important for electron beam therapy.

22. (Yes / No) The "effective SSD" for an electron beam depends only on the energy of the beam.

23. (Yes / No) The surface dose of the electron beams (in %) decreases with increasing electron energy.

24. (Yes / No) The electron beams penetrate to nearly the same extent in lung and unit density tissue.

25. (Yes / No) The peak depth R_{100} depends only on the energy of the electron beam.

26. (Yes / No) Film can be used for measuring the electron beam depth dose profile and its range parameters.

27. (Yes / No) A percentage depth ionization curve also represents a percentage depth dose curve for electron beams, measured using Farmer-type chamber.

28. (Yes / No) For measuring the range parameters of an electron beam, lateral scatter equilibrium is essential.

29. (Yes / No) For electron beams, output or PDD becomes field size dependent when there is no lateral scatter equilibrium.

30. (Yes / No) The backscattering of electrons increases with increasing electron energy or atomic number of the scatterer.

31. (Yes / No) Knowing the beam output at the normal treatment distance, FSD (or SSD), the output at an extended SSD can be calculated using the inverse square law.

32. (Yes / No) The position of a virtual electron source, along beam central axis, is independent of electron energy or field size.

Choose the right answer(s) (more than one may be correct):

33. Plane-parallel chambers are recommended for the dosimetry of low-energy electron beams (<10 MeV), in place of Farmer-type chambers, because _____.
    a. Farmer-type chambers are too large
    b. Farmer-type chambers cannot be used for electron dosimetry
    c. Farmer-type chambers give rise to significant electron fluence perturbation at low electron energies
    d. Farmer-type chambers are less accurate for low-energy electrons

34. The surface dose for electron beams _____.
    a. is much larger compared to photon beams
    b. for a given beam energy, is the same for all accelerators
    c. can be measured with an extrapolation chamber
    d. none of the above are true

35. Bremsstrahlung radiation is mainly produced by accelerator electrons _____.
    a. in the accelerator head
    b. in the patient
    c. in air
    d. only at very high electron energies

36. For an electron beam of energy E (in MeV), the PDD is dependent on field size W _____.
    a. for W less than E cm
    b. for W less than E/2 cm
    c. for any field size W
    d. only for W <10 cm

37. A scattering foil is used in the clinical accelerator _____.
    a. to reduce the electron energies
    b. to produce Bremsstrahlung
    c. to spread the accelerator "pencil beam" into a larger field for treating patients
    d. for none of the above

38. The electron beam incident on the patient exhibits _____.
    a. no spectrum (monoenergetic)
    b. a spectrum with a Gaussian distribution
    c. a spectrum of rectangular distribution
    d. a spectrum with a skewed distribution

39. The electron beam incident on a patient can be characterized by _____.
    a. a most probable energy
    b. a mean energy
    c. a maximum energy
    d. a single energy

40. The electron energy specified by the machine manufacturer refers to _____.
    a. the maximum energy of the electrons
    b. the mean energy of the electrons
    c. the most probable energy of the electrons
    d. none of the above

41. The mean energy of the electron beam, E_0, can be determined fairly accurately from the parameter _____.
    a. $2.33 R_{50}$
    b. $3 R_{50}$
    c. R_{50}
    d. $2 R_{50}$

42. The clinical electron beam incident on a patient _____.
    a. is monoenergetic
    b. exhibits a spectrum
    c. has a range parameter, R_{50}, that is of relevance in dosimetry
    d. has a most probable energy that is of relevance in dosimetry

43. Compared to a 10 MV photon beam, a 10 MeV electron beam _____.
    a. has more penetration
    b. exhibits more surface dose
    c. exhibits no scattering in the patient
    d. has a much flatter cross-beam profile

44. A clinical electron beam exhibits a narrow spectrum, which is characterized by a maximum energy, most probable energy, mean energy, and a spectral width.
    a. The most probable energy is related to the practical range of the electron beam.
    b. The mean energy is related to clinical dosimetry.
    c. The maximum energy is the same as the most probable energy.
    d. The spectral width is not affected by electron scattering in the patient.

45. The interaction of a pencil beam of electrons (accelerated in the wave guide) with the accelerator components (scattering foil, monitor chamber, collimators, etc.) and the patient results in _____.
    a. lateral spreading of the beam due to scattering
    b. production of low-energy electrons increasing the width of the spectrum
    c. production of Bremsstrahlung x-rays
    d. an increase in mean energy

46. The range of a 10 MeV electron beam in water is about _____.
     a. 10 mm
     b. 10 cm
     c. 5 cm
     d. none of the above

47. _____ is the rough energy of a 10 MeV electron beam incident on a water phantom at a depth of 2 cm.
     a. 10
     b. 8
     c. 6
     d. none of the above

48. For a clinical electron beam, _____.
     a. the skin dose increases with increasing electron energy
     b. there is significant skin sparing compared to photon beams
     c. $R_{50,D}$ (50% depth dose, in water, in cm) is related to mean energy of the beam
     d. the reference depth for beam calibration, as per the AAPM protocol (TG-51) is $0.6 R_{50,D} - 0.1$ cm

49. A clinical electron beam _____.
     a. is contaminated with Bremsstrahlung photons
     b. beyond the range, deposits no energy
     c. has a Bremsstrahlung tail of a few percent
     d. has none of the above characteristics

50. With electron beams, _____.
     a. high-Z materials are used as bolus
     b. Perspex, polystyrene, or any tissue-equivalent material is used as bolus
     c. bolus with a beveled edge must be used to decrease dose gradients below the edges in the medium
     d. bolus-like materials are also used to degrade the electron beam

51. The electron depth dose measurements in water must be carried out with _____.
     a. an ionization chamber
     b. silicon diode
     c. film
     d. TLD

52. For an electron beam, lateral scatter equilibrium exists when the field size is _____.
     a. roughly half the electron energy
     b. of the order of the electron energy
     c. much larger than the electron energy
     d. very much smaller than the electron energy

53. For a 10 MeV clinical electron beam incident on a patient, the required field size for lateral scatter equilibrium is about _____.
     a. 10 cm x 10 cm
     b. 15 cm x 15 cm
     c. 5 cm x 5 cm
     d. 3 cm x 3 cm

54. Electron beams of different field sizes are produced using _____.
     a. linac jaw settings as in the case of photon beams
     b. different applicator (called cone) attachments to the linac head
     c. different cutouts inside cones
     d. none of the above means

55. According to the AAPM TG-51 dosimetry protocol, the reference depth (in cm) for electron beam calibration is _____.
    a. $R_{50,D}$
    b. $0.5 R_{50,D}$
    c. $0.6 R_{50,D} - 0.1$
    d. R_p

56. The reference depth for monitor unit calculations in electron beam therapy is _____.
    a. d_{max}
    b. $R_{50,D}$
    c. $0.6 R_{50,D} - 0.1$
    d. none of the above

57. The pencil electron beam coming out of a wave guide is spread into larger field sizes, in a clinical accelerator using _____.
    a. a flattening filter
    b. scattering foils
    c. electromagnetic scanning
    d. none of the above

58. The "virtual source" of a clinical electron beam always lies at _____.
    a. the scattering foil position
    b. an exit window
    c. the flattening filter position
    d. an experimentally determined distance

59. Electron beam therapy is used for _____.
    a. deep-seated tumors
    b. superficial tumors
    c. all types of tumors
    d. none of the above

60. In electron beam therapy, the dose prescription point is generally _____.
    a. the patient's skin surface
    b. reference depth of calibration
    c. 90% isodose depth
    d. none of the points

61. In electron therapy, the electron applicator normally touches the patient's skin. If the electron beam applicator is kept at a height, h, above the skin, due to problems in positioning the applicator on the skin, then _____.
    a. the penumbra increases
    b. there is improved coverage of the target area
    c. the beam output, at reference depth, increases
    d. measurement at d_{max} follows an inverse square law

62. The output of a 7 MeV electron beam, 10 cm x 10 cm and 100 cm SSD field, at d_{max} is 1 cGy/MU. _____ is the MU required to deliver 250 cGy to the 90% treatment depth for a treatment field size of 12 cm x 12 cm. The output factor for the 12 cm x 12 cm field is 1.01 cGy/MU.
    a. $250 / (0.90 \times 1.01)$
    b. $250 \times 0.90 / 1.01$
    c. $250 \times 0.90 \times 1.01$
    d. none of the above

63. An electron beam is incident obliquely on a patient (e.g., breast irradiation). Compared to normal incidence situation, _____.
    a. skin dose increases
    b. skin dose decreases
    c. dose maximum shifts away from the surface
    d. therapeutic depth shifts toward the surface

64. When two adjacent electron fields abut on the skin or at a depth, there is a possibility of overdosing of regions around the junction because of _____.
    a. high beam output
    b. excess electron scattering in the patient
    c. difficulties in electron collimation
    d. none of the above

65. When cutouts are used to shape electron fields, cutout factors must be determined for each cutout for accurate dosimetry. However, the output variation is not very significant if the _____.
    a. open area or cutout area is still large enough for lateral electronic equilibrium
    b. blocking is more than about 25% of the applicator field area
    c. blocking is of high-Z material
    d. blocked field has a regular shape

66. In electron beam therapy, sometimes internal shielding is used (e.g., treatment of lip, eyelid, etc.) to reduce the transmitted dose to normal tissues beyond the treatment volume. If Pb or some high-Z material is used as an internal shield, one must take care of _____.
    a. chemical reactions of this material with tissues
    b. Bremsstrahlung production in the shield
    c. backscatter of electrons that may overdose normal tissues
    d. none of the above

67. Electron backscatter from internal shielding _____.
    a. increases with Z
    b. decreases with increase in energy, for any Z
    c. does not depend on Z
    d. is absorbed in a low-Z material lining so that they do not overdose normal tissues

68. Which of the electron beam characteristics listed below is true?
    a. High surface dose gives more uniform dose to the target in the buildup region.
    b. Rapid dose falloff reduces dosing of normal tissues beyond the target depth.
    c. There is a net increase in tumor control probability.
    d. There is a slight increase in normal tissue complications.
    e. All of the above are true.

69. _____ is the minimum field size that must be used for determining the beam quality or for beam calibration of the clinical electron beams.
    a. Any field size
    b. 10 cm x 10 cm
    c. Field size adequate to establish lateral scatter equilibrium
    d. Field size $>2 R_p$
    e. Minimum available field size

70. Which of the following is true regarding the use of extended SSD instead of the normal or calibration SSD in the dosimetry of clinical electron beams? Extended SSD distance _____.
    a. has minimal effect on the central-axis depth dose and off-axis ratios for the SSD changes involved in clinical practice
    b. has significant influence on output factors and beam penumbra
    c. d_{max} dose and normal SSD d_{max} dose are related by ISL, as in the case of photon beams
    d. cannot be used in electron therapy because of the complexity of electron dosimetry

71. The Bremsstrahlung contamination of the electron beam _____.
    a. can be determined from the tail of the electron beam PDD curve
    b. increases with increasing beam energy
    c. is less than 2% up to 20 MeV
    d. arises as a result of electron interactions with scattering foils, jaws, applicators, the patient, etc.
    e. depends on the accelerator design

72. PDD of electron beams can be measured in a water phantom using a _____.
    a. cylindrical therapy level chamber
    b. diode
    c. PP chamber
    d. radiographic film
    e. well-type chamber

73. _____ is the important correction factor that must be considered for converting the depth ionization curve into depth dose curve for the electron beams.
    a. The cavity perturbation
    b. Chamber wall correction
    c. Saturation correction
    d. Temperature/pressure correction
    e. SPR water to air

74. Which of the electron beam characteristics listed below is true?
    a. The 90% isodose is the only prescription isodose line and must enclose the tumor target.
    b. Low-energy isodose curves bulge out.
    c. High-energy isodose curves show constriction.
    d. Low-energy electrons do not produce any x-ray contamination.
    e. Electron beams can be combined with photon beams for treating some clinical cases.

75. The surface dose of the mega voltage electron beams is usually in the range of _____.
    a. 10% to 20%
    b. 30% to 40%
    c. 40% to 60%
    d. 75% to 95%
    e. 95% to 100%

76. The PDD at 6 cm for a 9 MeV electron beam will be _____.
    a. 100%
    b. 90%
    c. 75%
    d. a few percent
    e. 0%

77. Compared to 6 MeV electrons, 20 MeV electrons exhibit _____.
    a. larger surface dose
    b. sharper dose falloff beyond in the 80% to 20% region
    c. broader plateau
    d. lower Bremsstrahlung tail
    e. no field size dependence for any field size

78. Compared to scatter equilibrium field size, smaller field sizes exhibit _____.
    a. a decrease in d_{max}
    b. a decrease in d_{90} depth
    c. an increase in surface dose
    d. a decrease in R_p
    e. a decrease in PDD for a given depth

79. When a bolus is used, the depth dose curve _____.
    a. remains unaffected
    b. shifts upstream by bolus thickness
    c. shifts downstream by bolus thickness
    d. shifts by a fraction of bolus thickness upstream
    e. shifts by a fraction of bolus thickness downstream

80. Output of an electron beam depends on _____.
    a. energy
    b. applicator configuration
    c. field size
    d. SSD
    e. skin collimation
    f. all of the above

81. The output of a 6 MeV electron beam at $d_{max} = 1.5$ cm for the normal SSD of 100 cm is 1.05 cGy/MU. If a bolus of 1 cm thickness is placed on the skin, _____ is the output for the new SSD (in cGy/MU).
    a. 1.07
    b. 1.05
    c. 1.03
    d. 1.0
    e. 0.98

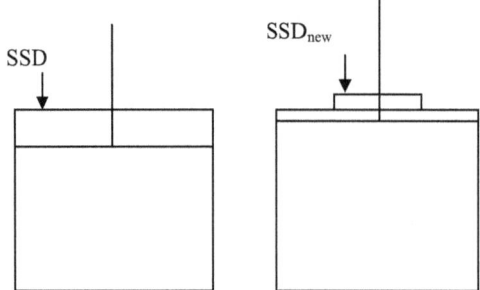

Bolus shifting the d_{max} point by bolus thickness toward the source

82. The output of a 6 MeV 10 x 10 field electron beam at d_{max} = 1.5 cm for the normal SSD of 100 cm is 1.02 cGy/ MU. If an air gap of 5 cm is used with respect to the skin, _____ is the output for the new SSD (in cGy/MU). The air gap correction factor (AGCF) for this case (6 MeV, 10 x 10, 100 SSD) is 0.989 for a 5 cm air gap.
    a. 1.24
    b. 1.17
    c. 1.06
    d. 0.92
    e. 0.74

83. We use a 1 cm thick bolus in a 6 MeV electron beam treatment. Which of the statements below are true?
    a. The d_{max} moves upstream by 1 cm.
    b. The dose at 2 cm depth decreases.
    c. The dose at d_{max} increases by 2%.
    d. The dose at d_{max} decreases by 2%.
    e. The dose at d_{max} increases or decreases depending on the beam energy a,b,c.

84. The following figure depicts the electron spectrum of a clinical linac incident on the patient. Match the numbers in the figure to the energy nomenclature.

Number		*Nomenclature*
a. 1	_____	i. E_{max}
b. 2	_____	ii. $\bar{E}_0$
c. 3	_____	iii. $\bar{E}_{p,0}$

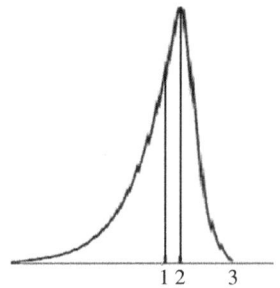

Electron beam spectrum of clinical linac

85. To deliver 200 cGy to 90% isodose, the dose at d_{max} must be _____ cGy.
    a. 300
    b. 222.2
    c. 200
    d. 156.6
    e. 98.4

86. The patient chest wall is being treated by an electron beam. The chest wall thickness requires a 9 MeV electron beam for treatment. Use of a 12 MeV electron beam, instead of 9 MeV, _____.
    a. would be advisable since the chest wall can be treated with 100% certainty
    b. would lead to unnecessary irradiation of the lung
    c. is not possible without irradiating the lung to high dose
    d. is possible without irradiating the lungs if proper thickness of bolus is used for this treatment
    e. would increase the tumor control probability

87. As the energy of the electron beam increases, _____.
    a. the percentage of surface dose increases
    b. R_{90} increases
    c. $R_{10} - R_{90}$ falloff increases
    d. R_p increases
    e. Bremsstrahlung contribution increases
    f. all of the above

88. PDD for a rectangular field, PDD(L x W), is given by _____.
    a. [PDD(L x L) + PDD(W x W)] / 2
    b. [PDD(L x L) – PDD(W x W)]
    c. [PDD(L x L) × PDD(W x W)]
    d. [PDD(L x L) × PDD(W x W)]$^{1/2}$
    e. [PDD(L x L) / PDD(W x W)]

89. As field size increases, _____.
    a. PDD increases marginally
    b. D_s increases marginally
    c. d_{max} increases
    d. output generally increases
    e. all of the above

90. A 12 MeV electron beam delivers a dose of 200 cGy at 90% isodose level. If the output of the configuration (i.e., the applicator with the insert) is 1.01 cGy/MU, the MUs to deliver the prescription dose are _____.
    a. 300.5
    b. 285.7
    c. 250.6
    d. 220
    e. 134.8

91. A lesion extending up to 3.5 cm is treated with a 12 MeV electron beam at 100 cm SSD. 1 cm of bolus is placed on the skin. If 200cGy is delivered to the distal edge of the tumor, _____ are the skin dose and maximum tissue dose, respectively, given D_{max} depth = 3 cm. The PDD for the treatment beam can be taken from the values given here:
    PDD (0 cm) = 82.7%; PDD (1 cm) = 91.3%; PDD (3.5 cm) = 97.5%
    PDD (4 cm) = 89.8%; PDD(4.5 cm) = 73.8%; PDD (5 cm) = 52.3%
    a. 288.6, 276.6
    b. 264, 260.7
    c. 247.4, 271
    d. 220, 232.4
    e. 200, 112.4

92. The output factor of a 9 MeV treatment field (15 x 15 applicator carrying a 4 x 4 insert) is 0.954. The output factor is normalized to 10 x 10 applicator field size. Treatment SSD = 100 cm. The beam calibration is given by D_{cal} (100, 10 x 10, 2.3 cm) = 1.03cGy/MU. The MU required to deliver 250 cGy at d_{max} = 2.3 cm for the treatment field are _____.
    a. 285.5
    b. 254.4
    c. 246.6
    d. 234.5
    e. 213.8

93. In the above problem, _____ MU are required to deliver 200 cGy to 95% isodose curve which is at a depth of 2.8 cm for the treatment field.
    a. 214.3
    b. 213.3
    c. 200.6
    d. 187.5
    e. 1879.3

94. In the above problem, _____ would be the MUs if the treatment had been delivered at 105 cm SSD. The SSD factor (inverse square factor × air gap correction factor) is 0.855.
    a. 288.7
    b. 250.6
    c. 234.2
    d. 212.1
    e. 187.8

95. The output factor of a 9 MeV 15 x 15 applicator carrying a 4 x 4 insert is 0.954. The output factor is normalized to 10 x 10 applicator field size. The beam calibration is given by D_{cal} (100, 10 x 10, 2.3 cm) = 1.03 cGy/MU. The MU required to deliver 250 cGy at d_{max} = 2.3 cm for the treatment field are _____.
    a. 284.5
    b. 267.7
    c. 254.4
    d. 223.4
    e. 198.3

96. The SSD correction factor consists of _____.
    a. an air gap correction factor
    b. an inverse square law correction factor
    c. an air gap correction factor and an inverse square law correction factor
    d. an air gap, inverse square law, and energy dependence correction factors
    e. an energy correction factor

97. _____ is the OF (output factor) for a 6 x 12 field size shaped in a 15 x 15 applicator. The OFs of a 6 x 6 field and a 12 x 12 field shaped in the same applicator are 1.008 and 1.001, respectively.
    a. 1.032
    b. 1.028
    c. 1.01
    d. 1.004
    e. 0.99

98. _____ is the AGCF (air gap correction factor) for a 6 x 12 field size shaped in a 15 x 15 applicator. The AGCFs of a 6 x 6 field and a 12 x 12 field shaped in the same applicator are 0.980 and 0.991, respectively.
    a. 1.023
    b. 1.010
    c. 0.985
    d. 0.95
    e. 0.78

99. _____ is the MU to deliver 180 cGy to 90% dose contour for a 6 MeV electron beam with a 6 x 12 treatment field size at 100 cm SSD, shaped in the 15 x 15 applicator. OF (6 x 12) = 0.985. The beam calibration is given by D_{cal} (100, 10 x 10, 2.3 cm) = 1.03cGy/MU.
    a. 185.2
    b. 180.5
    c. 179.7
    d. 178
    e. 177.4

100. In the above problem, _____ are the MUs if the treatment had been delivered at 110 cm SSD. The inverse square law correction factor is factor is 0.829 and the AGCF is 0.964.
    a. 245.2
    b. 238.3
    c. 234.4
    d. 222.05
    e. 178.8

E. Effects of Beam-modifying Devices

Circle the right answer (Yes or No):

1. (Yes / No) The skin dose can be reduced by using a bolus (of less than the buildup thickness) for treatment with accelerator photon beams.

2. (Yes / No) Wedge filters and tissue compensators must be mounted close to the patient to reduce electron contamination in the beam.

3. (Yes / No) The wedge factor is independent of field size for high-energy photon beams.

4. (Yes / No) When using a 60° wedge, a wedge of any angle <60° can be produced by combining the open beam and the wedged beam.

5. (Yes / No) The custom blocks, used for shielding critical structures in the treatment of MV photon beams must be diverging.

6. (Yes / No) MLCs used with various linacs are identical.

7. (Yes / No) One disadvantage of using an MLC compared to custom blocking is the reduced field shaping accuracy.

8. (Yes / No) Custom blocks are not kept too close to the patient skin because they increase the electron contamination reaching the patient.

9. (Yes / No) A straight-sided (non-diverging) block in the treatment field produces a transmission penumbra at the depth of interest and hence should never be used.

10. (Yes / No) The penumbra produced by a straight-sided block at the clinical depth is independent of its position.

11. (Yes / No) The thickness of the custom block depends on its distance from the source.

12. (Yes / No) The size of the custom block depends on its distance from the source.

13. (Yes / No)   The accuracy of custom blocks cannot be easily checked using the simulator.

14. (Yes / No)   The shielded area at the depth of interest in the phantom is better defined (i.e., without p.u.) with a diverging block.

15. (Yes / No)   Conformal treatment improves tumor control.

16. (Yes / No)   Custom blocking gives better conformity compared to treatment with MLCs.

17. (Yes / No)   Exact delineation of the target volume is not necessary for conformal planning.

18. (Yes / No)   IMRT treatment gives better conformity compared to 3D-CRT treatments.

19. (Yes / No)   Using a beam modifier as shielding material, the dose below the shaded region can be reduced to zero.

20. (Yes / No)   Beam modification by shielding is not very useful in radiotherapy since the dose in the shadow region can never be eliminated because of finite transmission of primary sand scatter.

21. (Yes / No)   Any tissue equivalent material can be used for photon shielding.

22. (Yes / No)   Shielding affects the isodose curves.

23. (Yes / No)   Compensators are designed to make the dose distribution uniform at a reference depth.

24. (Yes / No)   Shadow tray made of Perspex can also be used as a beam modification device.

25. (Yes / No)   The wedge angle is independent of depth.

26. (Yes / No)   A physical wedge affects the treatment duration.

27. (Yes / No)   Beam hardening is more pronounced in ^{60}Co beams.

28. (Yes / No)   A proper wedge angle uniquely determined by the hinge angle must be used during treatments.

29. (Yes / No)   Bolus should not be used in megavolt radiation therapy.

30. (Yes / No)   The linac jaws cannot be used as collimators for electron beams.

Choose the right answer(s) (more than one may be correct):

31. A flattening filter _____.
    a. preferentially reduces the photon intensity in the central region of the beam
    b. hardens the beam at off axis points compared to the central axis point
    c. broadens the photon beam to obtain larger field sizes
    d. positioning is not important as long as it is in the beam path.

32. Treatment fields can be shaped using _____.
    a. standard blocks
    b. custom made blocks
    c. a multileaf collimator
    d. none of the above

33. Disadvantages of customized blocking are _____.
    a. it is time consuming
    b. it must be made for each beam orientation
    c. it increases surface dose to the patient
    d. none of the above

34. Errors in custom field blocking can arise due to _____.
    a. incorrect source-to-film distance (SFD)
    b. incorrect source-to-tray distance
    c. incorrect positioning of the block on the tray
    d. none of the above

35. Beam shaping using custom blocks or MLCs _____.
    a. minimizes dose to normal tissues
    b. reduces tumor dose
    c. is needed for treatment planning purposes
    d. none of the above

36. The dose under a block in a patient is _____.
    a. 0%
    b. about 3% to 5%
    c. more than 3% to 5% and depends on how much patient scatter the shadow region of the patient receives
    d. none of the above

37. The thickness of blocks used for field blocking (in HVL) is _____.
    a. about 10
    b. about 5
    c. about 3
    d. about 1

38. To reduce the dose in the shielded region to <5% of the open field dose, the minimum shielding block thickness must be _____.
    a. >3 HVL
    b. >5 HVL
    c. >10 HVL
    d. >5 TVL

39. The shadow block _____.
    a. must always have sloping edges to avoid transmission penumbra
    b. can have simpler straight edges if there are no critical structures to be protected at the edges of the shadow region or open field scatter into the shadow region is significant
    c. is made of high-density material
    d. can affect treatment outcome if it is not positioned accurately

40. Disadvantages in using custom blocks are _____.
    a. a small error in block position can adversely affect treatment
    b. fabrication, alignment, etc., take lot of time
    c. it reduces accuracy of treatment compared to MLC blocking
    d. none of the above

41.   Advantages in using MLC (compared to custom blocking) for field shaping are _____.
       a.   treatment delivery is faster
       b.   it is equally convenient for any angled field
       c.   it gives better conformity to target shape
       d.   none of the above

42.   The tray transmission factor _____.
       a.   depends on beam quality
       b.   depends on tray thickness and composition
       c.   must be determined for every machine
       d.   is greater than 1

43.   The tray transmission factor is _____.
       a.   >1
       b.   <1
       c.   =1
       d.   none of the above

44.   The tray transmission factor for a ^{60}Co unit was measured as 0.96. The open beam output was 120 cGy/min. The beam output, in cGy/min, with the tray attached to the machine is _____.
       a.   115.2
       b.   120
       c.   125
       d.   none of the above

45.   Cerrobend™ _____.
       a.   is the trade name for lead
       b.   is a low-melting-point alloy used for making blocks
       c.   has a density the same as lead
       d.   has a density about 20% less than lead

46.   _____ are accessories used to shape electron fields.
       a.   Cutouts
       b.   Blocks
       c.   Compensators
       d.   Transmission filters

47.   Divergent blocks are used with photon beams to reduce _____.
       a.   geometric penumbra
       b.   transmission penumbra
       c.   block scatter
       d.   none of the above

48.   A patient is being treated with a collimator of 15 cm x 15 cm. To save the normal tissue around the target volume, the field is shaped to get an equivalent field of 12 cm x 12 cm. The treatment depth is 7 cm. The blocking _____.
       a.   reduces effective primary reaching the depth of interest along the central axis
       b.   reduces scatter reaching the depth of interest along the central axis
       c.   reduces the PDD or TAR values of the blocked field
       d.   increases the PDD or TAR values of the blocked field

49. The thickness of the lead cutout you would use with a 10 MeV electron beam is about _____.
    a. 5 mm
    b. 5 cm
    c. 10 mm
    d. 20 mm

50. The thickness of lead required for making a field block for a clinical photon beam is about 8 cm. If Cerrobend material is used, the required thickness would be _____.
    a. the same
    b. about 20% less
    c. about 20% more
    d. two times the lead thickness

51. Bolus is used in electron therapy _____.
    a. to enhance the skin dose
    b. for depth compensation
    c. to broaden the electron beam
    d. for none of the above

52. A wedge filter _____.
    a. is made of tissue-equivalent material
    b. tilts the isodose curve
    c. reduces the beam intensity
    d. alters the beam quality

53. The wedge angle is the angle of _____.
    a. the wedge
    b. tilt of the isodose curve with respect to the normal to central axis at a reference depth
    c. tilt of the gantry while using the wedge
    d. none of the above

54. Wedge angle is _____.
    a. the angle of tilt of isodose curve with respect to the beam central axis
    b. measured along the beam central axis at a reference depth in a phantom
    c. defined as the ratio of doses with and without a wedge
    d. specified for depth much beyond d_{max}

55. The dose under the shielding block (i.e., in the shadow region) depends on _____.
    a. field size
    b. size of the block
    c. position of the block
    d. all the above

56. The tray transmission factor for a shielding block is _____.
    a. <1
    b. 0
    c. 1
    d. >1

57.   The tray transmission factor for a clinical beam was determined to be 0.98. Compared to open field (i.e., without tray) conditions, the _____.
a.   reference dose rate (with tray) decreases by 2%
b.   reference dose increases by 2%
c.   treatment duration or the monitor units increase by 2%
d.   treatment duration or the monitor units decrease by 2%

58.   The wedge transmission factor (WTF) _____.
a.   varies with field size
b.   does not vary with field size
c.   varies with depth
d.   does not vary with depth

59.   A wedge filter in the beam path _____.
a.   leads to beam hardening
b.   leads to beam softening
c.   increases open field PDD
d.   decreases open field PDD

60.   The independent jaw movement feature of linacs helps in _____.
a.   field blocking
b.   field splitting
c.   field matching
d.   creation of dynamic wedge fields

61.   Compensating filters are generally used to _____.
a.   filter the electron contamination
b.   reduce the beam energy
c.   compensate for surface irregularity
d.   none of the above

62.   The use of bolus _____.
a.   compensates for surface irregularities
b.   increases the skin dose by eliminating skin sparing
c.   corrects for internal heterogeneity
d.   all of the above

63.   A compensator can be used to compensate for _____.
a.   sloping skin surface
b.   varying patient thickness over the target volume (or irregular patient surface)
c.   nonuniform energy fluence across the treatment field
d.   none of the above

64.   In the case of accelerator photon beams, a compensator _____.
a.   is placed on the skin surface for better accuracy
b.   is placed at a distance from the skin to reduce patient discomfort
c.   is placed at a distance from skin to reduce excess dose to the skin from electron contamination
d.   gives compensation in three dimensions

65.   A compensator _____.
a.   is usually made of aluminum, brass, or lead
b.   is made of a tissue-equivalent material
c.   is demagnified because it is kept closer to the source
d.   must be of the same size irrespective of its distance from the source

66. A wedge _____.
    a. often functions as a compensator for a sloping surface
    b. profile can be created only by a physical device
    c. is usually made of Perspex™
    d. is usually made of lead or some high-Z material

67. A universal wedge is a wedge _____.
    a. of the largest wedge angle in use
    b. that can be used with any accelerator
    c. that can be used for many field sizes
    d. none of the above

68. A motorized wedge _____.
    a. is an internal wedge of large wedge angle
    b. gives greater flexibility in choosing wedge angles
    c. is much smaller in size compared to a physical wedge
    d. is always present during treatment

69. In a linac that makes use of a motorized wedge, a wedge of any angle (less than the angle of the internal wedge) is produced by _____.
    a. changing the wedge mounted in the linac
    b. using wedges of different angles in the linac
    c. suitably combining the open field and wedge field exposures
    d. none of the above methods

70. Advantages of the universal wedge compared to an individualized wedge are _____.
    a. a single wedge works for many field sizes
    b. a reduction in treatment setup times
    c. less attenuation in the wedge
    d. it needs no wedge transmission factor

71. A dynamic wedge is _____.
    a. a physical wedge in motion during irradiation
    b. a technique of producing a wedged field without a physical wedge
    c. the technique of moving one of the collimator jaws across the field to create a wedged beam profile of any desired angle
    d. none of the above

72. Treatment accessories used in radiation therapy are _____.
    a. wedges
    b. compensators and bolus materials
    c. field-shaping blocks
    d. none of the above

73. A wedge transmission factor of 0.895 indicates a wedge attenuation of the incident beam by _____.
    a. 89.5%
    b. 10.5%
    c. 100%
    d. none of the above

74. The angle between the central axes of two overlapping wedge beams is known as the _____.
    a. wedge angle
    b. hinge angle
    c. obliquity angle
    d. none of the above

75. While treating many tumors that are non-centrally located, two fields from the same side are used. The figure below shows two fields incident on a patient. The dose in the overlap region is _____.

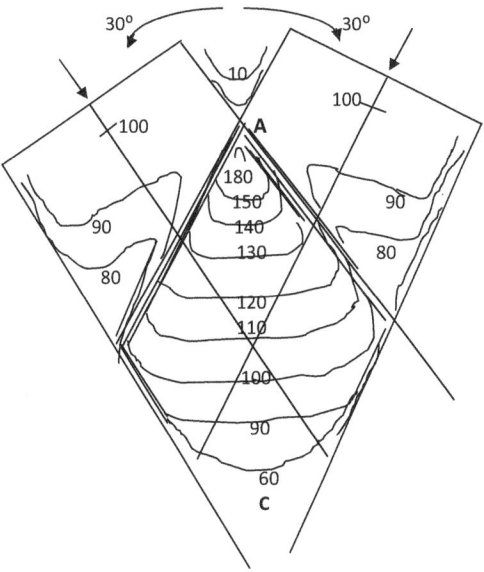

Two angled fields on the same side of the patient

    a. uniform
    b. nonuniform because of the contour of the patient
    c. nonuniform because points A and C are at different distances from the overlapping beams causing a large dose gradient
    d. not treatable for any existing tumor in this region (as per the plan in the figure)

76. The dose in the overlap region (in the figure above) can be made uniform by _____.
    a. wedging the two fields
    b. choosing the proper wedge angle that would make the isodose lines parallel to one another in the overlap region
    c. correcting for surface obliquity, if any, after wedging the fields
    d. none of the above

77. In a treatment involving a pair of wedged fields, the hinge angle is 90°. A wedge angle of _____ must be chosen to get uniform dose to the region of overlap.
    a. 90°
    b. 60°
    c. 45°
    d. none of the above

78. When using a pair of wedged treatment fields, the wedge angle is much smaller than the ideal one. The anterior region of overlap will lead to _____.
    a. high-dose region
    b. low-dose region
    c. no change in dose

79. When using a pair of wedged treatment fields, the wedge angle is much larger than the ideal one. The anterior region of overlap will lead to _____.
    a. high-dose region
    b. low-dose region
    c. no change in dose

80. For a treatment involving an internal wedge of wedge angle θ, a fraction of the tumor dose (f) is delivered by the wedged field and the rest without the wedge. The resultant or effective wedge angle produced by the treatment is approximately given by _____.
    a. $f\theta$
    b. $f^2\theta$
    c. θ/f
    d. none of the above

81. For a treatment involving a wedge pair, 30° wedges are required for obtaining uniform dose in the treatment volume. However, the hospital possesses only 45° wedge. _____ fraction of the tumor dose must be delivered with 45° wedges (and the rest with open fields) to restore dose uniformity in the treatment volume.
    a. 1/3
    b. 2/3
    c. 3/4
    d. none of the above

82. Beam modifiers are used for _____.
    a. shielding some regions from direct radiation
    b. compensating an oblique surface to make radiation incidence normal
    c. tilting the isodose distributions by a desired angle
    d. flattening the beam profile
    e. all of the above

83. A thickness of lead _____ is considered acceptable (<5% dose) for shielding purposes.
    a. equivalent to the range of the beam
    b. about 10 mm is adequate for high energy photon beams
    c. about 1 TVL
    d. about 3 TVL
    e. about 4.5 to 5 HVL

84. The advantages of Cerrobend as shielding material are its _____.
    a. high melting point
    b. high density
    c. non-toxicity
    d. all the above

85. The well known beam modifier in radiotherapy is the MLC of the linac head. The primary x-ray x-transmission of the MLC leaf (in %) is about _____.
    a. 15
    b. 10
    c. 5
    d. 2
    e. 0

86. Wedges are normally mounted at a distance of _____ cm from the skin surface.
    a. 0
    b. 5
    c. 10
    d. 15
    e. 30

87. The dose on the central axis for a 10 x 18.5 field ^{60}Co beam was measured to be 65 cGy/min. The wedge factor is 0.65. The dose at the same point in the wedged 10 W x 18.5 field is given by _____.
    a. 65
    b. 58.4
    c. 51.9
    d. 42.25
    e. 34.6

88. The dose on the central axis for a 12.5 W x 18.5 field ^{60}Co beam was measured to be 37.5 cGy/min. The wedge factor is 0.57. The dose at the same point in the open field is given by _____.
    a. 65.8
    b. 60
    c. 47.6
    d. 37.2
    e. 30.6

89. Which of the following statements are true in the use of wedge filters in radiotherapy?
    a. Wedge fields are generally used for very deep-seated tumors.
    b. The heel side of wedges should face each other during alignment.
    c. The hinge angle decides the wedge angle.
    d. The wedged fields are usually directed from the same side of the patient.
    e. All the above are true.

90. The beam modifiers in a linac are the _____.
    a. flattening filter
    b. secondary scattering foil (used in electron mode)
    c. dynamic wedge produced by the jaws
    d. multileaf collimator
    e. all of the above

91. Which of the following statements about flattening filters are true?
    a. Beam flatness is specified usually at 10 cm.
    b. The beam flatness should be within ±10%.
    c. The over flattening at shallower depths produces "horns" in dose profiles.
    d. The thinner outer rim causes more hardening on the periphery compared to the central region.
    e. All of the above are true.

92. If bolus of tissue-equivalent material is used in megavoltage therapy, _____ thickness must be used.
    a. any thickness
    b. about 5 to 10 mm
    c. depends on the beam energy
    d. buildup thickness
    e. MV/2 cm thickness

93. The purpose of a beam spoiler in a TBI irradiation using photon beams is _____.
    a. to increase the PDD at patient midplane
    b. to remove electron contamination
    c. to reduce lung dose
    d. to increase dose in the buildup region for better dose homogeneity
    e. for all the above reasons

94. _____ can be used as a tissue compensator.
    a. Bolus
    b. Wedge
    c. Cerrobend
    d. Dynamic MLC
    e. All of the above

95. The flattening filter is used in a linac beam to flatten the x-ray beam profile at 10 cm depth. This results in a beam that is _____.
    a. flat at all depths
    b. over flattened at shallower depths
    c. under flattened at deeper depths
    d. higher doses at field edges causing "horns" in the profile at d_{max}
    e. exhibiting all the above characteristics

96. The advantages of MLC beam shaping over Cerrobend beam shaping are _____.
    a. sharper penumbra
    b smoother beam shaping
    c. can be used for larger field sizes
    d. permits IMRT treatments
    e. low leakage
    f. all of the above

97. A beam spoiler is used in a high-energy photon beams to _____.
    a. reduce the energy of the photon beams
    b. increase the skin dose
    c. shift the PDD toward the source
    d. increase the dose in the buildup region by scattering electrons into the buildup region
    e. achieve all the above objectives

F. Irregular Field Calculations

Choose the right answer(s) (more than one may be correct):

1. The scatter-air ratio (SAR) _____.
    a. gives the scatter component of TAR
    b. is the difference between TAR and zero field TAR
    c. concept helps in separating the scatter and primary dose at a point in the patient
    d. none of the above

2. The equivalent square field of any irregular field for any dose function can be determined _____.
    a. by Sterling's formula
    b. by Clarkson's sector integration method
    c. only by Monte Carlo computation
    d. by none of the above

3. The scatter dose at any point in a phantom (patient) can be determined _____.
    a. by the evaluation of scatter-air ratio (SAR)
    b. by the evaluation of scatter maximum ratio (SMR)
    c. from the output of the beam for a standard field size
    d. by none of the above

4. Field blocking _____.
    a. reduces scatter volume
    b. reduces scatter dose at the depth of interest
    c. leaves the primary dose reaching the phantom along beam central axis relatively unaffected (assuming that blocking is neither along central axis nor a large fraction of open field)
    d. modifies the primary dose rate

5. The photon beam dose profile at depth of clinical interest is _____.
    a. fairly uniform across the field size for ^{60}Co and linac beams
    b. non-uniform across field size for linac photon beams due to flattening filter effects
    c. non-uniform across field size for linac photon beams, but the variation is the same at all depths
    d. uniform across the field size for all photon beams, all depths, and all beam qualities

6. SAR is _____.
    a. the scatter component of TMR
    b. obtained by subtracting zero field TAR from TAR of the given field
    c. not dependent on field size
    d. generally required for the dosimetry of irregular fields

7. SMR is _____.
    a. the scatter component of TMR
    b. obtained by subtracting zero field TMR from TMR of the given field
    c. not dependent on field size
    d. generally required for the dosimetry of irregular fields.

8. When a blocked field is not a significant fraction of an unblocked (or open) field, _____.
    a. the effective primary reaching the patient is altered to a negligible amount
    b. the collimator scatter factor, S_c, remains the same as that of the unblocked field
    c. the phantom scatter factor, S_p, remains the same as that of the unblocked field
    d. the output factor remains the same as that of the unblocked field

9. In a radiation therapy treatment, using ^{60}Co beam, the open field, 12 cm x 12 cm requires some blocking at the periphery to protect normal structures around PTV. The blocked field was 8 cm x 8 cm in size.
    The _____.
    a. S_c of blocked field = S_c of unblocked field
    b. S_c of blocked field > S_c of unblocked field
    c. S_c of blocked field < S_c of unblocked field
    d. S_{cp} of blocked field = S_{cp} of unblocked field

10. The figure below is an example of a mantle irregular field. The scatter dose contribution, SMR, reaching the point of interest Q in the phantom along the beam central axis, from sector 1 needs to be computed. The sector angle is 10° as shown in the figure. The SMR contribution at Q from sector 1 is given by _____.
    a. $(1/9) \times$ SMR contribution from the whole mantle field
    b. $(1/9) \times$ SMR contribution for the circular field with radius r_1
    c. $(1/36) \times$ SMR contribution for the circular field with radius r_1
    d. none of the above

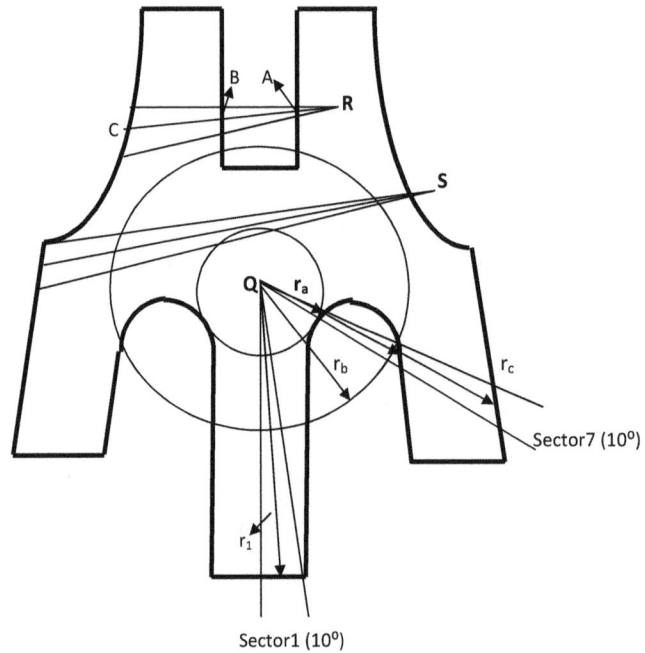

Illustration of Clarkson's method for mantle field dosimetry

11. In the figure above, the SMR contribution at Q due to sector 7 is given by _____.
    a. SMR contribution from the whole sector (of radius r_a)
    b. SMR contribution only from the sector of radius r_c
    c. SMR contribution only from parts of the sectors in the field
    d. none of the above

12. In the figure above, the SMR contribution at the off-axis point R due to sector of radius RC is given by _____.
    a. SMR contribution from the whole sector (of radius RC)
    b. SMR contribution only from the sector of radius RA
    c. SMR contributions from: sector of radius RA + (sector of radius RC − sector of radius RB)
    d. none of the above

13. In the figure above, the SMR contribution at the point S outside the field due to sector of radius SB is given by _____.
    a. zero since the point is outside the field
    b. SMR contribution only from the sector of radius SA
    c. SMR contribution from (sector of radius SB − sector of radius SA)
    d. none of the above

G. Special Calculations

Choose the right answer(s) (more than one may be correct):

1. A patient is to be treated using two adjacent fields that abut at 5 cm depth as shown in the following figure. SSD = 100 cm. The field widths at the abutting depth for the two fields are 24 and 26 cm, respectively. The gap between the fields on the skin is "g," which is _____ cm.
   a. 2.6
   b. 2.5
   c. 2.4
   d. none of the above

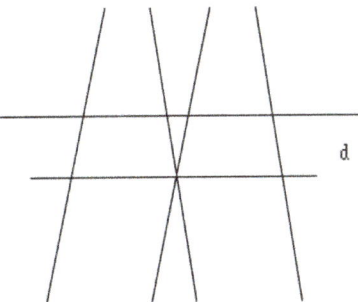

Abutting of fields at depth "d"

2. In the previous problem, if the gap on the skin is different from what is required for abutting fields at 5 cm depth, _____.
   a. the fields will abut at a different depth
   b. the junction at 5 cm depth will be underdosed or overdosed
   c. the dose uniformity at 5 cm will be maintained over the full length
   d. it will have no effect on the treatment outcome

3. Any error in the placement of adjacent fields will _____.
   a. be more serious with beams of smaller penumbra
   b. be more serious with beams of larger penumbra
   c. be equally serious for beams of any penumbra size
   d. in no way influence the treatment outcome

4. Sometimes adjacent POP treatment fields are needed. In this case, _____.
   a. there can be no overlap of three fields
   b. there is overlap of three fields if the pairs are of different field sizes
   c. three-field overlap cannot be avoided
   d. three-field overlap is of no consequence

5. While treating a patient for craniospinal irradiation (CSI), two orthogonal fields must abut at the desired depth in the patient. If both fields are divergent for the abutting of lateral and spinal fields, _____.
   a. only the collimator must be rotated through the angle of divergence of the spinal field
   b. only the couch must be turned through the angle of divergence of the cranial field
   c. both collimator and couch must be turned through the angle of divergence of abutting spinal and cranial fields, respectively
   d. neither must be rotated by any angle

6. In a CSI treatment, the cranial field on the abutting side is made nondiverging using a half beam block. The size of the abutting spinal field is 40 cm x 6 cm. SSD is 100 cm. If θ is the angle of rotation of the collimator required for the abutting of the cranial and spinal fields, tan θ is given by _____.
   a. 0
   b. 0.2
   c. 0.5
   d. 1

7. In a CSI treatment, the cranial field on the abutting side has a divergence ψ. The cranial field size at midline, at SAD = 100 cm, is 20 cm (length times width) x 18 cm. If ψ is the angle of rotation of the couch required for the abutting of the cranial and spinal fields, tan ψ is given by _____.
   a. 0
   b. 0.1
   c. 0.5
   d. 1

8. When two adjacent electron fields abut on the skin or at a depth, there is a possibility of overdosing of regions around the junction because of _____.
   a. error in exact matching of fields
   b. lateral spreading of low isodose lines
   c. difficulties in electron collimation
   d. the finite range of electrons

9. In a treatment involving adjacent photon fields abutting at depth of dose specification, the overlap region below the junction gets overdosed and the region above the junction gets underdosed. In order to make the dose more uniform, _____.
   a. the gap on the surface can be increased or decreased in successive fractions
   b. the two fields can be angled in opposite directions to make the beam edges parallel at the junction
   c. the two fields can be treated by half fields (using half beam blocks or asymmetric jaws), the central axes becoming the matching edges of the two adjacent fields
   d. matching pair of POP fields can be used

10. When using two adjacent fields in POP geometry, junctioning at midplane, the _____.
    a. dose is more uniform in the volume around midplane
    b. high-dose overlap regions (hot spots) are formed when field sizes of POP of fields are unequal
    c. hot spots are formed when SSDs are unequal (even if field sizes of the pairs are unequal)
    d. three-field overlap cannot occur

11. It is more difficult (compared to photons) to use abutting electron fields to treat large target volumes because _____.
    a. bulging of low-value isodose curves and constriction of high-value isodose curves make it very difficult to achieve dose uniformity
    b. any increase in required field gap can cause low-dose regions
    c. slight overlapping of fields causes hot spots
    d. electron beams cannot be collimated

12. While using adjacent photon fields for treatment, _____.
    a. it does not matter where the high-dose region occurs
    b. it does not matter where the low-dose region occurs
    c. dose nonuniformity due to error in abutting can be minimized by increasing the penumbra width at the junction
    d. dose uniformity can be improved by moving the junction or the gap during the course of treatment

13. In rotation therapy, _____.
    a. the machine rotates 360° about the patient
    b. average TMR or TAR has to be calculated for the entire rotation to determine the dose to the target center
    c. patients with centrally located tumors of reasonable size are generally treated
    d. the isodose contours are elliptical for a patient of circular cross section

14. Rotation therapy _____.
    a. is suitable for small tumors located centrally in the patient
    b. is suitable for large tumors
    c. is suitable for more superficial tumors
    d. offers slightly better dose distributions compared to multiple stationary field treatments

15. A patient is to be treated with 6 MV x-ray beam by rotation therapy with a field size of 7.5 cm x 7.5 cm at isocenter. The beam output is 1 cGy/MU at d_{max} depth for a 10 cm x 10 cm field. The calibration point is set at SAD = 100 cm. A dose of 200 cGy is to be delivered to the target center. OF (7.5 cm x 7.5 cm) = 0.973. <TMR> = 0.795.
    i. The dose rate at the target center (in cGy/MU) is given by _____.
       a. 0.774
       b. 0.973
       c. 1.22
       d. none of the above
    ii. The treatment duration (in monitor units) is given by _____.
       a. 258.4
       b. 154.8
       c. 194.6
       d. none of the above

16. In arc therapy, the starting and stopping field angles are given _____ weighting in computing TAR or TMR (see the figure below).
    a. 1
    b. 0.5
    c. 0
    d. none of the above

17. A 180° arc is used to treat the esophagus with a 20 cm x 6 cm ^{60}Co field (see the figure below). The dose delivered per fraction is 300 cGy for one half rotation. The fields are chosen at 10° intervals to determine the average TAR for the 18 depths of the arc angle. The average TAR is 0.600. The beam output (free space dose) at SAD for the given field is 100 cGy/min. The treatment time is given by _____.
    a. 0.6 min
    b. 1 min
    c. 5 min
    d. 8 min

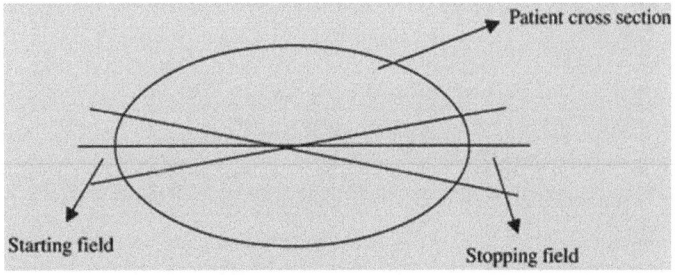

180° arc treatment

18. When a part of the open beam is blocked for reducing dose to an organ at risk, _____.
    a. primary dose in the patient under the unblocked region remains almost the same
    b. scatter dose in the patient under the unblocked region remains almost the same
    c. primary dose under the block is zero
    d. scatter dose under the block is zero

19. The total dose received by a point in the patient that lies outside the field is _____.
    a. zero
    b. the dose scattered by the irradiated patient volume
    c. the primary dose transmitted by the collimator and patient
    d. the sum of the transmitted primary dose and the dose scattered by the scatter volume

20. A patient is treated with a blocked field. The dose in the patient under the shadow region depends on _____.
    a. thickness of the block
    b. depth of the point of interest
    c. field size
    d. beam quality

21. Using Clarkson's method, _____.
    a. dose at any point in a patient can be computed for an irregular field
    b. scatter and primary components of dose at any point in a patient can be evaluated
    c. patient inhomogeneity corrections can be evaluated accurately
    d. errors in patient positioning can be minimized

22. Electron arc therapy _____.
    a. is used to treat superficial layers of skin that are not curved
    b. is used to treat superficial layers of skin that have a curvature, e.g., the chest wall
    c. requires proper dosimetric procedures for dosimetry
    d. can be used for any site

23. Total skin electron irradiation (TSEI) involves _____.
    a. treating large areas of skin and its underlying layers
    b. use of low-energy (about 3 to 6 MeV) electron beams
    c. measurement of beam uniformity and output for the treatment geometry
    d. an accuracy of about ±10%

24. TSEI usually involves _____.
    a. large SSDs
    b. beam angling to reduce dose from x-ray contamination
    c. measurement of electron energy incident on patient and skin dose
    d. none of the above

25. Total body irradiation (TBI) involves _____.
    a. irradiation of the whole body of the patient
    b. delivering dose in a single fraction or multiple fractions
    c. delivering the whole body dose that is uniform within 5%
    d. use of ^{60}Co beams or accelerator photon beams.

26. In TBI treatment, the _____.
    a. prescribed dose is about 1200 cGy, generally delivered using POP of fields
    b. dose is delivered as a single fraction
    c. dose is delivered as multiple fractions, 10 to 15 cGy per fraction
    d. dosimetry must be done for the treatment conditions

27. In the TSEI technique, the _____.
    a. dose is prescribed at the skin at umbilicus, along the central ray
    b. typical dose delivered is of the order of 4000 cGy and 200 cGy/fraction
    c. photon dose received by the patient can be neglected
    d. dosimetry is done at conventional SSD of 100 cm

28. Stereotactic radiotherapy is a treatment technique that involves _____.
    a. delivering dose to small stereotactically localized targets with high precision and accuracy
    b. target localization within about 1 mm
    c. accuracy of dose delivery within about 5%
    d. dose delivery in the range of 1000 to 5000 cGy depending on target size

29. The stereotactic radiotherapy technique _____.
    a. delivers dose as a single fraction
    b. treats functional disorders in brain
    c. treats small benign or malignant masses in brains
    d. involves patient simulation for precise localization of PTV; the patient must use stereotactic head frame for the precise localization and dose delivery

30. Stereotactic treatments are carried out with a _____.
    a. Gamma Knife
    b. ^{60}Co unit
    c. linear accelerator
    d. linear accelerator with suitable accessories and modifications

31. Stereotactic procedures _____.
    a. are generally used to treat brain lesions
    b. involve the same immobilization devices as in the treatment of head and neck cases
    c. involve accurate clinical target localization techniques
    d. are used only for the treatment of malignancies

32. In stereotactic treatment procedures, _____.
    a. the target volume can often be identified from conventional simulation and portal imaging procedures
    b. the setup accuracy is much more important compared to conventional treatment procedures
    c. noncoplanar beams are normally used
    d. avoiding critical structures bordering on the target volume is not of great importance

33. Stereotactic localization of an intracranial target involves _____.
    a. use of a rigid head frame to establish a fixed relationship between the target position and the coordinate system of the head frame
    b. angiography, CT, and MRI as diagnostic procedures for target localization
    c. a rigid head frame that is required during target localization
    d. the same stereotactic frame for all diagnostic procedures

34. Dosimetry of small stereotactic fields requires _____.
    a. the problem of lack of charged particle equilibrium be given special consideration
    b. that TMR and off-axis ratios (OFFARs) be measured for small fields
    c. the use of the same chamber used in conventional radiation therapy (0.6 cm^3 chamber) for the measurement of dosimetric functions like TMR, OFFAR, etc., of stereotactic fields
    d. significant correction for OFFAR with respect to depth

35. For a 6 MV photon beam used for stereotactic radiosurgery, the minimum beam size diameter (in mm) required for electronic equilibrium is about _____.
    a. 5 mm
    b. 10 mm
    c. 15 mm
    d. 30 mm

36. For optimum irradiation of stereotactic lesions, _____.
    a. POP fields must be made use of
    b. as many beams as possible are used isotropically in the hemisphere surrounding the lesion
    c. Gamma Knife source configuration is an example
    d. spherical dose distribution is necessary

37. In Gamma Knife radiosurgery, _____.
    a. 201 ^{60}Co sources distributed isotropically in a hemisphere are used
    b. dose distribution and dose to OAR can be altered by changing the activity of individual sources
    c. collimator diameters vary between 4 and 18 mm
    d. less expense is incurred compared to linac-based radiosurgery

38. The use of cylindrical tertiary collimators in stereotactic radiotherapy _____.
    a. provides more precise collimation
    b. gives rise to a spherical dose distribution surrounding the target volume
    c. gives sharper dose falloff compared to secondary collimators
    d. is ideal for the treatment of all lesions

39. The linac-based stereotactic treatment procedures are _____.
    a. single plane transverse rotation
    b. multiple noncoplanar converging arcs
    c. single arc dynamic rotation
    d. none of the above

40. Radiosurgery with single plane rotation _____.
    a. is similar to conventional rotation therapy but with smaller field and single fraction treatment of stereotactically located intracranial lesions
    b. gives rise to same dose falloff in all directions around the target
    c. gives better dose falloff compared to multiple noncoplanar converging arc technique
    d. involves a 360° rotation of the gantry

41. Stereotactic dynamic radiotherapy _____.
    a. involves simultaneous and continuous gantry and couch rotation during the treatment procedure
    b. involves a gantry angle greater than 180°
    c. gives dose falloff outside the target volume comparable to a Gamma Knife unit
    d. gives better isocenter accuracy compared to a Gamma Knife unit.

42. When treating acoustic neuroma with Gamma Knife, _____ is the typical prescription dose.
    a. 1.0 Gy
    b. 2.0 Gy
    c. 12 Gy
    d. 20 Gy
    e. 60 Gy

43. Treatment of AVMs with SRS is typically performed in _____.
    a. a single fraction
    b. 2 fractions with a one-day gap between fractions
    c. 5 daily fractions
    d. 8 daily fractions
    e. >10 fractions

44. When 16 Gy is prescribed to the 80% isodose line for a brain tumor, 10% dose isodose line would represents a dose of _____.
    a. 16 Gy
    b. 8.0 Gy
    c. 2 Gy
    d. 1.6 Gy
    e. none of the above

45. The order of dose prescribed in SRS treatment of brain metastases is about _____.
    a. 1 Gy
    b. 2 Gy
    c. 6 Gy
    d. 18 Gy
    e. 60 Gy

46. The order of dose prescribed in SRS treatment of trigeminal neuralgia is about _____.
    a. 2 Gy
    b. 10 Gy
    c. 30 Gy
    d. 60 Gy to 70 Gy
    e. 120 Gy

47. Typical single fraction dose in SRS delivery, except the treatment of trigenminal neuralgia, is in the range of _____.
    a. 1 to 2 Gy
    b. 5 to 10 Gy
    c. 15 to 25 Gy
    d. 30 to 40 Gy
    e. 100 to 120 Gy

48. A TBI patient is treated at an extended SSD of 400 cm. The linac beam calibration is given by D_{cal} (d_{max}, 100 cm SSD) = 1 cGy/MU. Approximately _____ MUs are required to deliver 100 cGy to the dose maximum depth at the extended SSD.
    a. 2865
    b. 2400
    c. 2000
    d. 1600
    e. 840

49. _____ is the collimator rotation (in degrees) for a 20 x 20 cranial field to abut with a vertical spinal field of length 36 cm in a 100 cm SSD treatment.
    a. 30°
    b. 20°
    c. 15°
    d. 10°
    e. 0°

50. A patient is treated with 6 MV abutting fields with a gap of 1.4 cm on the skin. If the same patient were treated by a 10 MV linac beam, the gap would _____.
    a. be greater than for 6 MV by (10/6)%
    b. be greater than that for 6 MV by (10/6) times
    c. remain the same
    d. be less than for 6 MV by (6/10)%
    e. be less than for 6 MV by (6/10) times

51. A TBI patient is treated with 6 MV linac at an SSD of 400 cm. _____ MUs are required to deliver 75 cGy at d_{max} using an AP field. D_{cal} (100, 40 x 40, d_{max}) = 1 cGy/MU.
    a. 1200
    b. 1000
    c. 956.4
    d. 870.8
    e. 820.2

52. The TBI patient is being treated with a linac. The patient's height is 170 cm. _____ is the minimum SSD (in cm) to cover the patient without collimator rotation. The collimator is set to 40 x 40 field size.
    a. 560
    b. 425
    c. 346.5
    d. 212
    e. 187

53. A patient is treated at an SSD of 4 m. A dose of 150 cGy is planned at 10 cm depth along the central axis. The machine output is given as 600 MU/min and TMR (10 cm) = 0.872. _____ is the machine ON time (in minutes). D_{cal} (100 cm, d_{max}) = 1.0 cGy/MU neglecting the OF variation with field size.
    a. 23.4
    b. 21.9
    c. 18.3
    d. 12.6
    e. 4.6

54. A pregnant woman is treated with 6 MV AP/PA fields to a total dose of 4200 cGy. If the fetus is at 10 cm distance from the field edge, _____ is the order of dose (in cGy) received by the fetus.
    a. 120
    b. 84
    c. 65
    d. 32
    e. 15

H. Corrections for Tissue Inhomogeneities and Surface Obliquities

Choose the right answer(s) (more than one may be correct):

1. The presence of lung in the beam path increases the dose behind the lung (per cm of lung tissue), for a 10 MV x-ray beam, by approximately _____.
    a. 1%
    b. 2%
    c. 5%
    d. 10%

2. The presence of lung in the beam path increases the dose behind the lung for a ^{60}Co beam by approximately _____.
    a. 1%
    b. 4%
    c. 8%
    d. 10%

3. A lung thickness of about 7 cm in the beam path increases the dose behind the lung for a 10 MV x-ray beam by approximately _____.
    a. 0%
    b. 7%
    c. 14%
    d. 21%

4. The PDD at 10 cm depth for a 15 x 15 field, 10 MV x-ray beam is 75.1%. The PDD at the same point for a lung of about 8 cm thickness above that point is obtained by applying a correction factor of _____.
    a. 1.0
    b. 1.08
    c. 1.16
    d. none of the above

5. The thickness of the lung in the ray path is 8 cm, 2 cm below the skin. The target center is at a depth of 5 cm from the ray exit point of the lung (see the figure below under the next question). The water equivalent depth of the target center (in cm) is _____.
    a. 14 cm
    b. 9.4 cm
    c. 6 cm
    d. none of the above

6. In the figure below, the lung correction for 6 MV x-rays is 2.5% per cm of lung. The dose behind the lung _____.
    a. increases by 2.5%
    b. increases by 20%
    c. decreases by 2.5%
    d. decreases by 20%

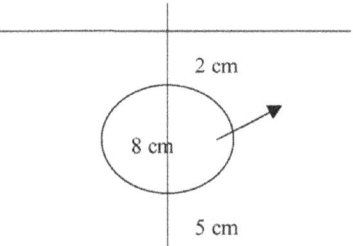

Concept of water equivalent depth

7. Two important effects due to presence of bone in the patient that must be taken into account in dose planning with electron beams are _____.
    a. reduction in the dose behind the bone
    b. an increase in dose behind the bone
    c. lateral scattering of electrons in bone giving rise to a high-dose region in the patient
    d. less scattering in bone (compared to water) giving rise to a low-dose region in the patient.

8. The following figure shows bone and air cavity inhomogeneities in a patient. Assume a bone density of 1.5 g/cm³.
   i.   The water equivalent depth to point P_1 (in cm) is given by _____.
        a.   10
        b.   13
        c.   14.5
        d.   none of the above
   ii. The water equivalent depth to P_2 (in cm) is given by _____.
        a.   10
        b.   13
        c.   14.5
        d.   none of the above

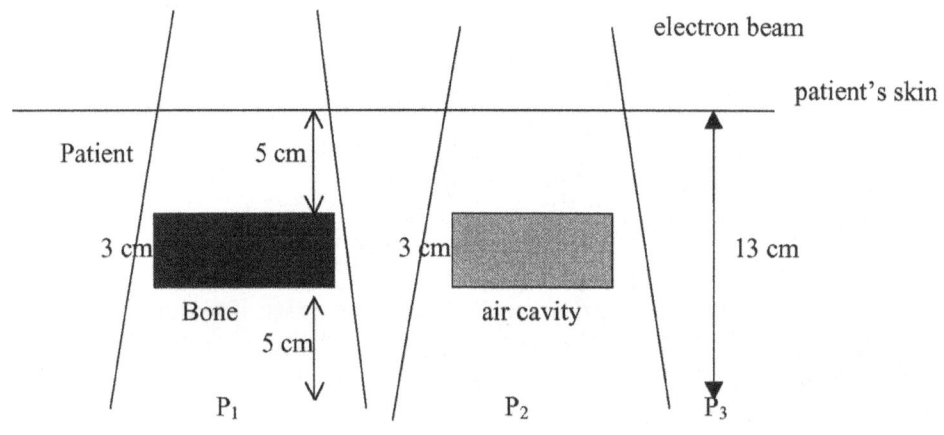

Inhomogeneity corrections

9. Presence of bone in the beam path decreases the dose behind bone (per cm of bone), for a 10 MV x-ray beam by about _____.
   a.   1%
   b.   2%
   c.   5%
   d.   10%

10. Presence of bone in the beam path decreases the dose behind bone for a ^{60}Co beam by about _____.
   a.   1%
   b.   3.5%
   c.   6%
   d.   10%

11. A bone thickness of about 7 cm decreases the dose behind bone for a 10 MV x-ray beam by about _____.
   a.   0%
   b.   7%
   c.   14%
   d.   none of the above

12. The PDD at 10 cm depth for a 15 x 15 field 10 MV x-ray beam is 75.1%. The PDD at the same point with bone of 5 cm thickness present above that point is obtained by applying a correction factor of _____.
   a.   1.0
   b.   0.92
   c.   0.90
   d.   none of the above

13.   In the isodose shift method used for correcting for inhomogeneity in a patient for a point lying beyond the inhomogeneity, the shift _____.
      a.   of isodose chart is toward the skin for bone
      b.   of isodose chart is away from the skin for lung
      c.   is the same for all inhomogeneities
      d.   none of the above statements are true

14.   The isodose shift method is used for correcting for 4 cm of lung that comes in the ray path in a treatment involving a ^{60}Co beam. The point of interest lies beyond lung. To determine the dose at this point, the isodose chart must be shifted away from the skin by _____.
      a.   0.4 cm
      b.   1.6 cm
      c.   4 cm
      d.   none of the above

15.   The isodose shift method is used for correcting for 2 cm of hard bone that comes in the ray path in a treatment involving a 4 MV beam. The point of interest lies beyond bone. To determine the dose at this point, the isodose chart must be shifted toward the skin by _____.
      a.   0.5 cm
      b.   1 cm
      c.   2 cm
      d.   none of the above

16.   When treating an inclined skin surface, _____.
      a.   the standard PDD or TAR data need to be corrected
      b.   the dose at any point in the patient is decreased by tissue deficit
      c.   correction for missing tissue can be approximately determined from tissue attenuation data for the treatment beam
      d.   CT patient data and TPS can be used for applying accurate corrections

17.   The increase in dose per cm of missing tissue is 5% for a ^{60}Co beam. The dose at a depth of 5 cm, along a ray line is to be determined. The missing tissue along the ray line is 3 cm. The increase in dose (in %) at the point of interest is given by _____.
      a.   0.15
      b.   1
      c.   1.15
      d.   15

18.   Some methods available for correcting for surface obliquity are the _____.
      a.   isodose shift method
      b.   effective SSD method
      c.   TAR method
      d.   Clarkson's method

19. In the following figure, "h" represents tissue excess and "k" tissue deficit along the ray lines AB and CD, respectively, in a treatment involving 6 MV x-ray beam. The isodose shift factor for the beam is 0.6. To correct for the surface obliquity, _____.
    a. the isodose point B must be shifted downward by 0.6 h
    b. the isodose point D must be shifted downward by 0.6 k
    c. the isodose point B must be shifted upward by 0.6 h
    d. no shift in standard isodose curve is necessary

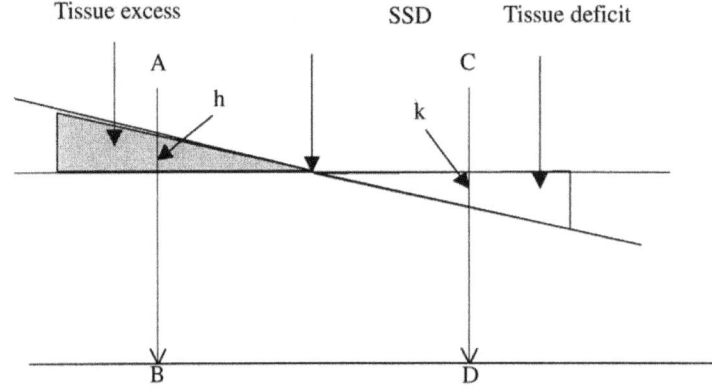

Corrections for body surface shape

20. The isodose shift method _____.
    a. is easiest of all methods used for surface obliquity correction
    b. shifts the isodose by the same factor for all beam energies
    c. can be used to correct a complete isodose curve
    d. is also applicable for electron beams.

21. The isodose shift factor _____.
    a. decreases with increasing beam energy
    b. increases with increasing beam energy
    c. shifts the isodose curve toward the surface in case of tissue excess
    d. shifts the isodose curve away from the surface in case of tissue excess

22. The isodose shift factor _____.
    a. increases with increasing beam energy
    b. is easier to apply to correct for surface irregularity
    c. is the most accurate method for correcting for patient contour
    d. varies with depth

23. The isodose shift method _____.
    a. can be applied for internal heterogeneities as well
    b. shifts the curve toward the surface in case of high-density inhomogeneity (e.g., bone)
    c. shifts the curve away from the surface in case of low-density inhomogeneity (e.g., bone)
    d. is the most difficult of all surface obliquity correction methods

24. The lung correction (% change in MU when heterogeneity correction is taken into account) depends on _____.
    a. beam energy
    b. field size
    c. electron density
    d. path length
    e. prescription dose
    f. all the above factors

25. Two plans are compared which deliver the prescription dose at the isocenter at 18 cm depth. The lung occupies a length of 10 cm in the first field and 5 cm in the second field. If lung correction is not made, the first field would deliver _____% more dose compared to the second field at the isocenter. Lung density is 0.3.
    a. 15
    b. 12
    c. 9
    d. 5
    e. 2

26. To apply heterogeneity correction to the CT data, one must know for each pixel the _____.
    a. physical density
    b. electron density
    c. proton density
    d. μ value
    e. Hounsfield number
    f. all the above values

27. A tumor situated adjacent to a 10 cm size lung in a 6 MV beam will increase the tumor dose by about _____.
    a. 5%
    b. 10%
    c. 14%
    d. 20%
    e. 24%

28. In an esophagus treatment using a 10 MV beam, the beam traverses a length of 10 cm in the lung. If there is no lung correction, the actual dose at the isocenter will be _____ compared to the calculated dose.
    a. 5% higher
    b. 10% lower
    c. 15% higher
    d. 15% lower
    e. 20% higher

29. With increasing photon energy, lung correction factor _____.
    a. increases
    b. decreases
    c. remains the same
    d. increases up to certain energy and then decreases

FS @ 120 ssd 8x12
what is it @ 100?

$$\frac{100}{120} \times W \qquad \frac{100}{120} \times L$$

46.5   34.5

15

15

100ssd

focal

46.5   34.5

□   □

15x15   @100

⑦

$$\frac{?}{15} = \frac{46.5}{100}$$

focal

46.5   34.5

15x15   100 cm

⑤

$$\frac{?}{15} = \frac{34.5}{100}$$

IV. Dose Calculation Methods ANSWERS

A. Applied Mathematics

1. b

2. e

3. d   Collimator field defines the field projection at SAD distance (100 cm)
       Field length at 100 cm = (100 / 120) × field length at 120 cm.

6.6   10

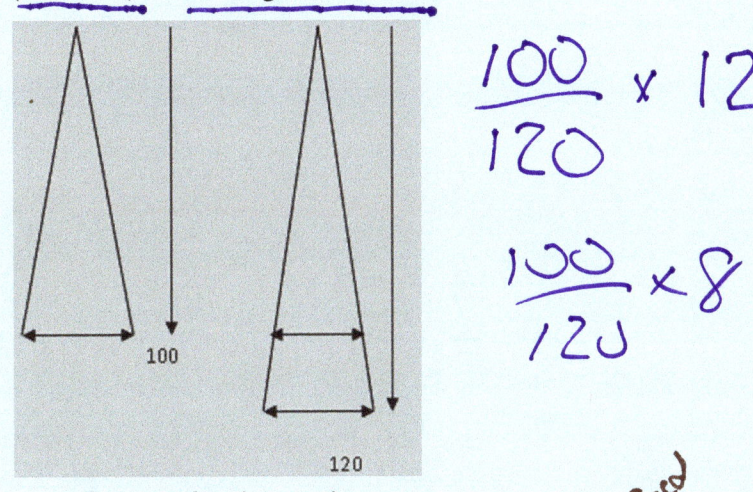

Concept of similar triangles

$$\frac{100}{120} \times 12$$

$$\frac{100}{120} \times 8$$

4. c   From the figure below, one can see a / b = d / 100, from which the jaw opening at "d" can be obtained.

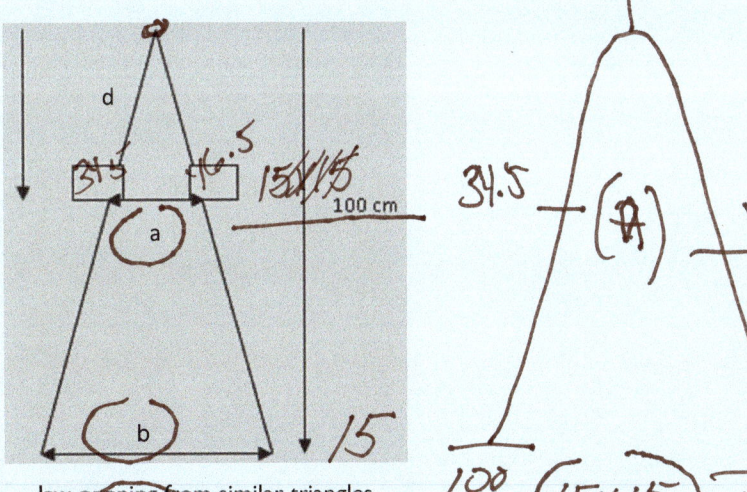

Jaw opening from similar triangles

focd

34.5 — (A) — 46.5

100 (15×15) 100

5. b

6. c

$$\frac{34.5}{15} = \frac{?}{100}$$

$$\frac{46.5}{15} = \frac{?}{100}$$

7. c   s / SDD = pu (penumbra width) / DSD for pu on surface
        s / SDD = pu(d) / (DSD + 10) for pu at depth 'd' = 10 cm

8. a   For a field length L α d; area α d^2 or $A_1 / (d_1)^2 = A_2 / (d_2)^2$. The divergence of the field is shown below, which will make the concept clear.

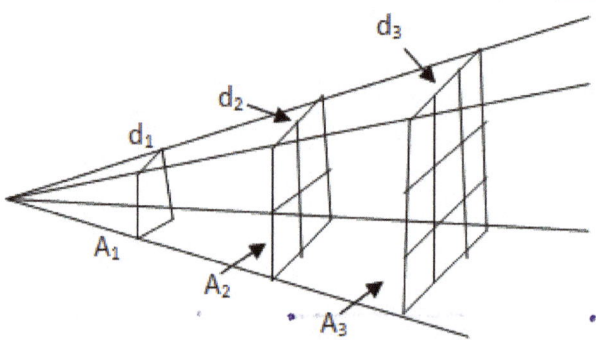

Concept of field divergence for photon radiation

9. d

10. e   AK varies as $(1/d^2)$ and area A varies as d^2 and so AK $\times$ area is constant.

11. d

12. d

13. a   The scatter volumes are a bit different since the field sizes are defined on surface and at d_{max} for the two geometries but the effect can be neglected for the typical SSDs involved in treatments.

14. c   The magnification factor (M) is "source-to-film distance (SFD) / source-to-object distance." The reverse gives the minification factor. Knowing M and object or image size the other parameter can be determined.

15. d

16. c   Minification = (40/100). Object size = minification factor $\times$ (image size / object size).

17. d

18. d   The two triangles ABC and CDE are similar triangles. So, BC is $(L_1 / 2)$ where L_1 is the field length. CD = $(g / 2)$ where 'g' is the field gap on the skin. From the two triangles, $(g / 2) / (L_1 / 2) = d / SSD$. Similarly, for the adjacent field, $(g/2) / (L_2 / 2) = d / SSD$. In this problem the two field lengths are equal, but can also be different.

19. c   A 2 cm field size difference gives rise to a difference of 10 in TMR, so one cm field difference will give a difference of 5. The TMR value is 778 + 5 = 783. With proper decimal place, it is 0.783 for an 11 x 11 field.

20. c   A 5 cm field size difference gives rise to a difference of 28 in TMR. So a 3 cm field difference will give a difference of 3/5 of 28 or 16.8. So TMR value is 720 + 17 = 737. With proper decimal place, it is 0.737 for a 13 x 13 field.

21. b

22. c

23. b   Dose rate × time = constant. $D_1 \times T_1 = D_2 \times T_2$; $T_2 = (D_1 \times T_1) / D_2$.

24. a

25. d   Dose / dose rate = time.

B. Beam Calibrations (Photon and Electron Beams)

1.   a, b   Low-energy x-rays (<100 kV), in-air measurements with PP chamber is recommended; medium-energy x-rays, air or water, with cylindrical chamber, is recommended; N_K calibration is recommended; a reference depth of 2 g/cm^2 is recommended for in-water measurements.

2.   b   Equation: $D_W(0 \text{ cm}) = [M_C \, N_K \, P_{stem,air}] \, B_W \, [(_m\mu_{en})_{W,air}]_{in \text{ air}}$
   You need to multiply only two large factors, M_C and N_K. Other factors are smaller. The nearest figure to this result is the correct answer, so this is the method to pick the correct answer when there is shortage of time. Of course, you must know the formula and know how to arrive at the correct result. When two or three answers are very close, you must actually check by multiplying all factors in the equation.

3.   d

4.   c   The chamber measures 20.215 cm. The ISL factor $(20.215)^2 / 20^2$ is the required correction factor to refer the measurement to 20 cm SSD.

5.   d

6.   b

7.   e

8.   c

9.   c

10. d

11. a, b, c

12. a

13. a, e

14. e

15. d

16. a, b, c, d

17. a. iv.   b. iii   c. ii.   d. i.   e. v.

18. a

19. d

20. b   Formula for $p_{pol} = |(M^+ - M^-) / 2 M^-|$. M^+ and M^- refer to the chamber readings for the positive and negative voltages applied to the chamber. M^- is the polarity used in the clinic for reference measurements.

21. a   Formula for $p_{ion} = \{1 - (V^H / V^L)\} / \{M^H / M^L) - (V^H / V^L)\}$.
M^H and M^L refer to the chamber readings for the high and low voltages (usually 300 V and 150 V) applied to the chamber.

22. b   The correction is obtained from the equation: $[273.2 + 20.7) / (273.2 + 22)] \times (760 / 750.1)$.

23. b   The calibration value has no dimensions. The actual charge is 1 nC, and the hospital electrometer must read the same, but it reads 0.997 nC. So, the calibration factor = 1.00 / 0.997 = 1.003.
Now 1.003×0.997 gives the correct value, namely 1 nC.

24. c   It is easy to guess the correct answer. Only three terms are important: chamber reading, chamber calibration factor, and k_Q, i.e., $4.073 \times 10^{-8} \times 4.533 \times 10^9 \times 0.971 = 179$. So the answer c is the nearest number. The multiplication of all other factors would amount to less than a few percent.
The equation for dose $= M \times (P_{ion} \times P_{pol} \times P_{TP} \times P_{elec}) \times N_{Dw,Co} \times k_Q$ cGy.

25. d

26. a

27. d

28. c

29. e

30. c

31. c

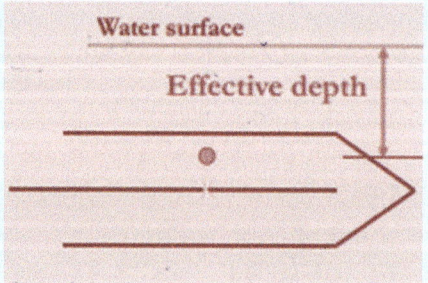

Effective depth of a cylindrical chamber

32. b

33. a, b, d

34. a. iv.   b. iii.   c. ii.   d. i.   e. v.

35. c   A 50% ionization depth can be taken to be R_{50} for this purpose avoiding the difficulty of converting depth ionization curve to depth dose curve and deriving R_{50} from the depth dose curve.

36. d

37. a

38. c $E_0 / 2.4 = R_{50}$

39. b   The approximate formula is $R_p = 1.271 \, R_{50} - 0.23$.

40. d $R_p \approx E_{p,0} \, (\text{MeV}) / 2$. The electron beam energies specified by the manufacturer are determined from practical ranges and, hence, refer to the most probable energies of the electron beams.

41. b R_p must be less than 6 cm. Use the relation given in the above example.

42. c, d   The lateral electron scatter must be retained with respect to the central axis. The maximum distance the electron can travel in lateral scattering is R_p and so the minimum field size is $2 \, R_p \approx \bar{E}_{p,0}$. Thus for a 10 MeV electron beam, a 10 x 10 field is the minimum required (so easy to remember!)

43. d   Since the dose fall is very steep for electron beams, the useful range of interest for treatment is the depth where the dose falls to 90% of the maximum dose, usually referred to as the therapeutic depth and denoted by R_{90} or d_{90}. If we treat beyond this depth, the dose will be very nonuniform and cause underdosing of the tumor. $\bar{E}_{p,0} \approx 4 \, R_{90}$. This is only an approximate formula but easy to remember. Hogstrom uses the expression $\bar{E}_{p,0} \approx 3.3 \, R_{90}$, but all equations are approximate.

44. d

45. b   The approximate rate of energy loss of electrons is about 2 MeV/cm in water. It can also be noticed that E(MeV) / 2 is approximately the practical range of the electron beam in water.

46. c $2 \text{ MeV/cm} \times 6 \text{ cm} = 12 \text{ MeV}$.

47. d   Harder relation $E(d) = \bar{E}_0 \, (1 - d/R_p)$ gives a value of 4.1, which almost agrees with the simpler relation of 2 MeV / cm energy loss.

48. b $200 / D_{cal}$ gives the MU required.

49. b, c, d

50. e

51. a   It is the same for all chambers of the same model. It can be measured for cylindrical chambers or a PP chamber. k_{ecal} is coming into the picture in the protocol only because the therapy chambers are calibrated at ^{60}Co beam quality that is available in all calibration laboratories. If the reference electron beam quality Q_{ecal} is available in a calibration lab, then one can directly obtain calibration at this beam quality and use $(k_{R50})'$ directly taking k_{ecal} as unity.

52. c   Equation for dose: $D_W(d_{ref}) = N_{Dw,Co} \, k_{ecal} \, M$.

53. d   Equation for dose: $D_W(d_{ref}) = N_{Dw,Co} \, k_{ecal} \, M \, K(T,P)$.
$K(T,P) = [(273.2 + 24) / (273.2 + 22)] \times (760 / 774)$

54. a, b, c, d
    Protocol provides a procedure for measuring this correction factor in the user electron beam.

55.   b   Dose at d_{ref} independent of the chamber used, i.e.,
$D_W(d_{ref}) = [N_{Dw,Co} \, k_{ecal} \, (k_{R50})' \, M \, P_{gr} \, K(T,P)]_{PTW} = [N_{Dw,Co} \, k_{ecal} \, (k_{R50})' \, M \, K(T,P)]_{NACP}$
$[N_{Dw,Co} \, k_{ecal}]_{NACP} = [N_{Dw,Co} \, k_{ecal} \, (k_{R50})' \, M \, P_{gr} \, K(T,P)]_{PTW} / M_{PTW}$
Temperature/pressure correction cancel out unless they vary during the two sets of measurements. This is a better method of determining $[N_{Dw,Co} \, k_{ecal}]$ factor of PP chamber instead of getting a calibration at ^{60}Co since PP chambers of any make do not behave in an identical fashion in a ^{60}Co beam.

56.   c   Dose at d_{ref} independent of the chamber used, i.e., $D_W(d_{ref}) = [N_{Dw,Co} \, M \, K(T,P)]_{NE}$.
$= [N_{Dw,Co} \, M \, K(T,P)]_{Roos}$
$[N_{Dw,Co}]_{Roos} = [N_{Dw,Co} \, M \, K(T,P)]_{NE} / [M \, K(T,P)]_{Roos}$ and $K(T,P) = 1$.

57.   e $D_W(d_{ref}) = [N_{Dw,Co} \, k_{ecal} \, (k_{R50})' \, M \, P_{gr} \, K(T,P) \, k_{sat} \, k_{pol}]$.

58.   a, c   According to TG-51, while PP chambers are recommended for electron beam dosimetry, cylindrical chambers are also acceptable. For very low-energy electron beams (<6 MeV or $R_{50} \leq 2.6$ cm) only PP chambers must be used. d_{ref} is not d_{max} but close to this point and is at a depth of $(0.6 \, R_{50} - 0.1)$ cm. The geometric center is at d_{ref} and the P_{gr} corrects the dose to refer it from d_{eff} (effective depth) to d_{ref}.

59.   a, c   For clinical calculations, the dose is referred to d_{max} point on central axis.

C. Basic External Beam Calculations (Photon Beams)

1.   Yes

2.   No

3.   No

4.   Yes

5.   No

6.   No   The field must be surrounded by 5 cm of water on all sides, so for smaller field sizes, smaller phantoms can be made use of.

7.   No   Often they have to be corrected for body curvature and internal heterogeneities since the beam data are acquired for a water phantom (homogeneous medium) for normal incidence.

8.   No

9.   Yes   The TAR concept involves equilibrium dose in air ("free space dose" as defined in Khan, 2010) which becomes more and more ambiguous and also difficult to evaluate. For further study, see Attix (1999) where the difficulty in the evaluation of this quantity for ^{60}Co beam is discussed. So the TMR concept replaces this quantity by a phantom-defined quantity that is less ambiguous.

10.   Yes

11.   No   They are.

12.   No

13.   Yes

14.   No   TAR is larger due to increased penetration.

15. Yes   Due to increased penetration of back scattered radiation.

16. No   Back scatter actually decreases and forward scatter increases.

17. No

18. No   TAR decreases with increased field asymmetry due to decrease in side scatter.

19. Yes

20. Yes

21. No   The actual dose will be less because of lack of backscatter thickness at patient exit point.

22. Yes

23. Yes

24. No   Scatter toward the central axis decreases as the beam quality increases; hence, the decrease in the field size dependence for higher-energy photon beams.

25. Yes

26. No   Photon emission from a ^{60}Co source is more isotropic compared to high-energy photon beams which are peaked in the direction of a central axis. So a flattening filter is not necessary.

27. Yes

28. Yes

29. No   A universal wedge is an internally mounted wedge that is usable for many different field sizes.

30. No   A dynamic wedge field is created by moving one jaw of the collimator toward the other, which creates the wedge field profile.

31. Yes   The dose measured at d_{max} point in a phantom, by varying the SSD, roughly follows (to better than 2%) the inverse square law for moderate changes in SSD (say within about 20 cm). This, however, must be verified for the treatment machine before putting it into clinical use.

32. No

33. Yes

34. Yes   This is because of electron and scatter photon contamination of the beam.

35. Yes   The entry window thickness becomes the thickness of the dead layer of the skin and the chamber would measure the dose just below the dead layer of the skin.

36. Yes   Because both parameters reduce the dose at a given depth.

37. No   It increases PDD since scatter generally increases dose on the central axis.

38. Yes

39. Yes   There is no physical wedge to differentially attenuate the photon beam.

40. Yes

41. Yes

42. No   It can be measured less frequently, but must be done more accurately.

43. No   The PDD used clinically in the department must be used.

44. No   A thin Perspex sleeve (≤1 mm) can be used with a non-waterproof chamber.

45. No   It depends on dose rate and not beam quality.

46. No   It is not applicable for kV x-rays. It is also not applicable for small field dosimetry since the beam calibration conditions (100 cm SSD, 10 x 10 field size) are not achievable for these fields (e.g., tomotherapy, stereotactic radiosurgery, etc.).

47. Yes

48. No   Wedge creates additional scatter which increases S_c factors by 5% to 10% compared to open beams, depending on the wedge angle and location.

49. Yes

50. Yes

51. No

52. Yes

53. Yes

54. No

55. Yes   The surface receives less scatter from the linac head.

56. No

57. Yes   Being a ratio, the distance corrections cancel out.

58. Yes

59. No   One should measure this factor for each linac. The differences in in-air output ratio for the same field size on different machines is primarily attributed to the difference in monitor back scatter.

60. No

61. No   It depends on how one derives the MUs from tumor dose to d_{max} dose or vice versa.
$D(d_{max}) = TD(d)$ / [PDD × wedge attenuation x tray attenuation
$TD(d) = D(d_{max})$ × PDD × [wedge attenuation × tray attenuation]
We always use the first equation since we derive d_{max} dose from tumor dose prescription, and we derive MUs from beam calibration.

62. No   Since there is no physical wedge present to harden the beam.

63. Yes

64. Yes

65. c

66. d

67. a

68. e

69. a

70. d

71. e

72. c

73. a, c, d

74. b, c

75. b

76. d

77. b $r = 5/2 = 2.5$ cm. $S_{eq} = 1.773 \times 2.5 = 4.43$ cm (Formula $S_{eq} = 1.773\ r$).

78. c   Open field area $= 16 \times 12 = 13.7 \times 13.7$ using 4 A/P formula.
Area of equivalent square $= 13.7 \times 13.7 = 187.69$ cm^2.
Blocked area of the open field $= 3 \times \{(8 \times 6) / 2\} = 72$ cm^2.
Unblocked field (i.e., treatment field after blocking) area $= 187.69 - 72 = 115.69$.
The side of the equivalent square of the blocked field $=$ root of $115.69 = 10.75$ cm.
Another method of determining the side of the equivalent square of the blocked field is to directly apply the 4 A/P formula to the treatment field, as shown in the next problem.

79. a

80. d   3.3 x 3.3 is the equivalent field of the blocked field. Note the area under the block is not the blocked field. The field with the blocking, or the open area of the field, is the blocked field (light shading).
Blocked field area $A_{bf} =$ Area of open field $A_{of} -$ Blocked area $= (5 \times 3) - (2 \times 1)$ cm^2
Blocked field perimeter $P_{bf} = 3+5+2+2+1+3$
Side of equivalent square of the blocked field $= S_{eq} = 4\ (A/P)_{bf} = 4\ (A_{bf} / P_{bf})$.

81. c $A_{bf} = 208 - 30 - 22.5 = 155.5$. $P_{bf} = 54.3$; S_{eq} of blocked field $= 4 \times A_{bf} / P_{bf}$.

82. a $OF_{bf} = OF_{of}\ [(S_{c.of}) \times S_{p,bf}]$ (or $NPSF_{bf}$); TMR is not required for the problem!

83. b   TD / TMR of equivalent field $\times OF_{eq.f} \times D_{cal}$; OF of bf = OF of its equivalent field, etc.
$MU = [200 / (0.728 \times 1.03 \times 1.01)]$

84. c $A_{of} = (14 \times 12) -$ Blocked area $= (6 \times 1) + (6 \times 7.5) / 2 + (4.5 \times 11) / 2 = A_{bf}$
$P_{bf} = 14 + 6 + 9.6 + 2 + 11.9 + 1$; $S_{eq,bf} = 4\ (A/P)$
$A =$ Area; $P =$ Perimeter; $S_{eq.bf} =$ Side of the equivalent square of blocked field.

85.   c   The electron contamination is unique to the linac and the clinical conditions which will affect the measurements at d_{max}. For $d > d_{te}$ (transient equilibrium depth) the influence of electron contamination is negligible, and the desired depth of measurement must be chosen in this region. Many recommendations suggest a reference depth of 10 cm for linac beams and 5 cm for ^{60}Co beams. These measurements are then referred back to d_{max} using PDD or TMR depending on the measurement setup (SSD or SAD).

           Since d_{max} is field size dependent, the PDD or TMR appropriate for the measurement field size must be used to refer the measurement **to its d_{max} depth**.

           This procedure must be adopted not only for output factor measurements, but also for beam calibrations where the protocols recommend measurement at stated transient equilibrium depth and referring the calibration to d_{max} depth. This is done in order to retain the definition of these quantities at d_{max} depth for the sake of convenience and continued clinical practice. We will illustrate these methods through some examples in this section.

86.   c   Use the formula $D_1 \times (d_1)^2 = D_2 \times (d_2)^2$.

87.   e

88.   b $103^2 / 100^2 \times 1.01$

89.   b

90.   a, b, c

91.   a, d, e   In "chamber in a mini phantom" geometry of irradiation, mini phantom functions as the buildup cap. Regular phantoms cannot be used for this measurement since phantom scatter contribution will complicate the measurements. This measurement is referred to as (OF) that includes both the collimator and phantom scatter. S_c, on the other hand, is measured in air, without the regular phantom that eliminates phantom scatter, and is referred to as $(OF)_{air}$. Please note that both measurements are relative measurements, normalized to a 10 x 10 field.

92.   a, b, c

93.   b

94.   d $OF = S_c \, S_p$; $(OF)_{air}$ is S_c. Other nomenclatures also exist in the literature.

95.   c   Field sizes are always in cm, so the dimension is not mentioned for convenience.

96.   a, c S_c is independent of SSD since both numerator and denominator vary according to ISL The mini phantom used for measuring S_c theoretically must be of transient equilibrium thickness to avoid the influence of contamination electrons which have ranges beyond d_{max}. A thickness of about 10 cm will be adequate for all photon beam quantities normally encountered in clinical practice. The dependence on collimator setting is obvious, and that is what S_c measures. Moderate blocking encountered in clinical practice changes phantom scatter seen by the measurement point and not collimator scatter, which forms part of the effective primary incident on the phantom.

97.   a, b, c

98.   b $(100 / 90) \times 3$ cm

99.   f

100. a

101. d

102. c

103. d

104. a, b, d

105. d   The irradiation geometry is symmetrical with respect to the midplane. So total MU = 200 / (0.683 × 1.022 × 1.01) = 283.68. Two thirds MU are delivered through the AP field and one third through the PA field as per the field weightings given.

106. e

107. c

108. b

109. a S_c (25 x 25) = (1.065 / 1.028). S_c × NPSF(20 x 20) gives OF(20 x 20) for the blocked field. S_c does not change since the collimator setting remains the same.

110. b   10 x 20 = Sq. field of size 13 cm. OF for this field is obtained from the OF table by interpolating between 10 and 15 cm fields, which is equal to 1.019. So MU = (130 / 1.019) cGy / 1.01 cGy/MU. The answer can be easily guessed. The output variation is just 2% and beam calibration is nearly 1. So the answer is very close to 130.

111. c   [200 / (0.701 × 1.053 × 1]

112. b

113. b   PDD(10) / PDD(6) = [D(10) / D(6)]

114. c

115. d

116. e

117. b   As SSD increases, PDD increases because of the ISL component. The correction due to ISL is given by $(106 / 101.6)^2 × (136.6 / 141)^2 = 1.022 ≈ 2\%$).

118. c   The maximum effect is due to output change with SSD. Output decreases by a factor of $(136.6)^2 / (101.6)^2 = 1.8$ or the MU increases by the same factor.

119. b   This involves two dimensional linear interpolations. You first determine PDD values for 7.5 cm depth for the given two field sizes (84.4 and 85.5) by interpolation. Then interpolate between these two values for the 6 x 6 field size. The same method applies for other dosimetric quantities as well e.g., TMR, TPR, etc.

120. a, b, c

121. c   The measurement at d_{max} is very similar to measurement in-air under equilibrium conditions, so the dose rate at 110 cm will go down according to ISL by a factor of $101.5^2 / 111.5^2$ for the given collimator setting. Note the field size on the phantom is not the same at 110 cm SSD. It diverges geometrically. (The scatter conditions are slightly different for the measurement point at the two SSDs since the surface field sizes are different, but this effect can be neglected for small changes in SSD.)

122. a   Applying ISL to dose at each point for both phantoms, one can show
$PDD_2 / PDD_1 = [(SSD_2 + 1.5)^2 / (SSD_2 + 8)^2] / [(SSD_1 + 1.5)^2 / (SSD_1 + 8)^2]$.
This factor is called the Mayneord factor.

123. a   Since field size is the same at SSD and SAD distances (see the figure in the question), the collimator setup is not disturbed, so the primary radiation incident on the phantom remains the same. Only the ISL correction changes the d_{max} dose at SAD setup.

124. b $[200 / (0.748 \times 1.030)]$ gives dose at d_{max} for a 10 x 10 field in the SAD setup. The ISL factor 0.970 refers the dose to calibration conditions. Dividing the result by 0.99 gives the required number of MUs.

125. d $0.99 \times [101.5^2 / 111.5^2] \times 1.051$

126. b

127. c $[200 / 0.758 \times 1.011 \times 1.013]$ gives the dose at d_{max} point in the treatment setup for a 10 x 10 field. We multiply this value by the ISL factor $(123)^2 / (103)^2$—note the value must increase—to refer it to d_{max} point under calibration conditions. Dividing by $D_{cal} = 1.005$ gives the required monitor units. Many authors come to treatment conditions from the calibration conditions. Though both are equivalent, we generally derive the monitor units from the tumor dose and treatment conditions, and it is desirable to do it the same way.

128. c

129. d $[200 / (0.619 \times 1.023 \times 1.020)]$ gives d_{max} dose for a 10 x 10 field. $K_{ISL} = (122.5 / 102.5)^2$ refers the dose to calibration conditions. $\{[200 / (0.619 \times 1.023 \times 1.020)] \times K_{ISL}\} / D_{cal}$ gives the MUs.

130. b

131. a   The dose in the SAD geometry will be more by the ISL factor $[(102.5)^2 / 100^2] \approx 1.05$. So the monitor units, in this case, will be less by about 5%, i.e., 219 MUs (answer in the last question) / 1.05 = 209.

132. c

133. e   Scatter increases with decreasing beam quality and increasing depth.

134. c

135. b   Magnification 1.4 means the film distance is 140 cm, so the size of the block at the tray $= 4 \times 65 / 140$ cm and the projection size on the skin is $4 \times 100 / 140$ cm.

136. c   Dose to be delivered = 150 cGy wedge field = 250 / 0.617 cGy open field dose. MU to be delivered = (250/0.617) cGy / 1.01 cGy/ MU gives the required MUs. If you have doubts regarding where the calibration factor goes (i.e., in the numerator or the denominator) just check the units as I have shown here. As for other beam-modifying factors, remember they all attenuate the beam, so the open field dose will always be higher, i.e., the correction factors must go in the denominator.

137. a

138. a   200 cGy at 12 cm for wedge field

= 200 / 0.759 cGy open field dose → [200 / (0.759 × 0.839)] cGy at d_{max}

→ [200 / (0.759 × 0.839 × 1.033)] cGy at d_{max} for 10 x 10 open field

→ [(200 / (0.759 × 0.839 × 1.033 × 0.99)] MU

The steps shown here explain how MUs are arrived at:

Step 1: Tumor dose corrected for beam modifiers attenuation.

Step 2: TMR to refer dose to d_{max} the calibration point.

Step 3: Field size correction to refer to calibration conditions.

Step 4: Divide by calibration dose rate to get the required MUs.

For problems involving SSD setup, the steps are the same except the use of PDD, instead of TMR, to refer the dose to the calibration depth.

139. a

140. c   (tan 40° / tan 60°) gives the dose fraction that must be delivered using the wedge.

141. a   0.48 × 200 cGy through wedge field and remaining from the open field

142. d   Dose fraction for wedge field = tan 30 / tan 60 = 0.33 = 66 cGy (i.e., 200 × 0.33).

$(MU)_{wf}$ = {66 / (0.410 × 0.641 × 1.030)} / 0.995 = 245

$(MU)_{of}$ = {134 / (0.624 × 1.030)} / 0.995 = 210

143. c   WA = 90° − (HA/2)

144. d

145. d   However, unlike the previous problem, the beam entry surface is not perpendicular to the beam incidence. Here the wedge angle may also take into account the surface obliquity and provide compensation in case a compensator has not been provided.

146. e   The wedge transmits 75% of the open field dose, so the wedged field dose (or the open field MUs) must be increased by 1 / 0.75 or by 1.33 times.

147. c   Since the fields are equally weighted (i.e., 1:1) each field delivers equal dose, i.e., 100 cGY.

$(MU)_{A/P}$ = {100 / (0.929 × 1.017) × 0.99}

$(MU)_{P/A}$ = {100 / (0.788 × 1.017) × 0.99}

148. e   Dose contribution by A/P field = (2/3) × 200 cGy; by P/A field = (1/3) × 200 cGy. Other factors in the denominator are the same.

149. b   The dose prescribed was to midline at 8.5 cm depth, but the dose delivered was, in fact, at 7.5 cm depth. The 6 MV beam attenuation in tissue is 1.5% per cm, so per field the dose escalation is about 1.5%. The total dose escalation is about 3% or 3 × 36 cGy more.

150. d   This depth is generally less than d_{max} but depends on the field size and field separation, so it must be determined for each case.

151. b

152. d   Dose at d_{max} / dose at d = 8 cm

= [TMR(d_{max}) / (SAD − 8 + 1.6)^2] / [TMR(d_{8cm} / (SAD)^2]

= {1.00 / 93.6^2} / {0.861 / 100^2}

153. b

154. d

155. a

156. e

157. e

158. e

159. a

160. b Each field contribution at midline = 100 cGy. 100 / PDD(10, 8 x 8, 100) gives entry dose = 149.925 cGy. Entry dose × PDD(18.5, 8 x 8,100) gives exit dose; entry dose × PDD(17, 8 x 8, 100) gives the cord dose due to A/P field; entry dose × PDD(3, 8 x 8, 100) gives the cord dose due to the P/A field. Add contributions of both the A/P and the P/A fields to get the total dose.

161. c Fields are equally weighted at the isocenter, implying a contribution of 100 cGy by each field. The rest of the calculations are straightforward, i.e., 100 / (TMR × OF × D_{cal}), etc.

162. a AP/PA and LL/RL field contributions are 250 × (2/3) and 250 × (1/3) respectively. Again, AP/PA are equally weighted, so each field contributes half of [250 × (2/3)]. Again, LL and RL are equally weighted, as can be seen from the figure.

163. c

164. d

165. c Fields are equally weighted and the midplane is at equal depth. Hence the fields will deliver equal dose or 200 / 2 = 100 cGy per field.

166. d MU per field = 100 / [PDD (7, 12 x 12, 100) × OF (12 x 12) × D_{cal}(1.6, 10 x 10)]
TMU delivered = 2 × (MU/field)

167. b When the fields are equally weighted, the dose delivered by RL and LL are (1/2) × 247 and (1/2) × 247 MUs, respectively.
When these two fields are weighted 2:1, the dose delivered by RL and LL are (2/3) × 247 and (1/3) × 247 MUs, respectively.

168. a In the above problem the tumor dose due to each field is unknown, but total PDD at the treatment plane = TPDD = (2/3) × PDD_{RL} + (1/3) × PDD_{LL} due to field weighting, which gives the total tumor dose, i.e., 100% dose = 200 × 100 / TPDD.
Due to field weighting, $D(d_{max})_{RL}$ = (2/3) × 100% and $D(d_{max})_{LL}$ = (1/3) × 100%.

169. d $TD_{RL} = D(d_{max})_{RL} \times PDD_{RL}$ and $TD_{LL} = D(d_{max})_{LL} \times PDD_{LL}$

170. c $ED_{RL} = D(d_{max})_{RL}$; $(E \times D)_{RL} = D(d_{max})_{RL} \times PDD(12.4, 12 \times 12, 100)$

171. b $ED_{LL} = D(d_{max})_{LL}$; $(E \times D)_{LL} = D(d_{max})_{LL} \times PDD(12.4, 12 \times 12, 100)$

172. e On the right-hand side of the patient, the entrance dose is due to the RL field, and the exit dose is due to the LL field. Beam weightage has already been applied to determine $D(d_{max})_{LL}$ and, hence, is not applied again for the PDD, i.e.,
$D(d_{max})_{RL} + D(d_{max})_{LL} \times PDD(12.4, 12 \times 12, 100)$

173. c On the left-hand side of the patient, the entrance dose is due to the LL field. and the exit dose is due to the RL field, i.e.
$D(d_{max})_{LL} + D(d_{max})_{RL} \times PDD(12.4, 12 \times 12, 100)$
The problems show a gradient of dose across the patient.

174. a $MU_{RL} = D(d_{max})_{RL} / \{OF (12 \times 12) \times D_{cal}\}; MU_{LL} = D(d_{max})_{LL} / \{OF (12 \times 12) \times D_{cal}\}$.

175. c The weightages for AP, PA, RL, and LL are 2:2:1:1 for a total of 400% for the four fields. $(2/6) \times 400$, $(2/6) \times 400$, $(1/6) \times 400$, and $(1/6) \times 400$ are the individual weighted percentages for the four fields, respectively, at the respective d_{max} points.

176. a The dose at d_{max} point for AP, PA, RL, and LL fields (in %) are 133.3, 133.3, 66.7, and 66.7, respectively. $210\% = 200$ cGy, and so $(200 / 210) \times 133.3$ and $(200 / 210) \times 66.7$ give the doses to the AP/PA d_{max} points and RL / LL d_{max} points, respectively.

177. e $\{D(d_{max})\}_{AP} = / OF(15 \times 15) \times D_{cal}$ would give the MUs for AP field, etc.

178. d $(200 / 230) \times 200$ gives dose to AP field; $(200 / 230) \times 100$ gives the dose to the RL and LL fields. The d_{max} doses can also be renormalized to a total d_{max} dose of 100% (as against 400% in the above problem). This will reduce each percentage by a factor of four. In this case, the weightages will be $(200 / 4)$, $(100 / 4)$, and $(100 / 4)$ for the three fields. The isodose line also becomes $230 / 4 = 57\%$. If the total d_{max} dose is normalized to 200%, we reduce all the percentages by 2. All will lead to the same results, which can be easily verified.

179. b $[D(d_{max})]_{AP} / [OF (12.5 \times 12.5) \times D_{cal}]$ MUs; $[D(d_{max})]_{RL \, or \, LL} / [OF (9 \times 9) \times D_{cal}]$ MUs.

180. d The fields are weighted at the isocenter in SAD techniques. Since they are equally weighted, they contribute equal doses at the isocenter (i.e., 100, 100 cGy).

181. d Each field delivers 100 cGy, so $MU = 100 / (TMR \times OF \times D_{cal})$.

182. a RL delivers $(2/3) \times 200$ cGy and LL delivers $(1/3) \times 200$ cGy at the isocenter. Dividing this by $(TMR \times OF \times D_{cal})$ gives the MU delivered by each field.

183. c $(2/3) \times 200 / (TMR \times OF \times D_{cal})$ and $(1/3) \times 200 / (TMR \times OF \times D_{cal})$ given the contribution of the RL and the LL fields, respectively.

184. c Dose at "A" is higher and dose at "B" is lower than the dose at the isocenter by the ISL, i.e., by $(100 / 94.6)^2$ at point "A" and by $(100 / 105.4)^2$ at point "B" for the RL field. D_{cal} is approximately the same at "A" and "B" due to the canceling effects of scatter and field size changes at these points, though it strictly applies to 100 cm SAD distance.
Dose at A and B due to RL field
Dose at A= $MU_{RL} \times D_{cal} \times TMR(d_{max}) \times K_{ISL}$; Dose at B= $MU_{RL} \times D_{cal} \times TMR(d_B) \times K_{ISL}$
Dose at A and B due to LL field
Dose at A= $MU_{LL} \times D_{cal} \times TMR(d_A) \times K_{ISL}$; Dose at B = $MU_{LL} \times D_{cal} \times TMR(d_{max}) \times K_{ISL}$
Summation gives the total dose at A and B due to the POP of fields.

185. e Fields are equally weighted at the isocenter, and the total contribution is 100% (see the figure next to question 185), so the contribution per field = 25%.
$95\% = 200$ cGy; $25\% = (200/95) \times 25 = 52.6$ cGy
MU per field = $52.6 / (TMR \times OF \times D_{cal})$ using the appropriate TMR values.

186. b   When the fields are equally weighted at the isocenter and the total contribution is 100% (see the figure next to question 185), the contribution per field = 25%.
95% = 200 cGy; 25% = (200/95) × 25 = 52.6 cGy
Here, fields are weighted 2:1 for AP/PA and RL/LL fields at the isocenter, and the total contribution is 100% (see the figure above).
AP or PA field contribution = (2/6) of 100% = 33.3%
RL or LL field contribution = (1/6) of 100% = 16.7%
95% = 200 cGy; 33.3% = (200/95) × 33.3 = 69.5 cGy; 16.7% = (200/95) × 16.7 = 35.2 cGy
MU = dose for the field of interest / (TMR × OF × D_{cal}) using the appropriate TMR values.

187. c

188. a

189. b, d

190. c, d

191. d

192. c, d

193. c

194. d

195. a

196. b

197. d   Calculate $D(d_{max})$ for treatment geometry for 10 x 10 field. = (TD × 100 / PDD × {$S_C(r_C)$ $S_p(r)$.
(Collimator field size determines S_C and field size on phantom determines S_p.)
Calculate $D(d_{max})$ at calibration geometry – K_{ISL} correction = $(101.6/121.6)^2$.
Note the calibration point is nearer the source, so the dose at calibration depth will be higher. So if you use the factor in the numerator, it must be >1, and if you use it in the denominator, it must be <1.
Use D_{cal} to calculate MUs
= (TD × 100 / PDD × {$S_C(r_C)$ $S_p(r)$ × K_{ISL} × D_{cal}]
= (100 × 100) / [73.1 × 1.007 × 1.009 × 0.698 × 0.99]

198. e   Calculate $D(d_{max})$ for treatment geometry for 10 x 10 field.
= (TD / TMR × {$S_C(r_C)$ $S_p(r)$
(Collimator field size determines S_C and field size r_d at 10 cm depth determines S_p.)
Calculate $D(d_{max})$ at calibration geometry – K_{ISL} correction = $(101.6 / 121.6)^2$.
Use D_{cal} to calculate MUs
= (TD / TMR × {$S_C(r_C)$ $S_p(r_d)$ × K_{ISL} × D_{cal}]
= 100 / [0.824 × 1.007 × 1.012 × 0.698 × 0.99]

199. d   Let P represent the primary dose at P and S_1, S_2, S_3 the scatter contribution from the field areas shown as 1, 2, and 3 in the figure for question number 199.

$D(P, 10, 12 \times 12) = P + S_1 + S_2 + S_3$

$D(P, 12 \times 4) = P + S_2$

$D(P, 10, 12 \times 12) - D(P, 12 \times 4) = S_1 + S_3$

$[D(P)]_{bl.fd} = P + S_1 + S_3 = P + [D(P, 10, 12 \times 12) - D(P, 12 \times 4)]$; $S_2 \approx 0$ below the block

$P \approx T \times D(P, 12 \times 4)$ where T = transmission coefficient $\approx 3\%$

$[D(P)]_{bl.fd} = P + [D(P, 10, 12 \times 12) - D(P, 12 \times 4)] = [D(P, 10, 12 \times 12) - D(P, 12 \times 4) \times (1 - T)$

$D(P) = 100 \times D_{cal} \times OF \times PDD$ for the field under consideration = D(P)

200. c

201. a

202. a

203. a, c, d

204. a, b, c

205. a, c, d

206. a, b, c

207. c

208. a, b, c

209. a, b, c

210. a

211. a, b, c

212. a, d

213. a, b

214. b, c

215. a, c

216. c

217. a

218. b

219. d   The calibration depth must be much beyond the depth of dose maximum, i.e., the transient equilibrium region, where the electron contamination does not reach. A depth of 10 cm will satisfy this criteria for the whole range of clinically used photon beam qualities. Earlier recommendations gave depths of 5 cm, 7 cm, and 10 cm for different beam qualities, but recent protocols recommend a single depth of 10 cm for all beam qualities.

220. a

221. d

222. b   The earlier protocols recommended that an effective center or the geometric center of the chamber might be placed at the calibration depth. However, it is much easier to place chambers in a water phantom with respect to their geometric center, so AAPM TG-51 recommends this approach.

223. c

224. a, c

225. a. ii.   b. i.   c. iv.   d. iii.

226. a. iii.   b. i.   c. ii.

227. a. ii.   b. i.   c. iii.

228. a, b, c, d

229. b

230. c

231. c

232. b

233. b

234. c

235. a

236. a

237. a

238. b

239. a

240. a, b

241. a, b

242. c

243. c

244. a

245. c

246. a

247. b

248. b

249. c

250. a

251. a

252. c

253. d

254. b

255. c

256. a

257. a

258. b

259. a, b

260. a, b, c

261. c

262. c D_W(1.5, 12.5 x 12.5, 125) and D_W (1.5, 12.5 x 12.5, 100) are related by the inverse square law. D_W (1.5, 12.5 x 12.5, 100) and D_{cal} (1.5, 10 x 10, 100) are related by the output factor, OF.

263. a

264. a, b

265. b, d $(2/3) \times 300$ cGy and $(1/3) \times 300$ cGy, respectively, for the anterior and posterior fields.

266. a, c, d

267. b

D. Basic External Beam Calculations (Electron Beams)

1.   No   In the AAPM TG-51 protocol, the electron beam quality is specified by the beam penetration parameter, $R_{50,D}$, the depth where the electron beam depth dose drops to 50% of the peak value. This is more accurate than the parameter, E_0, the mean energy of the electron beam used in the earlier protocols (as derived from R_{50} using an empirical equation).

2.   Yes

3.   No   It is the dose produced by the Bremsstrahlung radiation or the x-ray background.

4. No   The dose measured at d_{max} point in a phantom, by varying the SSD, roughly follows (to better than 2%) the inverse square law for small changes in distance (say, within about 5 cm), but the ISL is followed not with respect to the target position, but with respect to a point much closer than the target position. This apparent distance, which is also a function of field size, must be experimentally determined for every electron beam quality of the machine. See Khan (2010) for more details.

5. Yes

6. Yes

7. No   They very much do.

8. Yes

9. Yes

10. Yes

11. Yes   The stopping power ratio water to Si is independent of energy (and hence depth), so it will get cancelled while taking the ratio.

12. No   The "water /air" stopping power ratio is depth (energy) dependent. Hence, the ionization ratio needs different stopping power values in the numerator and the denominator to get the depth dose curve.

13. No   It does, and especially for large obliquity angles (>45°) the PDD curves are significantly different. Compensating the patient surface obliquities is essential for using the measured PDD curves, which are typically obtained for normal incidence.

14. No   For low-energy electron beams (i.e., <6 MeV) d_{ref} can be in the buildup region where the dose is increasing with depth and $P_{gr} > 1$.

15. No   For ≤6 MeV electron beams, TG-51 requires the use of PP chambers for electron dosimetry.

16. Yes

17. Yes   Because of isodose constrictions that occur with lower-energy electron beams, the choice of the field size must be strictly based on target coverage.

18. Yes

19. Yes   The x-ray contamination is defined as dose at $R_p + 2$ cm.

20. No   The electrons scatter more compared to photons in air (and also inside the linac head), so the geometric spread of the beam with respect to the focus is a rather accurate assumption for x-rays, but not electron beams. The electron beams "virtually" spread from a focus in space which does not coincide with the scattering foil, which is the point of electron production. So the "effective SSD" of the electron beam must be experimentally determined.

21. No   The concept is important because it is difficult to place the electron applicator on the patient surface due to the irregular surface of the patient. So treatments are often carried out at SSDs slightly different from the nominal SSD. This requires an inverse square law correction to the beam output as in the photon beams. The "virtual source position" helps in applying this correction.

22. No   The effective energy depends not only on the beam energy, but on the field size and the manner of collimation for a given beam energy. So it must be determined for every linac for its energy and field size configuration.

23. No   Unlike photon beams, it increases with increasing electron beam energy.

24. No   The penetration length can be 3 to 4 times larger compared to unit density tissue, depending on the lung density. If the electron energy penetrates even about 1 cm beyond the chest wall in a unit density medium, it will amount to about 4 cm penetration in the lung.

25. No   It depends also on the field size, since scatter influences the dose in the buildup region and the peak depth.

26. Yes   But a lot of precautions have to be taken while using films. Film linearity, film processor QA, and film alignment with respect to the phantom edge have to be carefully taken into account while using films for relative dosimetry. See Khan (2010).

27. No   This is true for photon beams, but not for electron beams while using a Farmer-type chamber. The stopping power ratio water to air and the fluence perturbation factor change with beam quality (depth). So the depth ionization curve requires corrections to convert to a depth dose curve. It is important to ensure that RFA systems used for collecting beam PDD data take these corrections into account. However, if Si diode dosimeter is used, as in the case of some commercial RFAs, the fluence correction factor is not very significantly depth dependent, and the Si-to-water stopping power ratio is practically energy independent. The depth ionization curve, in this case, can be taken to be the depth dose curve.

28. Yes

29. Yes

30. No   For any given material, back scatter decreases with increase in electron energy. For any given energy, back scatter increases with increase in atomic number. See Khan (2010).

31. No   The electron beam output does not vary by inverse square law with respect to the accelerator target position or the nominal SSD (as in the case of photon beams) but does with respect to an effective SSD (SSD_{eff}), which must be experimentally determined. SSD_{eff} is also a function of beam quality and field size (especially for smaller fields).

32. No   For further study, see Khan (2010).

33. c, d

34. a, c

35. a, b

36. a

37. c

38. d

39. a, b, c

40. c   The manufacturer usually measures the practical range of the electron beam, which is related to the most probable energy of the electron beam.

41. a

42. b, c   The clinical electron beam incident on the patient exhibits a skewed spectrum, characterized by maximum energy, most probable energy, mean energy, and a spectral width. The most probable energy, E_p, is related to the practical range of the beam. The beam energy specified by the manufacturer refers to most probable energy since it is determined from practical range. The mean energy, E_0, is related to and derived from $R_{50,D}$ the depth in water where PDD falls to 50% of the peak value. The dosimetry parameters—stopping power ratio, perturbation correction, etc.—are related to E_0. The AAPM protocol, TG-51, directly makes use of $R_{50,D}$ as beam quality parameter instead of using E_0. This is because the relationship used to derive E_0 from $R_{50,D}$ in earlier protocols did not give the true value of E_0. Read the AAPM TG-51 protocol or the IAEA TRS398 protocols for more information.

43. b

44. a, b   Spectral width increases with increasing electron interactions because of the increasing number of lower-energy electrons in the spectrum. So d is wrong. The whole spectrum shifts toward lower energies because of the degradation in electron energies. For further study, see Paliwal (1982).

45. a, b, c   Mean energy decreases. d is wrong since the spectrum as a whole shifts toward lower energies.

46. c

47. c   Electron energy loss in unit density medium: about 2 MeV/cm.

48. a, c, d   Unlike photon beams, buildup for electron beams is not very significant. Electron buildup occurs because of the production of knock on electrons (i.e., δ rays having finite range) and increase in fluence as a result of scattering. δ ray buildup is not very pronounced. Fluence buildup is somewhat pronounced for low-energy electron beams because of higher scattering. This buildup also becomes insignificant for higher-energy electron beams. For further study, see Khan (2010).

49. a, b, c   Electrons exhibit finite range and, hence, deposit no energy beyond the range. However, Bremsstrahlung produced by electrons can deposit energy in deeper layers and appear as a background tail of the electron beam depth dose curve.

50. b, c, d   See the figure below which illustrates the point mentioned in c.

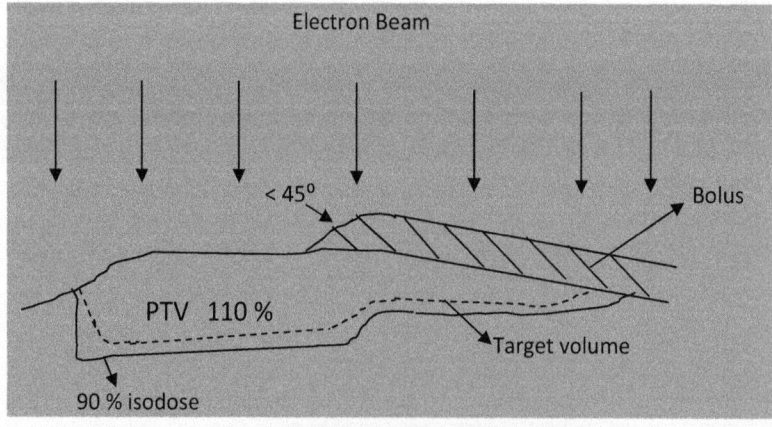

Use of bolus for depth compensation

51. a, b

52. b

53. a

54. b, c

55. c

56. a

57. b, c

58. d

59. b

60. c

61. a   Electron scatter in the "air gap" increases the penumbra and constricts the 80% and 90% isodose curves (refer to the figure under the answer to Question 67 below). The larger field needs to be used for target coverage, so b is wrong. The beam output at calibration depth decreases, so c is wrong. The inverse square law can be used to determine the output at d_{max}, but effective SSD must be used.

62. a   90% dose is 250 cGy; 100% dose is (250/0.9); 100% dose rate is 1.01 cGy/MU.

63. a, d   Increased obliquity increases skin dose and shifts d_{max} and therapeutic depth toward the patient surface. For further study, see Khan (2010).

64. d

65. a

66. c   To reduce dose due to backscatter, the internal shield is usually surrounded by a material to absorb back scatter electrons.

67. a, b, d

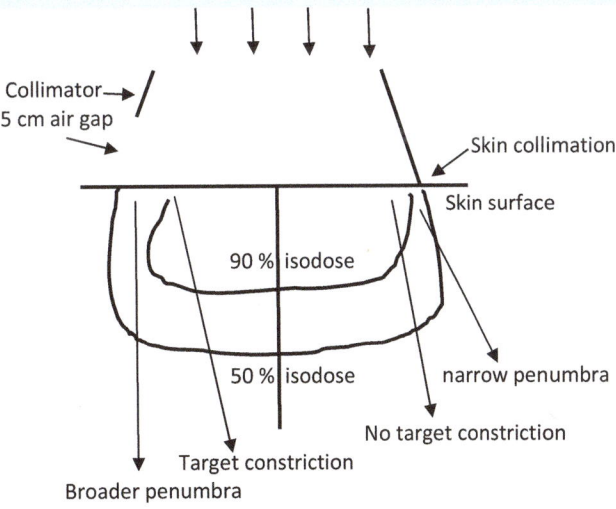

Effect of increased SSD on electron beam isodose profile

68. a, b, c

69. c, d   The lateral electron scatter must be retained with respect to the central axis. The maximum distance the electron can travel in lateral scattering is R_p, and so the minimum field size is $2 R_p \approx \bar{E}_0$. Thus for a 10 MeV electron beam, a 10 x 10 field is the minimum required (so easy to remember!).

70.   a, b   Though the first statement is true in most of the situations, it may not be applicable to all models of linacs since it depends on how the homogeneous fields are generated. So it needs to be verified for the user's linac.

          The ISL does not strictly apply to the electron beams since the divergence is because of scattering from the scattering foils and applicator walls. However, experiments show that for any given electron beam and field size, the beam diverges with respect to an effective source position, displaced from the focal point of the target, according to the ISL, and a correction factor can be applied to determine dose at d_{max} for the new SSD.

71.   a, b, d   Bremsstrahlung is typically less than 1% for 4 MeV electrons and varies by about 1% for every 4 MeV change in electron energy. For 20 MeV it is around 6%, but it very much depends on the linac and the construction details of the head.

72.   a, b, c

73.   e

74.   b, c

75.   d

76.   d

77.   a, c

78.   a, b, c, e

          Since smaller field sizes affect d_{90}, the electron energies must be properly chosen so that the tumor periphery is not underdosed. The change in d_{90}, however, is not significant at lower energies. R_p remains essentially the same and is not affected much.

79.   b

80.   f

81.   a   The modifying factor is the inverse square law correction = output for SSD × [(SSD + d_{max}) / (SSD + d_{max} – bolus thickness)]^2. Bolus shifts the depth dose curve upstream by a distance equal to bolus thickness.

82.   d   In the case of photon beams, the SSD factor would have been simply [(SSD$_{nominal}$ + d_{max}) / (SSD$_{new}$ + d_{max})]^2. However, the electron beams do not follow purely geometric divergence like the photon beam, so it requires a further correction to the inverse square law factor, which must be experimentally determined. This factor depends on electron energy, field size, and the air gap.

          The modifying factor due to inverse square law correction
= Output for SSD × [(SSD$_{nom}$ + d_{max}) / (SSD$_{new}$ + d_{max})]^2
Remember that the output decreases due to increased SSD if we go from 100 cm SSD to 105 cm SSD. This gives us the clue as to how the correction factor must appear in our equation.
Output for new SSD = output for old SSD × ISL factor × AGCF.

83.   a, b, c

84.   a1. ii. b2. iii. c3. i.

85.   b

86. d

87. f

88. d

89. e   PDD increases only slightly because the ranges of electrons scattered toward the axis are $\ll R_p$
          Increased scatter seen by the surface increases D_s. Lateral scatter of lower energies reduce effective
          energy and hence d_{max}. Increased scatter increases output, though it depends on the design of the
          collimating system of the linac.

90. d   200 / (0.90 × 1.01) cGy. Here the output of the actual treatment field is given.

91. c   Skin dose = 200 × {PDD (1 cm) / PDD (4.5 cm)} = 200 × (91.3 / 73.8)
          Maximum tissue dose = 200 × {PDD (3 cm) / PDD (4.5 cm)} = 200 × (100 / 73.8)

92. b   250 / [OF of treatment field × D_{cal}] gives the required MUs = 250 / [0.954 × 1.03] MU.
          Here the cone factor and the insert factor are combined into a single factor. Normally the insert factors
          are normalized to the respective cone or applicator, and the cone or application factors are normalized
          to 10 x 10 standard applicator. In our problems, all OFs are normalized to calibration field size of
          10 x 10, unless otherwise specified.

93. a   200 / 0.95 = dose at d_{max} for the treatment field in cGy = $[D_w(d_{max})]_{tr.\ fd}$
          $[D_w(d_{max})]_{tr.\ fd}$ / 0.954 = dose for the calibration field in cGy = $[D_w(d_{max})]_{cal.\ fd}$
          $[D_w(d_{max})]_{cal.\ fd}$ / D_{cal} in (cGy/MU) gives MUs for treatment.
          MUs = [200 / (0.95 × 0.954)] / 1.03 = 200 / (0.95 × 0.954 × 1.03)

94. b   i.e., MUs = 200 / [(0.95 × 0.954 × 1.03) × 0.855]

95. c   250 / [OF of treatment field × D_{cal}] gives the required MUs
          = 250 / [0.954 × 1.03] MU

96. c

97. d   OF (6 x 12 in 15 x 15) = [OF (6 x 6 in 15 x 15) × OF (12 x 12 in 15 x 15)]$^{1/2}$
          = (1.008 × 1.001)$^{1/2}$ = 1.004

98. c   AGCF (6 x 12 in 15 x 15) = [AGCF (6 x 6 in 15 x 15) × AGCF (12 x 12 in 15 x 15)]$^{1/2}$
          = (0.980 × 0.991)$^{1/2}$ = 0.985

99. e   = 180 / [0.985 × 1.03] MU

100. d   SSD factor = ISLF × AGCF = 0.829 × 0.964 = 0.799

E. Effects of Beam Modifying Devices

1.   No   The buildup would shift to the bolus and the skin would receive a higher dose.

2.   No   They must be at least about 15 to 20 cm from the patient surface for the electron contamination to
            reduce by scattering.

3.   No   There is a field size dependence that must be investigated for the treatment machine in use.

4.   Yes   A motorized wedge incorporated into the treatment head, as in the case of the Philips accelerator, can generate various wedge angles.

5.   Yes   In the case of MV photon beams (see answer to question 9 below).

6.   No   They differ in their design and radiation leakage characteristics and the way in which they have been incorporated into the treatment unit.

7.   Yes   But it is not a great disadvantage. Moreover, MLCs not only serve the purpose of beam shaping, but also increase reproducibility, save time and labor, and help to automate treatment dose delivery by IMRT.

8.   Yes   But there are also other important points to be considered in the use of custom blocks. The nearer the block is to the patient, the larger the size and weight of the block needed to shield the same volume because of the beam divergence. On the other hand, the transmission penumbra would be larger for a larger block-to-skin distance. A block-to-skin distance of about 15 to 20 cm is a good compromise for positioning custom blocks in patient treatment.

9.   No   A straight-edged block does give rise to transmission penumbra due to the beam divergence. However, two more points need to be considered here. For larger source size (e.g., ^{60}Co) the geometric penumbra would make the dose penumbra much less sharp, even for a diverging block. Secondly, the influence of the scatter dose received by the shielded region is also a parameter to be considered. Since scatter is more in the case of the ^{60}Co beam, there is no great advantage in using a divergent block. There is more justification, however, for using a divergent block in MV photon beams.

10.   No   It is not. The transmission penumbra is proportional to the block-to-skin distance.

11.   No   It depends only on the beam quality.

12.   Yes   The size and weight of the block increases with decreasing block-to-skin distance because of the beam divergence.

13.   No   The accuracy of custom block can be checked using the simulator light field and simulator film with blocking area marked on it.

14.   Yes   (See the answer to question 9 above.)

15.   Yes   It helps to escalate dose to clinical target.

16.   Yes   In principle, yes, but in practice it makes very little difference if the MLC leaf size is compatible with the size of the clinical target, i.e., for smaller target volumes, micro-multileaf may be necessary.

17.   No   Conformal technique aims to spare more normal tissues (compared to conventional technique) and also the critical targets bordering on the PTV. So exact delineation of target volume is very essential for this technique.

18.   Yes

19.   No

20.   No

21.   No

22.   Yes

23. Yes

24. Yes

25. No

26. Yes

27. No

28. No

29. No

30. No

31. a   The beam is hardened in the central region, so b is wrong. The beam cannot be broadened by multiple scattering as in the case of electron beams, so c is wrong. Positioning is critical because of the beam flattening being effected by the cone shaped flattener, so d is wrong.

32. a, b, c

33. a, b   Blocks are normally not very close to patient. Answer c is wrong.

34. a, b, c   Incorrect SFD will change the size of block fabricated; incorrect source to tray distance or incorrect positioning of block on tray will influence the region blocked in the patient.

35. a

36. c

37. b

38. b

39. a, b, c, d

40. a, b

41. a, b

42. a, b, c

43. b

44. a

45. b, d

46. a

47. b

48. b, c   When blocking is a minor percentage of the open field, say less than one third, the central region is still able to receive full collimator scatter. Answer a is wrong because as scatter volume decreases, so the scatter dose in the phantom along the central axis decreases and, hence, the PDD or TAR. This, of course, assumes that there is no blocking along the beam central axis.

49. a   Electron absorption is roughly 2 MeV/cm of unit density material.

50. c

51. a, b   Bolus shifts the buildup to the bolus material to extend high dose up to the skin surface. There could be variation in chest wall thickness or the clinical target thickness could be nonuniform. In this case, bolus in part of the treatment field will shift the therapeutic depth closer to the surface. See the figure on p. 316.

52. b, c, d

53. b

54. b, d

55. e

56. a

57. a, c

58. a, c   WTF does vary with field size and depth, but how significant the variation is needs to be investigated for the beam qualities of the machine in clinical use.

59. a, c   Wedge filter causes slight beam hardening and, hence, influences PDD, etc., but the wedge-hardening effect may not significantly alter relative values like PDD, TMR, etc., especially for the smaller depths involved in wedge beam treatments. However, it is important to confirm this with the machine in clinical use.

60. a, b, c, d

61. c

62. a, b

63. a, b

64. c, d

65. a, c   Size of compensator decreases with decreasing source-to-compensator distance due to beam divergence.

66. a, d   Wedge profile can also be created by dynamic wedge technique, without a physical wedge.

67. c A single wedge can be used for several field sizes (see the figure below). However, there is additional beam attenuation due to the larger thickness of the universal wedge. It is used when the beam output is large (e.g., with accelerator beams).

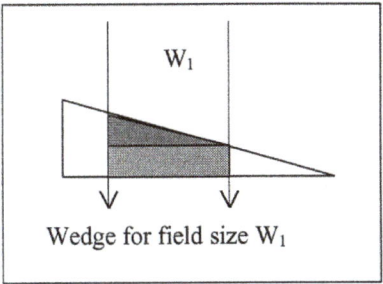 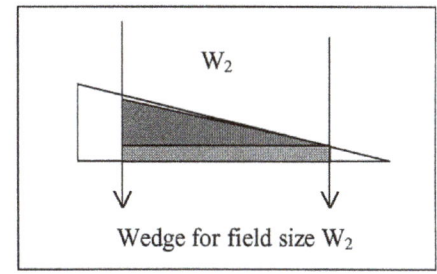

Concept of universal wedge

68. a, b, c

69. c

70. a

71. b, c

72. a, b, c

73. b

74. b

75. c, d

76. a, b, c For a given hinge angle, there is a wedge angle for a wedge pair of fields that would make the dose uniform in the beam overlap region. This assumes that the incident surface is normal to the central axis. If the patient surface is sloping, either compensator should be made use of or the wedge angle, derived from hinge angle, should be adjusted to take into account the surface irregularity. For further study, see Khan (2010).

77. c

78. a

79. b

80. a

81. b

82. e

83. e

84. b, c

85. d

86. d

87. d

88. a

89. b, c, d

90. e

91. a, c

92. c, d

93. d

94. e   Bolus is use with kV x-ray beams and electron beams. Wedges are used as simple compensators in the case of uniformly sloping surfaces. Beam Cerrobend is used for fabricating 2D tissue compensator. Dynamic MLC can modify the beam without the need for a physical 2D compensator.

95. e

96. d   Cerrobend shaping allows sharper penumbra, lower leakage, and larger field sizes compared to MLC. In fact, there is no dosimetric advantage in using MLC except ease of use and IMRT dose delivery.

97. d   Higher energies are used for better dose homogeneity, but they have pronounced skin sparing unlike the lower-energy beams. So to retain both characteristics, one uses high-energy photon beams with a beam spoiler.

F. Irregular Field Calculations

1.   a, b, c

2.   b

3.   a, b

4.   a, b, c

5.   a, b

6.   b, d   SAR depends on field size since scatter volume increases with an increase in field size.

7.   a, d   SMR is not TMR, zero field TMR. For further study, see Khan (2010).

8.   a, b   Blocking changes scatter volume and, hence, Sp, so c and d are wrong.

9.   a

10. c

11. c

12. c

13. c

G. Special Calculations

1.   d

2.   a, b

3.   a

4.   b   Larger field diverges into smaller field. Three-field overlap increases the dose in the overlap region, which is of concern if any OAR is present at that position. This overlap can be avoided by several means. See Khan (2010).

5.   c   See the figure below.

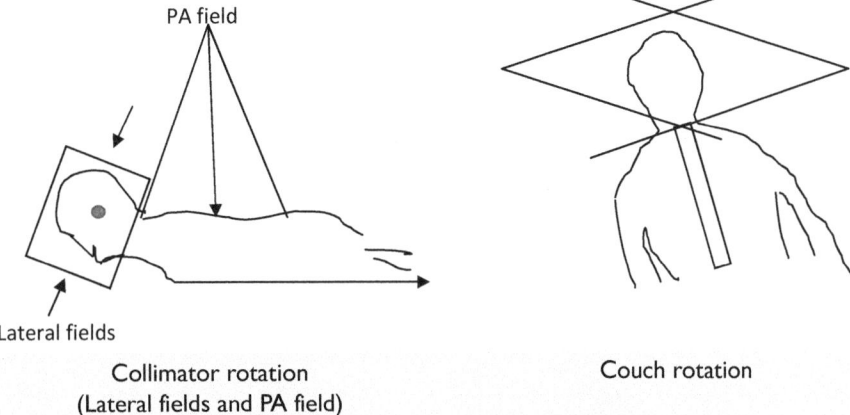

Collimator rotation   Couch rotation
(Lateral fields and PA field)

Cerebral spinal irradiation (divergent fields)

6.   b   Here the abutting border is having no divergence. So collimator rotation to match the divergence of the spinal field would be enough to match the two orthogonal fields. The inferior border of the brain field, in this case, will not diverge into the abutting border of spine field. Compare the figure below with figure (b) above.

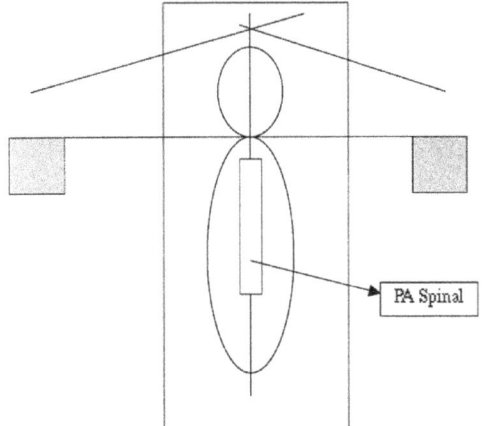

Cerebral spinal irradiation (non-divergent brain fields)

7.   b

8.   a, b

9.   a, b, c, d

Adjacent fields abut at the depth where the dose is required to be uniform (over the full length of the two abutting fields). This, however, creates high- and low-dose regions around the abutting junction (see the figure below). The high-dose region is created by the adjacent fields diverging into each other. In order to accept this plan for treatment, the low- and high-dose regions must be free of tumor and organ at risk, respectively. The "hot" and "cold" regions can be avoided by the various methods mentioned in this question, namely making the abutting field borders parallel to each other.

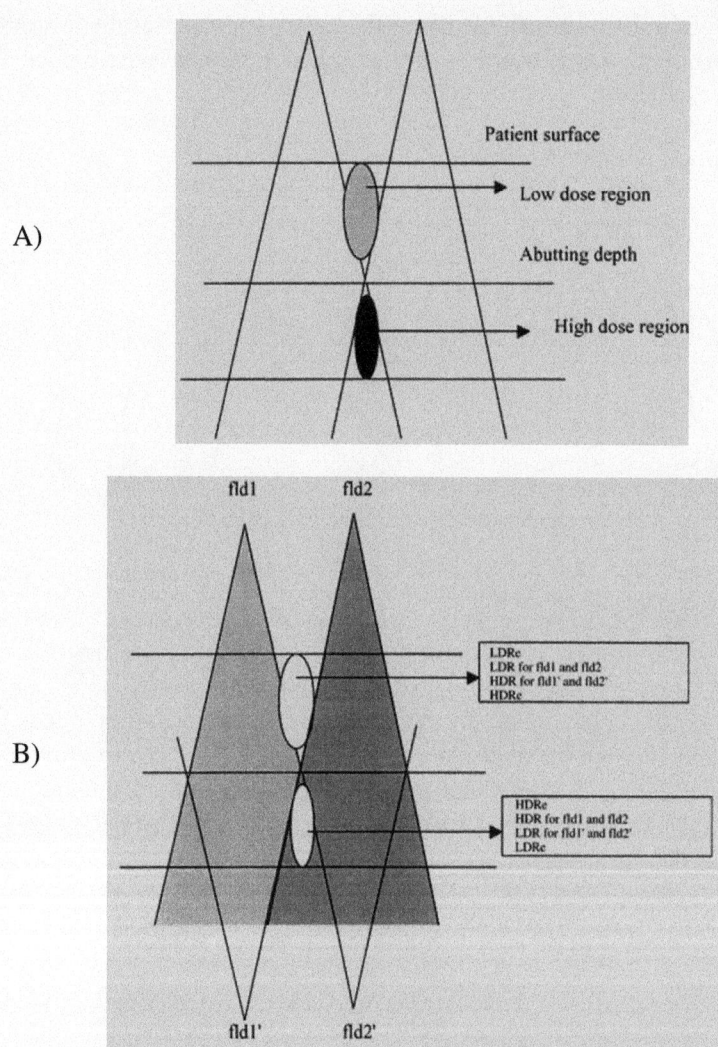

Abutting of adjacent fields at desired depth

10.   a, b   While using a matched pair of adjacent fields from the opposite direction, the low-dose and high-dose regions become high- and low-dose regions, thus making the dose more uniform around the midplane (see figure B above). However, if the divergences do not match for the POP of fields (e.g., when the fields are unequal), the larger fields would diverge into opposing smaller fields, causing a "hot spot" of the three-field overlap region. In such cases, the divergences can be matched by adjusting the SSDs accordingly. For further study, see Khan (2010) and Bentel et al. (1991).

11.   a, b, c   For further study, see Bentel et al. (1991).

12. c, d   See the answer to question 9 above; a and b are therefore wrong. The penumbra is sharp; a small error in abutting can cause large dose variations. This error can be minimized by increasing the penumbra width of the beam using a suitable wedge at the junction. Dose uniformity can be improved and the effect of cold and hot regions reduced by moving the junction along the "gap." For further study, see Wright (1983) and Shahabi (1989).

13. a, b, c   For a circular-field treatment, isodose contours are circular since the beam attenuation is the same in all directions; so d is wrong. However, for a patient of elliptical cross section, the isodose contours are ellipsoidal but normal to the major axis of the patient. For further study, see Leung (1994).

14. a   Off-center tumors, when treated by rotation therapy, will irradiate a large volume of normal tissues and better treatment procedures are available for these types of tumors. With the availability of high dose rates and higher penetration of linac beams, three or four fields can treat the majority of tumors, so rotation therapy is rarely practiced these days.

15. i. a.   ii. a.
    Calibration dose rate at $d_{max} \times OF \times TMR$ = dose rate at target center. Dose to target is not required, in this case.

16. b   Since the leading edge does not cover the two fields. Just compare the coverage of the starting or stopping fields with the coverage of any intermediate field, which will make this weightage clear. For further study, see Leung (1994).

17. c   "Free space dose rate" × treatment time × average TAR = tumor dose rate. For further study, see Khan (2010) on free space dose.

18. a   Moderate blocking (i.e., less than about 25% of open field) does not affect the effective primary reaching the phantom. Scatter dose under the unblocked region decreases since the shadow region would scatter very little radiation. The dose under the block comprises the primary leakage radiation and scatter going into the shadow region from unblocked regions.

19. d

20. a, b, c, d

21. a, b

22. a, b, c

23. a, b, c, d

24. a, b, c

25. a, b, d   Delivered dose is uniform to within about ±10%.

26. a, b, c, d
    High-dose TBI is delivered in a single fraction or in small number of fractions (200 cGy/fraction, 6 fractions, 1200 cGy, total dose). Low-dose TBI is delivered in 10 to 15 fractions, 1 to 15 cGy/fraction. For further study, see Podgorsak (2004).

27. a, b   Photon dose received by the patient cannot be neglected—c is wrong—and must be estimated carefully. Dosimetry must be carried out under treatment conditions.

28. a, b, c, d

Stereotactic treatments involve PTVs of about 1 to 35 cm^3 volume, dose prescription of 1000 to 5000 cGy, delivered in single or multiple fractions with a positional accuracy of better than about 1 mm and a dose delivery accuracy of better than 5%. For further study, see Podgorsak (2004).

29. b, c

30. a, d

For stereotactic treatments, a linac requires additional collimation facilities (e.g., MLCs or circular collimators), devices for immobilization of patients, stereotactic frames during dose delivery, and couch controls for movement or arresting movement with high precision.

31. a, c

32. b, c

33. a, c

34. a, b

35. d

36. b, c

37. a, c

38. a, b, c

39. a, b, c

40. a, b, d

In this technique, dose fall is much sharper in the direction perpendicular to the plane of rotation and more gradual in the plane of rotation because of the parallel opposed beams in all directions in the plane, so b is wrong. By avoiding beams in parallel opposed directions, the dose falloff around the target can be made more uniform. The multiple noncoplanar converging arc technique (see figure B above) was developed for this purpose.

41. a, b, c

A Gamma Knife unit gives better positional accuracy (±0.3 mm) compared to a linac since no dynamic movement is involved.

42. c

43. a

44. c

45. d

46. d

47. c

48. d

49. d

Divergence of the vertical field edge = θ; tan θ = (half width / SSD) = (18/100)
θ = tan^{-1}(0.18) = 10°

50. c

51. a At d_{max} we must give 75 cGy at 400 cm. At 100 cm corresponding dose will be 16 times higher, i.e., $75 \times 16 = 1200$ cGy. D_{cal} is 1 cG/MU or 1200 MUs.

52. b From similar triangles (see the figure below), $SSD_{min} / 85 = 100 / 20$ or $SSD_{min} = 425$ cm.

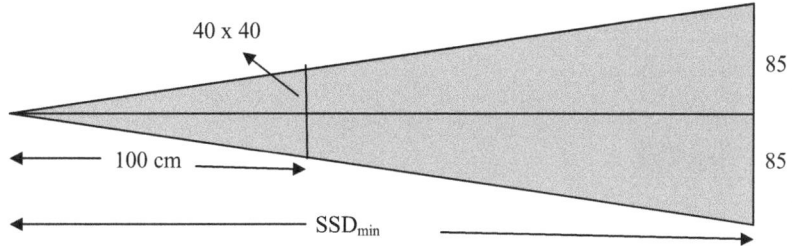

Similar triangles to compute field size at any distance

53. e [150 / TMR(10 cm) dose at d_{max} × ISL factor] MUs / 600 MU / min
        = [150 / {0.872 × (1 / 16) × 1.0}] MU / 600 MU/min.

54. b At 10 cm depth, the dose between 10 cm to 20 cm from the field edge the dose varies by 2% to 0.6%. So at 10 cm distance, the fetal dose is about $4200 \times (2/100)$ cGy.

H. Corrections for Tissue Inhomogeneities and Surface Obliquities

1. b

2. b

3. c

4. c

5. b

6. b

7.　a, c　　See the figure below.

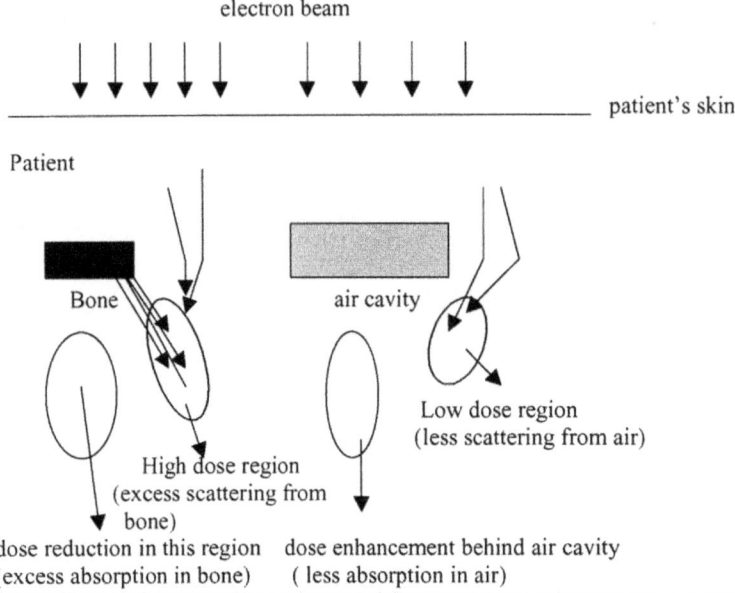

Perturbation due to inhomogeneities in a patient for electron beam

8.　i. c.　ii. a.

9.　b

10.　b

11.　c

12.　c　　Note: The corrections discussed in the above problems are only approximate and are evaluated using the Tissue-Air Ratio (TAR) method. They apply to points beyond the inhomogeneity. The correction increases the dose beyond lung and decreases the dose beyond bone. The percentage correction per cm of lung is 4 (^{60}Co), 3 (4 MV), 2 (10 MV), and 1 (20 MV). The percent correction per cm of bone is 3.5 (^{60}Co), 3 (4 MV), and 2 (10 MV). See Khan (2010).

　　　　　The actual correction will depend on field size, spatial extent of the inhomogeneity, location of the point with respect to inhomogeneity, and other scatter conditions. More accurate correction factors are provided by treatment planning systems.

13.　a, b

14.　b

15.　b　　Note: The isodose shift method is applied here to determine the shift along a ray line that passes through the inhomogeneity. Using this method for an isodose curve, the whole curve can be shifted, correcting for the inhomogeneity. The method requires experimentally determined shift (per cm of inhomogeneity) for various beam qualities. Khan (2010) tabulates approximate shift factors determined for bone and lung for ^{60}Co and 4 MV x-ray beams.

16.　a, b, c, d

17.　c

18.　a, b, c

19. b, c

20. a, c

21. a, c

22. b

23. a, b, c

24. a, b, c, d

25. c $11 \times 0.3 + 8 = 11.3$ cm tissue for field 1
             $6 \times 0.3 + 13 = 14.8$ cm tissue for field 2
             Path length differ by 3. 5 cm tissue
             Tissue attenuation per cm $\approx 2.5\%$, i.e., $3.5 \times 2.5 \approx 8.8\%$

26. b   The attenuation of each pixel depends on the electron density. The CT yields the Hounsfield numbers, which are converted to electron density by a special calibration.

27. e   10 cm lung $= 10 \times 0.3 = 3$ cm tissue, i.e., 7 cm tissue is missing, and tissue attenuation is about $7 \times (3.5\% / \mathrm{cm}) \approx 24\%$ neglecting scatter effects.

28. c   10 cm lung amounts to 3 cm tissue length or 7 cm missing tissue. At 10 MV, the tissue attenuation is roughly 2.5% per cm, so it amounts to about 18% increased dose due to missing tissue. But lung also scatters less, so the actual dose will be less. 15% is the correct answer.

29. b   At 6 MV the correction is $\approx 3.5\% / \mathrm{cm}$, but at 10 MV it is $\approx 2.5\% / \mathrm{cm}$.

V. Brachytherapy

A. Basic Concepts

Circle the right answer (Yes or No):

1. (Yes / No) ^{226}Ra sources were only used in the form of needles.

2. (Yes / No) ^{222}Rn seeds used in the earlier days were radon gas sealed in Au tubes.

3. (Yes / No) ^{226}Ra and ^{222}Rn sources are still in use in brachytherapy.

4. (Yes / No) ^{226}Ra sources were used in intracavitary and interstitial treatments.

5. (Yes / No)   Remote afterloading units are becoming more and more widespread in brachytherapy use.

6. (Yes / No)   For ^{125}I or ^{103}Pd sources embedded in a medium, the dose fall with respect to distance significantly deviates from the inverse square law.

7. (Yes / No)   All brachytherapy sources are bare sources and must be handled with care to avoid any escape of radioactivity.

8. (Yes / No) ^{192}Ir sources are available both as wires and as seeds for brachytherapy use.

9. (Yes / No) ^{192}Ir wire sources are wires made of pure Ir encapsulated in a Pt sheath of 1 mm thickness.

10. (Yes / No)   The air kerma rate (AKR) at 1 m from a 1 mg Ra source (0.5 mm Pt filtration) is 7.25 μGy m^2/h. From the definition of mg-Ra-eq, 1 mg-Ra-eq of any source would produce the same AKR at 1 m distance as the 1 mg Ra source (0.5 mm Pt).

11. (Yes / No)   The active length of a brachytherapy source is the same as its physical length.

12. (Yes / No)   The exposure rate around a line source (no self attenuation) is determined using the sievert integral.

13. (Yes / No)   Mean basal dose rate is the dose rate of an implant in the Paris system of dosimetry.

14. (Yes / No)   In the treatment of gynecological cancers, dose to the organ at risk, namely the rectum, cannot be measured.

15. (Yes / No)   In a particular intracavitary treatment of cancer of the cervix, the "point-A" dose rate is 100 cGy/hr. The rectum and bladder dose rate is 60 cGy/hr. The clinician wants to prescribe 30 Gy to "point A" and a maximum of 18 Gy to the rectum. In this situation, do you think bladder dose can become a limitation?

333

16. (Yes / No) The concept of PTV is as important in brachytherapy as in external beam therapy.

17. (Yes / No) The concept of GTV is NOT as important in brachytherapy as in external beam therapy.

18. (Yes / No) MRI is superior to CT in the delineation of GTVs in gynecological cancers.

19. (Yes / No) LDR treatments can be simulated by delivering higher dose rates in pulsed mode instead of continuous irradiation.

20. (Yes / No) 3D dose optimization and conventional x-ray based planning lead to same probability of tumor control or OAR sparing in brachytherapy.

21. (Yes / No) The institutions are responsible for the verification of the certificate values of the manufacturer calibration of brachytherapy sources.

22. (Yes / No) Gold is used to make eye plaques since it is very effective in shielding low-energy photons.

23. (Yes / No) A lead apron must be used by the physician interacting with a patient with ^{137}Cs insertion, to reduce his exposure, since ^{137}Cs energies are quite penetrating.

24. (Yes / No) The AKS of ^{192}Ir HDR source must be measured every day before treatment starts, using a well chamber, and the value entered into the TPS.

25. (Yes / No) All the brachytherapy sources in the hospital must be leak tested every 6 months even if they are not in use.

26. (Yes / No) In Paris system of dosimetry the prescription dose need NOT encompass the target volume.

27. (Yes / No) In Paris system of interstitial dosimetry the source strength is more important than the linear source strength.

28. (Yes / No) Paris system is applicable only to low activity sources.

29. (Yes / No) If the Paris system rules are correctly implemented the reference isodose surface, should enclose the target volume.

30. (Yes / No) Dose rate constant is unique for a given radioisotope.

31. (Yes / No) Interstitial implants can be template based rigid needle implants or flexible catheter implants.

32. (Yes / No) According to TG-40 recommendations, only 10% of the seeds must be assayed before using them in implants.

33. (Yes / No) The anisotropy correction factor is negligible for ^{125}I and ^{203}Pd pellet sources due to their small size.

34. (Yes / No) For all the brachytherapy gamma sources the ratio $(\mu_{tr})_{water,air}$ is 1.11.

35. (Yes / No) The dose near (i.e., up to 5 cm) higher energy brachytherapy source pellets, in a tissue medium, can be easily calculated by applying just the inverse square law.

36. (Yes / No) The ^{137}Cs sources used for brachytherapy and ^{226}Ra sources have very similar dose distributions in a tissue medium.

37. (Yes / No) Manchester system was developed to treat skin cancers and interstitial tumors using radium sources.

38.   (Yes / No)   Manchester system of dosimetry cannot be applied for the modern brachytherapy.

39.   (Yes / No)   In Manchester system of planar implants, the dosimetry plane is sandwiched between two source planes.

40.   (Yes / No)   Point A has not been formally defined in the Manchester system even though Point A dose has been mentioned.

41.   (Yes / No)   In the Manchester system of intracavitary therapy, the treatment is specified in terms of fixed implant time for a given tumor volume and not in terms of mg-hr.

42.   (Yes / No)   Uterine perforation cannot occur during tandem insertion during brachytherapy treatment of cervix cancer.

43.   (Yes / No)   The dose in tissue at 1 cm from the vaginal cylinder is independent of the ovoid diameter.

44.   (Yes / No)   ICRU 38 recommends tumor dose to be prescribed as an isodose surface that surrounds the target volume (uterus, cervix) rather than at single point like Point A to avoid underdosing or overdosing of the target volume.

45.   (Yes / No)   3D planning of brachytherapy shows ICRU 38 specifications of bladder and rectal doses are very conservative estimate of doses to OARs.

46.   (Yes / No)   Implant containing two or more planes is known as multi-plane implants.

47.   (Yes / No)   Interstitial brachytherapy is only used in palliative cases.

48.   The advantages of brachytherapy over external beam therapy are _____.
    a.   localized therapy
    b.   more conformal
    c.   sharp falloff at target periphery
    d.   any part of the body can be treated
    e.   any size of tumor can be treated

49.   The disadvantages of brachytherapy are that this modality _____.
    a.   can treat only well-localized tumors
    b.   can treat only small-sized tumors
    c.   can treat only body cavities (lung, rectum, vagina, etc.)
    d.   is more laborious
    e.   involves great radiation hazard

50.   _____ sources have been used in brachytherapy treatments.
    a..   Photon
    b.   Beta
    c.   Neutron
    d.   Alpha
    e.   All of the above

51.   The criteria for treatment accuracy in brachytherapy are accurate _____.
    a.   planning of brachytherapy rooms
    b.   source positioning
    c.   source calibration
    d.   dosimetric model for dose computations
    e.   choice of patients

52. The main types of brachytherapy implants are _____.
    a. intracavitary
    b. interstitial
    c. intraluminal
    d. intravascular
    e. surface plaque
    f. intracranial

53. Due to its relatively long half life, high average energy, and radiation hazard, radium's use in brachytherapy has been discontinued. The isotope _____ is most commonly used as a radium substitute in intracavitary gynecological implants.
    a. ^{60}Co
    b. ^{137}Cs
    c. ^{192}Ir
    d. ^{198}Au
    e. ^{125}I

54. Implants that are removed from the patient once the prescription dose has been delivered are generally known as _____.
    a. permanent implants
    b. removable implants
    c. short term implants
    d. temporary implants
    e. unimportant implants

55. The _____ parameter is defined in free space.
    a. air kerma strength
    b. dose rate constant
    c. geometry factor
    d. anisotropy function
    e. source activity

56. Shielding of the _____ is provided in Fletcher-Suit and Henschke applicators.
    a. prostate
    b. cervix
    c. bladder
    d. rectum
    e. all of the above

57. The early empirical radium systems for cervix cancer treatment are the _____.
    a. Manchester system
    b. Stockholm system
    c. Paris system
    d. U.S. system
    e. British system

58. In the Manchester system of intracavitary therapy, Point B _____.
    a. is also defined in addition to Point A
    b. is 3 cm lateral to Point A on pelvic wall
    c. moves if Point A moves
    d. is location of the obturator nodes in pelvis
    e. dose is the same as dose to cervix

59. Paterson-Parker tables _____.
    a. are no longer useful for brachytherapy because newer types of sources have come into clinical use
    b. are applicable to sources of photon energies comparable to Ra
    c. can be planned with sources of equal source strengths
    d. are identical to the Paris system developed by Dureix and others in France

60. The widely used systems for interstitial brachytherapy are the _____.
    a. Manchester system
    b. Quimby system
    c. Paris system
    d. Swedish system
    e. brachy system

61. The AKR at 10 cm from a source container was measured to be 180 μGy/hr. One should stand _____ (in cm) from the container so that the radiation level at that place would be about 20 μGy/hr.
    a. 20
    b. 30
    c. 50
    d. 70
    e. 100

62. The ^{192}Ir HDR afterloading systems that are in common clinical use are _____.
    a. GammaMed 12i
    b. VariSource
    c. HDR microSelectron
    d. Curietron
    e. Selectron

63. Advantages of HDR treatment are _____.
    a. negligible radiation hazard
    b. placement accuracy
    c. possibility of dose optimization
    d. best radiobiological advantage compared to LDR
    e. less traumatic

64. Advantages of manual afterloading technique are _____.
    a. there is very low radiation hazard compared to preloading technique
    b. applicator position verification is possible before source loading
    c. there is much less duration of treatment compared to preloading technique
    d. very little clinical experience is required to practice this technique
    e. radiobiology of low-dose-rate treatments is unknown

B. Brachytherapy Source Characteristics

Circle the right answer (Yes or No):

1. (Yes / No) ^{125}I decaying by electron capture goes to the excited state of ^{125}Te.

2. (Yes / No) Electron capture decay of ^{125}I does not change the identity of the radionuclide.

3. (Yes / No) All ^{125}I sources used in brachytherapy will have identical source characteristics.

4. (Yes / No) Permanent implants are used only for boosting dose in the central region of the tumor following an external beam treatment.

5. (Yes / No) If the implantation is not according to the preplan, the sources have to be removed and implanted again as per the preplan.

6. (Yes / No) If the radiation dose (measured in air kerma) from an ^{125}I source is <1 μGy/h at 1 m from the patient, then no restrictions apply for visitors and relatives near him.

7. (Yes / No) ^{125}I seed source (model 6711) is made of thin ^{125}I wire encapsulated in titanium.

8. (Yes / No) The exposure rate constant or air kerma rate constant of a brachytherapy source helps to compute the radiation output of a brachytherapy source from its activity.

9. (Yes / No) The Total Reference Air Kerma (TRAK) is the sum of the products of the Reference Air Kerma Rate and the irradiation time for each source.

10. (Yes / No) For distances larger than, say, five times the implantation volume, the center of the implant can be treated as the source point, and the total strength of the implant as the strength of a single source at the source point.

11. (Yes / No) Brachytherapy sources are now specified in terms of their activity instead of in older units of mg-Ra-eq.

12. (Yes / No) Integrated Reference air Kerma (IRAK or K_{ref}) is defined as the product of Air Kerma Strength and the duration of exposure.

13. (Yes / No) Dose rate constant is proportional to AKS for a given radioisotope.

14. (Yes / No) Sievert integral gives rise to large errors when applied to lower-energy sources like ^{192}Ir, ^{169}Yb, and ^{125}I rather than higher-energy ones (e.g., ^{137}Cs).

Choose the right answer(s) (more than one may be correct):

15. A brachytherapy source is encapsulated to _____.
    a. produce Bremsstrahlung in the encapsulation wall
    b. reduce the intensity of the brachytherapy source
    c. increase the useful life of the source
    d. contain radioactivity and stop particles other than gamma in delivering tumor dose

16. According to recent AAPM recommendations, a brachytherapy source must be specified in terms of _____.
    a. mg-Ra-equivalent
    b. effective activity
    c. air kerma strength (AKS)
    d. reference air kerma rate (RAKR)

17. 1 mg Ra (0.5 mm Pt filtration) gives an exposure rate of 8.25 R/h at 1 cm. The strength of a Ra substitute is 2 mg radium equivalent (mg-Ra-eq). It gives an exposure rate at 1 cm of _____.
    a. 8.25 R/h
    b. 16.5 R/h
    c. 4.12 R/h
    d. none of the above

18. An AKS of 10 U refers to an AKS of _____.
    a. 10 cGy cm^2/h
    b. 10 µGy m^2/h
    c. 1 Gy m^2/h
    d. none of the above

19. The strength of an ^{192}Ir HDR source in free space at a distance of 50 cm from the source was measured as 80 mGy m^2/h. The AKS of the source (in mGy m^2/h) is _____.
    a. 20
    b. 40
    c. 640
    d. none of the above

20. The low-dose-rate treatments with Ra were prescribed in terms of mg-hr and with Ra substitutes, mg-RaEq-hr. When the sources are expressed in terms of AKS, the dose prescription must be expressed by _____.
    a. Gy m^2/h
    b. Gy m^2
    c. R cm^2
    d. none of the above

21. The exposure rate constant of a ^{125}I source is 1.45 R cm^2/mCi.hr. The AKS of a 1 mCi (apparent activity) ^{125}I seed ("air kerma/exposure" conversion factor is 0.876 cGy/R) is given by _____.
    a. 1.45×0.876
    b. $1.45 / 0.876$
    c. $(1.45 / 0.876)^2$
    d. none of the above

22. Match the exposure rate constant (expressed in R cm^2/mCi.h) to the brachytherapy radionuclide.

Exposure Rate Constant		*Radionuclide*
a. 3.28	_____	i. ^{192}Ir
b. 4.69	_____	ii. ^{60}Co
c. 13.07	_____	iii. ^{125}I
d. 1.45	_____	iv. ^{137}Cs

23. All the recent recommendations state that brachytherapy sources, for the purposes of dosimetry, must be specified only in terms of AKS. However, source strength is still being specified in other units. In order to implement the recommendation, one must _____.
    a. continue to specify brachytherapy source strength in different units
    b. henceforth specify the source only in terms of AKS for dosimetry purposes
    c. specify AKS and also give conversion factors to express strength in terms of other units
    d. specify in old units (e.g., mg-Ra-eq) and give a conversion factor to obtain AKS of the source

24. The clinical usefulness of a brachytherapy source depends mainly on its _____.
    a. specific activity
    b. half life
    c. photon energy
    d. atomic number

25. The AKS of a brachytherapy source is defined as _____.
    a. air kerma rate, specified in free space at 1 m from the perpendicular bisector of the source
    b. air kerma rate, specified at 1 m from the source
    c. air kerma rate, measured in free space at 1 m from the perpendicular bisector of the source
    d. in none of the above ways

26. The AKS of a brachytherapy source has units of _____.
    a. $Gy\ m^2\ hr^{-1}$
    b. $R\ m^2\ hr^{-1}$
    c. $Sv\ m^2\ hr^{-1}$
    d. Ci/hr
    e. none of the above

27. The AKS of a brachytherapy source is 15 U. This refers to an AKS of _____.
    a. $15\ Gy\ m^2\ hr^{-1}$
    b. $15\ cGy\ cm^2\ hr^{-1}$
    c. $15\ \mu Gy\ m^2\ hr^{-1}$
    d. none of the above

28. _____ is the apparent activity (in mCi) of a ^{137}Cs source whose strength has been quoted by the manufacturer as 20 mg-Ra-eq. [Exposure rate constant (ERC) of 137Cs and ^{226}Ra (0.5 mm Pt filtration) are $3.26\ R\ cm^2/mCi.h$ and $8.25\ R\ cm^2/mg.h$, respectively.]
    a. $20 \times (8.25\ /\ 3.26)$
    b. $20 \times (3.26 \times 8.25)$
    c. 3.26×8.25
    d. 20×3.26
    e. none of the above

29. For a typical brachytherapy source, the exposure rate at any point around it is influenced by _____.
    a. distribution of radioactivity
    b. self absorption in the source
    c. attenuation characteristics of the encapsulation
    d. medium perturbation
    e. distance from the source
    f. all of the above

30. _____ is the equivalent continuous length for a train source made of 5 seeds with 0.8 cm distance between the source centers?
    a. 6.2 cm
    b. 4 cm
    c. 3.2 cm
    d. 2.8 cm

31. Match the brachytherapy source to half life.

Source		Half Life	
a. ^{137}Cs	_____	i.	5.26 years
b. ^{192}Ir	_____	ii.	30 years
c. ^{60}Co	_____	iii.	59.4 days
d. ^{125}I	_____	iv.	17 days
e. ^{103}Pd	_____	v.	74.2 days

32. Match the brachytherapy source to its average energy of emission.

Source		Average Energy	
a. ^{137}Cs	_____	i.	21 keV
b. ^{192}Ir	_____	ii.	28 keV
c. ^{60}Co	_____	iii.	0.662 MeV
d. ^{125}I	_____	iv.	1.25 MeV (1.17, 1.33 energies)
e. ^{103}Pd	_____	v.	370 keV

33. Match the brachytherapy source to its mode of decay.

    Source   *Mode of Decay*
    a. ^{137}Cs   _____ i.   electron capture
    b. ^{192}Ir   _____ ii. β^- / γ
    c. ^{60}Co
    d. ^{125}I
    e. ^{103}Pd

34. Radioactive nuclide ^{125}I, decaying by electron capture, produces _____.
    a. neutron emissions
    b. proton emissions
    c. characteristic x-rays
    d. gamma emissions
    e. all the above

35. A brachytherapy source is specially constructed to _____.
    a. contain the radioactivity inside the source
    b. stop betas from getting out (in the case of beta-gamma radionuclides)
    c. identify the source through imaging
    d. reduce anisotropy
    e. modify its emission spectrum

36. The clinically useful radiations produced by brachytherapy sources are _____.
    a. gamma rays
    b. beta rays
    c. characteristic x-rays
    d. Bremsstrahlung from the encapsulation
    e. neutrinos

37. Match source configurations to the sources listed below (one source may have more than one configuration).

    Sources   *Source Configurations*
    a. ^{137}Cs   _____ i.   wire
    b. ^{60}Co   _____ ii. pellet / miniature cylinder
    c. ^{192}Ir   _____ iii. tube
    d. ^{125}I   _____ iv. needle
    e. ^{203}Pd   _____ v.   seeds

38. Due to its relatively long half life, high average energy, and radiation hazard, _____ isotope is most commonly used as a radium substitute in intracavitary gynecological implants.
    a. ^{60}Co
    b. ^{137}Cs
    c. ^{192}Ir
    d. ^{198}Au
    e. ^{125}I

39. The AKS of a ^{192}Ir HDR source is 40 Gy cm^2/hr. The radiation level at 1 m from the source in 10 minutes is given (in µGy) is _____.
    a. 40
    b. 6.7
    c. 4
    d. 0.67
    e. 0.34

40. The palladium source is able to be visualized radiographically by _____.
    a. a lead marker between graphite pellets
    b. radiation coated onto a silver rod (acts as an x-ray marker)
    c. a tungsten marker between graphite pellets
    d. radiation coated on a lead rod (acts as x-ray marker)

41. The important characteristics of a permanent implant source are _____.
    a. energy
    b. half life
    c. implantation technique
    d. size
    e. decay mode

42. _____ many mCi of ^{192}Ir are in a 0.75 mg-Ra-eq seed ($\Gamma_{x,Ra}$ = 8.25 and $\Gamma_{x,Ir}$ = 4.69).
    a. 6.6
    b. 4.29
    c. 3.76
    d. 1.31
    e. 0.469

43. _____ mCi of ^{137}Cs are in a 15 mg-Ra-eq ($\Gamma_{\gamma,Cs}$ = 3.26).
    a. 124
    b. 98
    c. 38
    d. 24
    e. 64

44. 611 mg h (radium) is equal to _____ mCi hr of ^{192}Ir.
    a. 611
    b. $611 \times (8.25 \times 4.69)$
    c. $611 / (8.25 \times 4.69)$
    d. $611 \times (8.25 / 4.69)$
    e. $611 \times (8.25 / 3.26)$

45. _____ mg-Ra-eq of ^{137}Cs are in a 38 mCi source.
    a. 124
    b. 38
    c. 15
    d. 24
    e. 16.7

46. One becquerel is equal to _____ Ci.
    a. 1
    b. $10^{10} / 3.7$
    c. $1 / (3.7 \times 10^{10})$
    d. 3.7×10^{-10}
    e. $1 / (3.7 \times 10^{-10})$

47. One mg h corresponds to a total reference air kerma (TRAK) of _____ µGy/h @ 1 m.
     a. 1
     b. 8.25
     c. 7.22 / 4.69
     d. 7.22
     e. 8.5 × 3.2

48. At a point 5 cm from a source, the exposure rate is 3.3 R/hr. _____ is the mg-Ra-eq of this source.
     a. 26
     b. 13
     c. 10
     d. 8.25
     e. 4.26

49. 30 mCi of ^{137}Cs is expressed as _____ mg-Ra-eq.
     a. 11.85
     b. 37
     c. 3.2
     d. 0.469
     e. 0.038

50. The mg-Ra-eq exposure rate converts to _____ µGy in SI units.
     a. 8.25
     b. 7.64
     c. 7.22
     d. 3.22
     e. 0.87

51. To convert exposure rate constant (in R) to air kerma rate constant (in cGy), one must use a conversion factor of _____.
     a. 6.25
     b. 0.95
     c. 0.876
     d. 0.765
     e. 0.4

52. The Γ_x factor is expressed in units of _____.
     a. (R-cm^2) / mCi-hr
     b. cGy/mCi-hr @ 1 cm
     c. R/mCi-hr @ 1 cm
     d. (cGy-cm^2) (mCi-hr)
     e. µGy/mCi-hr @ 1 m

53. The air kerma rate constant _____.
     a. is unique for each radioactive isotope
     b. can be used to calculate the air kerma rate output of a point source of known activity
     c. can be used to calculate the air kerma rate output of an encapsulated spherical source of known apparent activity
     d. can be used to calculate half life
     e. is stated in units of roentgens

54. The exposure rate at 50 cm from a 30 Ci ^{137}Cs source (in R/hr) is given by _____.
    a. 1.98
    b. 0.04
    c. 40
    d. 80
    e. 137

55. The exposure rate at 30 cm from a 10 Ci ^{192}Ir HDR source is given by _____.
    a. 80.2 R.hr
    b. 20.6 R/hr
    c. 400 mR/hr
    d. 52 m R/hr
    e. 12 mR/hr

56. _____ is the exposure rate (in mR/hr) at 30 cm from a 3 mCi ^{125}I source ($\Gamma_{x,I}$ = 1.46 R cm^2/ hr mCi).
    a. 49.6
    b. 4.9
    c. 4.9×10^3
    d. 4.9×10^3
    e. 8.25×10^{-3}

57. _____ are the units used for denoting air kerma strength.
    a. U
    b. mGy
    c. Ci
    d. Bq
    e. Sv a

58. Expressing in terms of mg-Ra-eq, 1 U = _____ mg-Ra-eq.
    a. 1
    b. 1 / 8.25
    c. 7.227
    d. 1 / 7.227
    e. 8.25 / 7.227

59. One U is equal to _____ mCi for ^{138}Cs.
    a. 3.26 / 8.25
    b. 8.25 / 3.16
    c. (8.25 / 3.26) / 0.138
    d. (8.25 / 3.26) × 0.138

60. The units of TRAK are _____.
    a. µGy at 1m
    b. µGy / hr at 1m
    c. µGy m^2 / hr at 1m
    d. µGy m^2
    e. R m^2

61. Regarding the specification of gamma ray sources, ICRU 38 and ICRU 58 say brachytherapy gamma ray sources must be specified in terms of _____.
    a. air kerma strength
    b. reference air kerma rate
    c. exposure strength
    d. dose in water at 1 cm from the source center
    e. becquerels

62. Reference air kerma rate (RAKR) is defined as the air kerma rate _____.
    a. in a reference material
    b for a reference source
    c. in air at a reference distance $d_{ref} = 1$ m (in the absence of attenuation and scattering)
    d. in air at 1 cm distance
    e. in air at 30 cm distance

63. Air kerma strength is defined as the measured air kerma rate of a source _____.
    a. at a reference distance of 25 cm in air
    b at 1 cm in water medium
    c. at 1 m in tissue medium
    d. in air, at any convenient distance, not necessarily at 1 m, corrected for attenuation and scattering, and specified at 1 m distance.

64. The units of integrated reference air kerma (IRAK) are _____.
    a. μGy at 1 m
    b. μGy / hr at 1 m
    c. μGy m^2 / hr at 1 m
    d. μGy m^2
    e. the same as total reference air kerma (TRAK)

65. IRAK is equal to _____.
    a. mg-Ra-eq /hr $\times$ 8.25
    b. mg-Ra-eq-hr $\times$ 7.22
    c. mCi $\times$ 7.22
    d. mCi / 7.22
    e. Bq $\times$ 7.22

66. Calculate the air kerma strength of a ^{192}Ir source in terms of U given the following data:
    Linear activity = 1 mCi/cm; Source length = 3 cm; AKS to mCi conversion factor = 4.2 U/mCi.
    a. 25
    b. 12.6
    c. 9
    d. 4.2
    e. 3

67. For most radionuclides used in brachytherapy with photon energies above 200 keV, the ratio $(\mu_{tr})_{water,air}$ is
    _____.
    a. 1.11
    b. 1.00
    c. 0.965
    d. 0.873
    e. 0.248

68. Beta sources used for endovascular brachytherapy must be specified in terms of _____.
    a. reference air kerma rate
    b. air kerma strength
    c. becquerels
    d. reference absorbed dose rate at a distance of 2 mm from the source center (axis)
    e. apparent activity

69. The mean life of a brachytherapy source is defined as the average time _____.
    a. for the seeds to remain in the patient
    b. for the complete decay of radioactive atoms at the same initial rate
    c. for the dose to patient to become zero
    d. to reduce the implant activity to zero

C. Dose Distributions

Choose the right answer(s) (more than one may be correct):

1. For a brachytherapy source embedded in a medium, the factors that influence the relative dose in the medium are _____.
   a. distance from source
   b. self attenuation in source
   c. medium attenuation
   d. scatter buildup.

2. For a brachytherapy seed embedded in a tissue medium, the predominant factor that influences the dose falloff with respect to distance is given by _____.
   a. self attenuation in the source encapsulation
   b. attenuation of primary in the medium
   c. scatter buildup in the medium
   d. inverse square fall in energy fluence

3. For a pellet source of ^{137}Cs or ^{192}Ir embedded in a tissue medium, the dose variation over a short distance (1 to 5 cm) _____.
   a. roughly (say, within 5%) follows the inverse square law
   b. does not follow the inverse square law due to medium attenuation and scattering
   c. is nearly constant since attenuation and scatter balance each other
   d. is much steeper than the inverse square law

4. For a pellet source of ^{103}Pd or ^{125}I embedded in a tissue medium, the dose falloff over a short distance (1 to 5 cm) roughly (say, within 5%) _____.
   a. follows inverse square law
   b. does not follow inverse square law due to medium attenuation and scattering not compensating each other
   c. is nearly constant since attenuation and scatter balance each other
   d. is much steeper than the inverse square law.

5. For a point source of known AKS embedded in a water medium, the dose at a distance "d" from the source is influenced by _____.
   a. the inverse square law
   b. medium scatter
   c. the attenuation coefficient of water
   d. none of the above

6. A ^{137}Cs source of AKS 10 U (1 U = 1 μGy m^2/hr) is embedded in a water medium. _____ is the dose rate at 5 cm from the source. Medium perturbation factor is 0.951 and the mass energy absorption coefficient, water to air, is 1.11.
   a. $10 \times (5 \times 10^{-2})^2 / (0.951 \times 1.11)$
   b. $10 \times 0.951 / \{1.11 \times (5 \times 10^{-2})^2\}$
   c. $10 \times 1.11 / \{0.951 \times (5 \times 10^{-2})\}$
   d. $10 \times 0.951 \times 1.11 / (5 \times 10^{-2})^2$

7. A ^{137}Cs source of AKS 10 U is embedded in a water medium. _____ is the dose rate at 3 cm from the source. Medium perturbation factor is 0.971 and the mass energy absorption coefficient, water to air, is 1.11.
   a. $10 \times 0.971 \times 1.11 / (3 \times 10^{-2})^2$
   b. $10 \times 0.971 / \{1.11 \times (3 \times 10^{-2})^2\}$
   c. $10 \times 1.11 / \{0.971 \times (3 \times 10^{-2})\}$
   d. $10 \times (3 \times 10^{-2})^2 / (0.971 \times 1.11)$

8. A ^{125}I source of AKS 10 U is embedded in a water medium. _____ is the dose rate at 3 cm from the source. Medium perturbation factor is 0.526 and the mass energy absorption coefficient, water to air, is 1.01.
   a. $10 \times 0.526 / \{1.01 \times (3 \times 10^{-2})^2\}$
   b. $10 \times 0.526 \times 1.01 / (3 \times 10^{-2})^2$
   c. $10 \times 1.01 / \{0.526 \times (3 \times 10^{-2})\}$
   d. $10 \times (3 \times 10^{-2})^2 / (0.526 \times 1.01)$

9. In the Manchester system, a source plane with distribution of sources according to the Manchester rules delivers a dose _____.
   a. of 6000 R in 6 days to the target volume
   b. that is 10% higher than the minimum dose to the target
   c. that is uniform (to within 5%) in the target treated
   d. not as per the above statements

10. The Manchester system of interstitial implantation technique _____.
    a. was originally developed for ^{226}Ra sources
    b. was originally developed for ^{137}Cs and ^{60}Co sources
    c. is applicable to all sources that are in use in brachytherapy
    d. is not suitable for brachytherapy

11. For a single plane implant in the Manchester system, the treatment plane is parallel to the source or implant plane at a distance of _____.
    a. 5 mm on either side of source plane
    b. 5 mm on one side of source plane
    c. 10 mm on either side of source plane
    d. 10 mm on one side of source plane

12. The basic needle arrangement for a single plane implant in the Manchester system consists of _____.
    a. a parallel array of needles across the area of implant
    b. 1 cm spacing between adjacent needles
    c. crossing of needles at ends
    d. no correction for rectangular implants

13. The surface area of an interstitial implant using the Manchester system _____.
    a. is defined by the active lengths of peripheral needles and the distance between them
    b. is always equal to the treatment surface area
    c. treats less treatment area if the needles are not crossed
    d. treats any volume
    e. activity distribution is uniform in the implanted area

14. The Manchester system of dosimetry _____.
    a. gives the amount of activity required to deliver prescribed dose for the target area or target volume in question
    b. gives the rules for distribution of this activity across the implant surface or volume
    c. gives a dose uniformity within ±10% of the stated dose
    d. is not applicable to ^{60}Co or ^{137}Cs sources

15. The Paris system was developed for _____.
    a. all the Ra substitutes for which the Manchester system was inapplicable
    b. dosimetry of intracavitary applications
    c. temporary implants of ^{192}Ir wire
    d. permanent implants

16. The Paris system of dosimetry can be used for interstitial implants that utilize _____.
    a. ^{192}Ir wires
    b. ^{192}Ir ribbons
    c. ^{192}Ir HDR source
    d. none of the above sources

17. The Paris system of interstitial implant makes use of ^{192}Ir wires _____.
    a. of uniform linear exposure or air kerma rate
    b. of equal length and uniform spacing
    c. parallel to one another
    d. in one or more parallel (uniformly spaced) planes

18. In the Paris system of volume implant, the source points in the central plane form _____.
    a. equilateral triangles
    b. right angle triangles
    c. squares
    d. rectangles

19. In the Paris system, the basal dose rates (BDRs) are defined in the central plane for _____.
    a. midpoints of adjacent source points, for single plane implant
    b. source points in multiplanar implants
    c. centroids of squares or equilateral triangles formed by source points in multiplanar implants
    d. none of the above

20. The reference dose rate (RDR) in the Paris system of dosimetry is _____.
    a. mean basal dose rate
    b. 90% of mean basal dose rate
    c. 85% of mean basal dose rate
    d. none of these values

21. The reference isodose surface in the Paris system refers to _____.
    a. dose rate at a reference depth from any one of the sources in the implant
    b. the dose rate of the implant
    c. the isodose surface along which the dose rate is RDR
    d. the isodose surface that just encloses the PTV

22. According ICRU-38, the reference point of rectum for dose limiting or prescription is _____.
    a. at 0.75 cm posterior to posterior vaginal wall
    b. at 1.0 cm posterior to posterior vaginal wall
    c. at 0.5 cm posterior to posterior vaginal wall

23. Dose prescription in the classical Manchester system of dosage in gynecological treatment is _____.
    a. 8000 cGy delivered over 8 days
    b. 7000 cGy delivered over 7 days
    c. 6000 cGy delivered over 7 days
    d. 7000 cGy delivered over 6 days

24. The basal dose rate in the Paris system represents the dose rate in the central plane of the implanted volume and is defined as the _____.
    a. dose rate at a point in the center of a group of adjacent source points where the dose rate is highest
    b. dose rate at a point in the central region of a group of source points where the dose rate is lowest
    c. dose rate at a point that is outside the implant volume by 0.5 cm
    d. none of the above

25. U-shaped hairpin bend implants are done in _____.
    a. Ca-prostate
    b. Ca-breast
    c. Ca-base of tongue/floor of mouth
    d. brain tumors

26. The one specification that defines dose rates for LDR, MDR, and HDR units in brachytherapy is _____.
    a. 0.4–3.0 Gy/hr = LDR, 2–15 Gy/hr = MDR, HDR is >10 Gy/hr
    b. 0.4–1 Gy/hr = LDR, 1–10 Gy/hr = MDR, HDR is >10 Gy/hr
    c. 0.4–2 Gy/hr = LDR, 2–12 Gy/hr = MDR, HDR is >12 Gy/hr
    d. 0.4–5Gy/hr = LDR, 5–15 Gy/hr = MDR, HDR is >15 Gy/hr

27. According to the Paris system of implant dosimetry, the reference dose rate for the purpose of treatment dose prescription is _____.
    a. 80% of mean basal dose rate
    b. 85% of mean basal dose rate
    c. 90% of mean basal dose rate
    d. 75% of mean basal dose rate

28. The clinical definition of "point A" is _____.
    a. it represents ureter's tolerance
    b. it represents tolerance of uterus
    c. the area where the uterine vessel crosses the ureter, which is the limiting factor for dose prescription
    d. none of the above

29. The dose rate at "point B" indicates _____.
    a. dose to rectum
    b. limiting dose to bladder
    c. the dose to pelvic wall and obturator node
    d. none of the above

30. Which one of the statements regarding the Quimby system is correct?
    a. It tells that the implant delivers more radiation (mg-Ra-Eq-hr) per unit stated dose than an equivalent Manchester implant.
    b. It recommends more mg-Ra-Eq-hr/1000R than the Manchester system.
    c. The Quimby system is only for volume implant.
    d. The Quimby system is only for planar implant.

31. The dose distributions around brachytherapy sources are influenced by _____.
    a. emitted radiation
    b. embedded medium
    c. source construction details
    d. source
    e. all the above

32. The air kerma rate at a point "r" from a point source in free space varies _____.
    a. as $1/r$
    b. as $1/r^2$
    c. as $1/r^3$
    d. with the angular position of the measurement point

33. _____ are the factors that determine the dose at a point $P(r,\theta)$ for an encapsulated line source.
    a. air kerma strength
    b. dose rate constant
    c. radial dose function
    d. anisotropy function
    e. atomic number of the source
    f. all of the above

34. _____ accounts for the effects of absorption and scatter in the embedded medium along the transverse axis of the source.
    a. Radial isotropy function
    b. Radial dose function
    c. Geometry factor
    d. Air kerma strength
    e. Buildup function

35. The revision of the definition of AKS in the TG-43 update introduced which of the following modifications?
    a. Photons of energy $<\delta$ keV are excluded from the definition of AKS.
    b. Measurements must be made in "free space" or in vacuum.
    c. Measurements must be made at 1 cm distance.
    d. Measurements must be made in water medium.
    e. All of the above are true.

36. A point source approximation can be applied to a linear source if the dose point of interest is _____.
    a. very close to the source
    b. along the axis of the source
    c. along the perpendicular bisector of the source
    d. at a distance d >> source length
    e. anywhere provided the radioactivity is uniformly distributed across the source d

37. The dose rate constant of a seed source is defined at 1 cm distance from the source along its _____.
    a. longitudinal axis, in a water phantom
    b. longitudinal axis, in air
    c. transverse axis, in a water phantom
    d. transverse axis, in air
    e. oblique axis at 120 degrees, in a water phantom

38. The dose rate constant is dependent on _____.
    a. source encapsulation material and thickness
    b. spatial distribution of radioactivity in the source
    c. self attenuation in the source
    d. medium attenuation and scattering
    e. angular distribution of source output

39. _____ is the dose rate in water at 1 cm from a ^{192}Ir wire given the following data:
    Wire length = 5 cm; LAKS = 4.2 U/cm; Γ = 0.521 (cGy/hr)/U.
    a. 26.84
    b. 13.69
    c. 10.94
    d. 8.0
    e. 4.97

40. The geometry factor for a point source is given by _____.
    a. $1 / r^2$
    b. $\beta / (L \times r \times \sin \theta)$
    c. $1 / r^3$
    d. r^2

41. _____ is the ratio of geometric factors for a 3 cm long source at distances of 0.5 cm and 1.0 cm.
    a. $0.5 / 1$
    b. $(0.5 / 1)^2$
    c. $(1 / 0.5)$
    d. $(1 / 0.5)^2$
    e. $<(1 / 0.5)^2$

42. Which of the following is true about anisotropy?
    a. Point sources are also anisotropic.
    b. It exists for all brachytherapy sources.
    c. It is caused by source shape and construction characteristics.
    d. It is more pronounced at the top ends of a linear source.
    e. It does not influence dosimetry.

43. The radial dose function for a brachytherapy source used in TG-43 formalism _____.
    a. accounts for attenuation and scattering in the medium along the perpendicular bisector of the medium
    b. is analogous to Meisberger correction in the conventional formalism
    c. gives the dose variation around the source
    d. is actually the dose rate constant
    e. is "air kerma to dose in the medium" conversion factor

44. The value of the anisotropy constant is _____.
     a. >1
     b. =1
     c. <1
     e. >1 for higher energy sources and <1 for lower energy sources

D. Dose Calculations

Choose the right answer(s) (more than one may be correct):

1. Using the TG-43 formalism, _____ is the dose rate in tissue (in cGy/h) to a point 5 cm from a 0.5 mg Ra-eq ^{192}Ir seed on a line perpendicular to the long axis of the seed, using the following data: Source strength = 0.5 mg Ra-eq; 1 mg Ra-eq = 7.22 U; Dose rate constant = 1.12; Radial dose function = 0.996; anisotropy factor = 0.98.
    a. 2.0
    b. 1.23
    c. 0.74
    d. 0.16
    e. 0.04

2. Using conventional formalism, _____ is the dose rate in tissue (in cGy/h) to a point 5 cm from a 0.5 mg Ra-eq ^{192}Ir seed on a line perpendicular to the long axis of the seed using the following data: Source strength = 0.5 mg Ra-eq; $\Gamma_{x,Ra}$ = 8.25 R cm^2/ mg-Ra-hr; f_{med} = 0.957; Meisberger coefficient for ^{192}Ir at 5 cm M(5) = 1.006.
    a. 2.4
    b. 1.74
    c. 0.16
    d. 0. 08
    e. 0.006

3. The following figure shows the air kerma rate at a point near a typical brachytherapy source in air. Water kerma at the same point P (δm of air replaced by δm of water) is given by [Kw(r, θ)] air. [Kw(r, θ)]air / [Ka(r, θ)]air is given by _____.
    a. $(\mu_w) / (\mu_{air})$
    b. $(\mu_{tr})_w / (\mu_{tr})_{air}$
    c. $(\mu_{en})_w / (\mu_{en})_{air}$
    d. $(\mu_{tr})_{air} / (\mu_{tr})_w$
    e. $(\mu_{en})_{air} / (\mu_{en})_w$

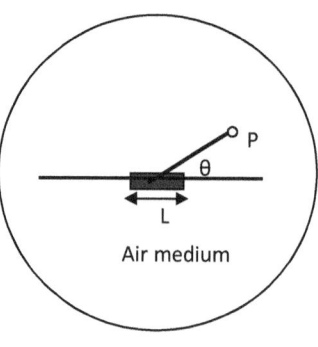

$[K_a(r, \theta)]_{air}$

4. The following figure shows the air kerma rate at a point near a typical brachytherapy source in air. The air medium is replaced by water medium (in figure B) except for the δm of air giving air kerma in water medium, $[K_a(r, \theta)]_w$. $[K_a(r, \theta)]_w / [K_a(r, \theta)]_{air}$ is given by _____.
   a. $(\mu_w) / (\mu_{air})$
   b. $(\mu_{tr})_w / (\mu_{tr})_{air}$
   c. Meisberger function in the conventional dosimetry formalism
   d. a factor that corrects for the attenuation and scatter introduced by the water medium
   e. $(\mu_{en})_{air} / (\mu_{en})_w$

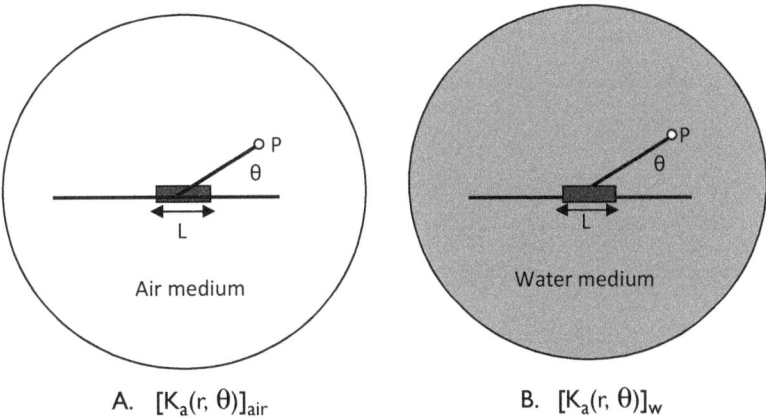

A. $[K_a(r, \theta)]_{air}$ B. $[K_a(r, \theta)]_w$

5. The dose rate at P, D (r, θ) from a brachytherapy source seed in a water medium is given by _____.
   a. S_K
   b. $S_K \Lambda$
   c. $S_K \Lambda / r^2$
   d. $S_K \Lambda G(r, \theta) g(r) F(r, \theta)$
   e. $S_K \Lambda [G(r, \theta) / G(r_0, \theta_0)] g(r) F(r, \theta)$

E. Source Localization

Circle the right answer (Yes or No):

1. (Yes / No)   Source localization refers to the determination of the three-dimensional coordinate and the orientation of each source relative to the patient anatomy.

2. (Yes / No)   CT based source localization is inferior to orthogonal films method.

3. (Yes / No)   Postoperative imaging can often help to improve dosimetry in brachytherapy.

4. (Yes / No)   An A/P or lateral film with known magnification factor for the film is enough to reconstruct the 3D coordinates of the seeds of a brachytherapy implant.

5. (Yes / No)   3D dose optimization and conventional x-ray based planning lead to same probability of tumor control or OAR sparing in brachytherapy.

6. (Yes / No)   MRI is superior to CT in the delineation of GTVs in gynecological cancers.

Choose the right answer(s) (more than one may be correct):

7. To calculate isodose distributions in brachytherapy treatment, _____.
   a. the geometry (position and orientation) of the implant sources must be known
   b. a minimum of two radiographs of the site at different angles is required
   c. a treatment planning system is required
   d. delineation of PTV is necessary

8. The geometric reconstruction of source positions generally involves _____.
   a. obtaining two orthogonal radiographs by isocentric method
   b. obtaining source positions from the image positions on radiographs by similar triangles (geometric reconstruction)
   c. excellent results even if the patient moves between the two radiographs
   d. sometimes problems in identification of sources on radiographs

9. Source localization gives _____.
   a. 3D coordinates of the source end points
   b. source orientation
   c. image magnification
   d. none of the above

10. Source localization in cervix treatment requires _____.
    a. at least one localization film
    b. at least two films, not necessarily orthogonal
    c. at least three films
    d. none of the above

11. In source localization techniques using films, _____.
    a. orthogonal films offer the best accuracy
    b. stereo-shift films offer the best accuracy
    c. non-orthogonal films offer the best accuracy
    d. target area can also be identified

12. CT-based source localization would help in _____.
    a. target volume identification
    b. source localization
    c. dose planning
    d. improved accuracy in treatment planning
    e. none of the above

13. A palladium source is able to be visualized radiographically by _____.
    a. a lead marker between graphite pellets
    b. radiation coated onto a silver rod (acts as an x-ray marker)
    c. a tungsten marker between graphite pellets
    d. radiation coated on a lead rod (acts as an x-ray marker)

14. The imaging methods used for imaging permanent implants are _____.
    a. spectroscopy
    b. MRI
    c. PET
    d. port films
    e. infrared imaging

15. The _____ imaging modality is limited to only axial slices.
    a. CT
    b. x-ray film
    c. ultrasound
    d. MRI

16. The _____ imaging modality was used to take this image of a permanent implant.
    a. x-ray machine
    b. ultrasound
    c. MRI
    d. PET
    e. CT

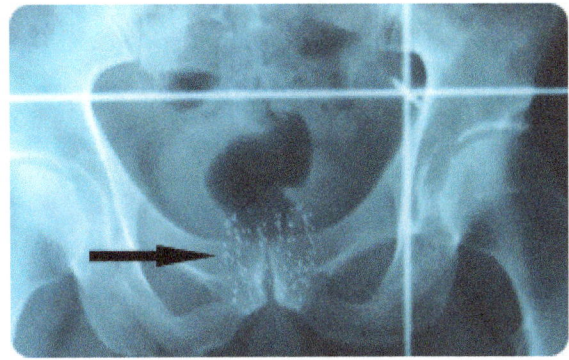

17. The _____ imaging modality was used to take this image which shows Point A and target dose for an intracavitary treatment.
    a. x-ray machine
    b. ultrasound
    c. MRI
    d. PET
    e. CT

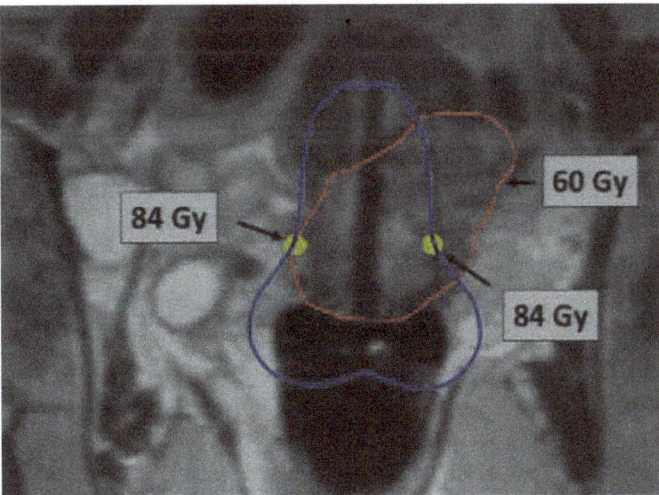

18. The _____ imaging modality was used to take this image which shows an interstitial multicatheter brachytherapy implant.
    a. x-ray machine
    b. ultrasound
    c. MRI
    d. PET
    e. CT

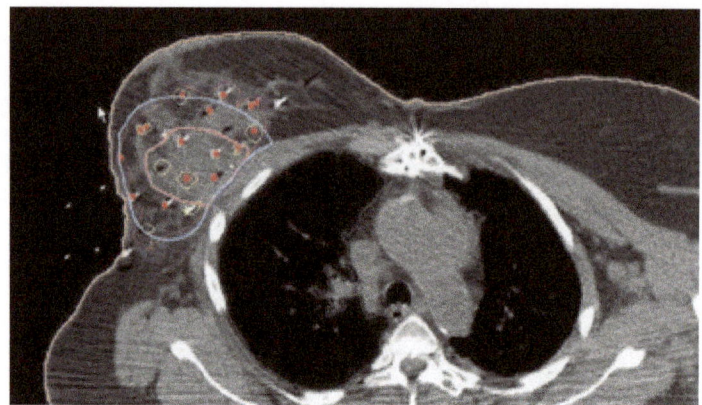

19. The _____ imaging modality was used to take this image which shows a scan of the prostate gland.
    a. CT
    b. x-ray films
    c. ultrasound
    d. MRI
    e. PET

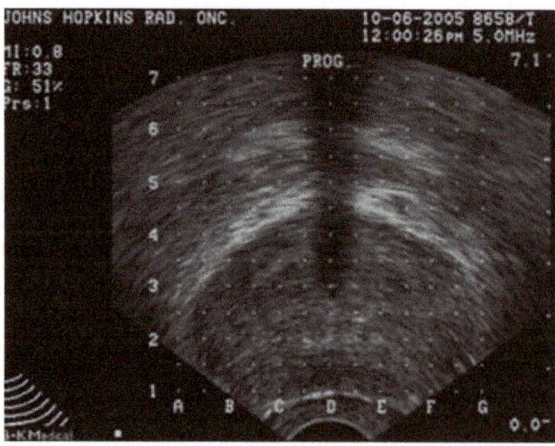

20. The imaging modalities that show high temporal and contrast resolution are _____.
    a. CT
    b. x-ray films
    c. ultrasound
    d. MRI
    e. PET

21. The imaging modalities used for real-time imaging include _____.
    a. CT
    b. x-ray films
    c. ultrasound
    d. MRI
    e. PET

22. The imaging modalities most limited by the expertise of the user include _____.
    a. CT
    b. x-ray films
    c. ultrasound
    d. MRI
    e. PET

23. The imaging modalities capable of visualizing zonal anatomy of the prostate include _____.
    a. CT
    b. x-ray films
    c. ultrasound
    d. MRI

24. Most of the brachytherapy centers still make use of brachytherapy planning using _____.
    a. plain x-ray films
    b. ultrasound
    c. CT
    d. MRI
    e. CT + MRI

25. The imaging modality normally used during an implant is _____.
    a. CT
    b. fluoroscopy
    c. ultrasound
    d. MRI

26. _____ is normally used for preplanning.
    a. CT
    b. Fluoroscopy
    c. Ultrasound
    d. MRI

27. _____ is normally used for post-planning.
    a. CT
    b. Fluoroscopy
    c. Ultrasound
    d. MRI

28. In a pair of orthogonal radiographs of a gynecological insertion, the coordinates of two points on the radiographs were determined as follows:

Points	Coordinates		
	X	Y	Z
Point 1	1.5	0	2
Point 2	0	3	4.5

    The distance between the two points (in cm) is _____.
    a. 17.5
    b. 11.0
    c. 7.0
    d. 4.2
    e. 3

29. Implant geometry can be reconstructed by _____.
    a. single x-ray film from any angle
    b. MRI imaging
    c. orthogonal films
    d. stereo-shift films
    e. CT imaging

30. Match the numbered locations to the organs/sources from the figure which represents treatment of cervix cancer.

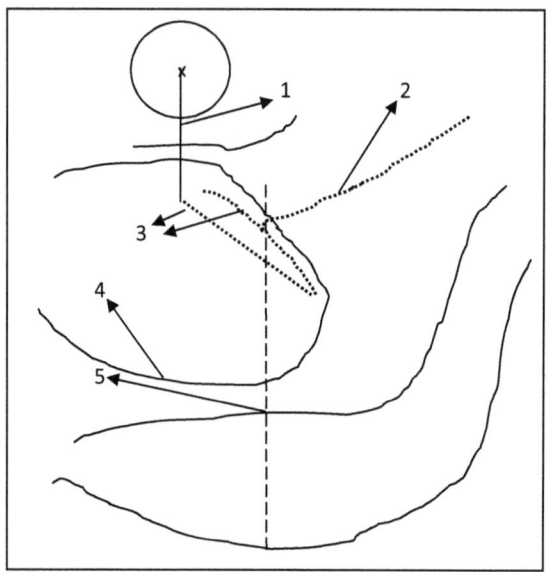

Intracavitary brachytherapy for cervix cancer

Location	*Organs*
a. Point 1	_____ i. intrauterine tube
b. Point 2	_____ ii. intravaginal sources
c. Point 3	_____ iii. rectum reference point (ICRU 38)
d. Point 4	_____ iv. bladder reference point (ICRU 38)
e. Point 5	_____ v. vaginal posterior wall

31. Drawbacks of ICRU reference point doses for OARs include _____.
    a. there are no disadvantages
    b. due to steep dose gradients, point doses are not representative of whole organ doses
    c. ICRU points do not provide information on the volumes of OARs that receive significant doses
    d. 2D imaging cannot provide much of the required information and 3D treatment planning is required
    e. reference points chosen elsewhere (e.g., at the center of bladder and rectum) would have provided better control over reducing complications

F. Interstitial Dosimetry

Choose the right answer(s) (more than one may be correct):

1. Brachytherapy techniques can be classified as _____.
    a. surface mold treatments
    b. interstitial treatments
    c. intralumenal therapy
    d. intracavitary therapy

2. The length of the target volume to be treated with an interstitial implant of iridium wires is 4.5 cm. _____ is the active length that should be used.
    a. 3.15 cm
    b. 4.5 cm
    c. 6.4 cm
    d. 2.25 cm

3. The dose delivered by a brachytherapy source implant depends on _____.
    a. total air kerma strength
    b. the duration of implant
    c. perturbation by the implant medium
    d. none of the above

4. The parameter that most significantly influences the dose near a brachytherapy source in an implant medium is _____.
    a. the encapsulation
    b. the energy
    c. the inverse square law fall in photon energy fluence
    d. none of the above

5. For an ^{125}I source (model 6711 seed), the dose distribution in the source plane is not isotropic because of _____.
    a. oblique filtration in source encapsulation
    b. excessive attenuation along the axis due to source encapsulation welding
    c. implant medium
    d. none of the above

6. For brachytherapy sources with an average photon energy greater than about 200 keV, the dose fall with respect to distance (>1 cm) from the source, in the embedded medium, roughly follows inverse square law because of _____.
    a. large half life
    b. no attenuation and scattering in the embedded medium
    c. attenuation and scatter buildup roughly compensating each other
    d. none of the above

7. Sources presently in use for permanent implants are _____.
    a. ^{192}Ir seeds
    b. ^{137}Cs seeds
    c. ^{125}I seeds
    d. ^{103}Pd seeds

8. Brachytherapy is practiced in cases of _____.
    a. skin cancers
    b. interstitial tumors
    c. tumors in accessible body cavities
    d. brain tumors only

9. In the treatment of skin cancer, the sources are kept _____.
    a. directly on the skin
    b. at a distance of 5 mm or 1 cm from the skin
    c. on a metallic spacer to maintain the source-to-skin distance
    d. on a tissue-equivalent spacer to maintain distance

10. Skin cancers can be treated by _____.
    a. brachytherapy sources
    b. kV x-rays
    c. low-energy electron beams
    d. ^{60}Co beams

11. Sources used in interstitial therapy are in the form of _____.
    a. tubes
    b. spheres
    c. needles
    d. wires
    e. ribbons
    f. seeds

12. Interstitial implants _____.
    a. are only temporary implants
    b. can be temporary or permanent implants
    c. involve direct insertion of sources into the tumor
    d. involve placement of applicators in the tumor volume before source insertion

13. The advantages of manual afterloading (MAL) procedure are _____.
    a. the elimination of exposure received by radiation oncologist during placement of afterloading catheters
    b. better accuracy (any error in catheter placement can be rectified during localization)
    c. higher activity sources can be made use of
    d. nursing staff also receives no exposure

14. A 7 cm^2 surface area is to be treated using Ra (0.5 mm Pt filtration) sources placed on an equal area 1.0 cm away. To deliver 60 Gy in 5 days, _____ activity of radium is required. The Manchester table gives the dose delivered by this surface mould for the skin surface at 1 cm treatment distance as 382 mg h / 10 Gy.
    a. 60×382
    b. $10 \times 382 / (5 \times 24)$
    c. $60 \times 382 \times (5 \times 24)$
    d. $6 \times 382 / (5 \times 24)$
    e. $6 \times (5 \times 24) / 382$

15. An implant containing two or more sources which lie in the same plane is known as a _____.
    a. brachy implant
    b. temporary implant
    c. single-plane implant
    d. nonplanar implant
    e. volume implant

16. An implant containing two planes which are generally parallel to each other is known as a _____.
    a. two-plane implant
    b. temporary implant
    c. single-plane implant
    d. nonplanar implant
    e. volume implant

17. If the calibration certificate of an HDR brachytherapy source quotes the calibration in apparent activity, and if the hospital well chamber is calibrated in AKS, how will you verify the manufacturer calibration?
    a. There is no way it can be verified since the calibrations are in two different units.
    b. The manufacturer must be asked to recalibrate the sources in AKS.
    c. Knowing the air kerma rate constant (AKRC) of the ^{192}Ir source, the apparent activity can be calculated.
    d. The AKS is numerically equal to apparent activity to within 3%, so verification is possible.
    e. The hospital can easily calibrate the well chamber in apparent activity and then verify the certificate value.

18. The most common methods for ^{192}Ir HDR source calibration are _____.
    a. a well-type chamber
    b. a 0.6 cc chamber and calibration jig for positioning chamber at 10 cm from source
    c. a parallel plate chamber
    d. an extrapolation chamber
    e. all of the above

19. A well-type chamber _____.
    a. has sensitivity that is dependent on the source depth in the well
    b. must be calibrated with respect to its peak response depth
    c. must be calibrated only for one type of source
    d. must be calibrated in terms of source apparent activity
    e. cannot be calibrated for wire-type sources

20. A ^{192}Ir ribbon can be treated as identical to a ^{192}Ir wire if the spacing between the seeds is _____ where the active length of seed is assumed to be L.
    a. equal to L
    b. >1.5 L
    c. ≤1.5 L
    d. ≥0.5 L
    e. <0.5 L

21. If the length of an implant is 7 cm, a _____ cm source length must be used to give uniform dose to the target volume.
    a. 14
    b. 10
    c. 7
    d. 4.9
    e. 3

22. Match thickness of implant to the implant configuration.

Implant Thickness	*Implant Configuration*
a. 0.5 source separation	_____ i. two-plane triangular
b. 1.2 source separation	_____ ii. two-plane square
c. 1.5 source separation	_____ iii. single-plane triangular

23. In a single-plane Paris system of implant, the intersource distance is 14 mm. The implant thickness (in mm) is given by _____.
    a. 21
    b. 16.8
    c. 14
    d. 7
    e. 0.5

24. For single-plane Paris system of implant, the source separation _____.
    a. is related to target thickness
    b. for a two-wire implant is "2T," with T being the target thickness
    c. for >2 wire implants is "1.7T"
    d. can increase to any extent

25. In a two-plane triangular implant (Paris system) the intersource distance is 16 mm. The implant thickness (in mm) is given by _____.
    a. 32
    b. 24
    c. 19.2
    d. 14
    e. 7

26. In a two-plane square implant (Paris system) the intersource distance is 12 mm. The implant thickness (in mm) is given by _____.
    a. 36
    b. 24
    c. 18
    d. 12
    e. 6

27. In Paris system of dosimetry, the maximum tissue thickness (in mm) allowed for a single-plan implant is _____.
    a. 15
    b. 12
    c. 10
    d. 8
    e. 5

28. The prescription dose rate in a Paris system of interstitial implant is typically 60 cGy/hr. _____ is the typical basal dose rate of the implant in cGy/hr.
    a. 96.74
    b. 70.5
    c. 60.0
    d. 51.0
    e. 30.47

29. Often the Paris system of implant is not accomplished strictly according to the rules of the system. Such plans are considered satisfactory and taken as acceptable provided the basal dose rates are within ±_____% of the mean basal dose rate of the implant.
    a. 10
    b. 13.6
    c. 15
    d. 25
    e. 100

30. Match the implants from the following figure to the quality in the Paris system.

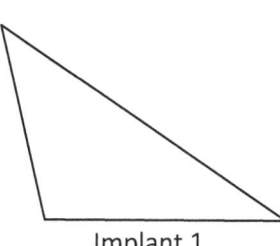 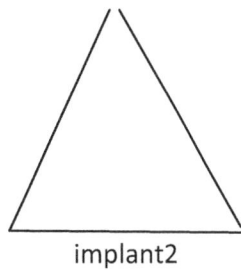 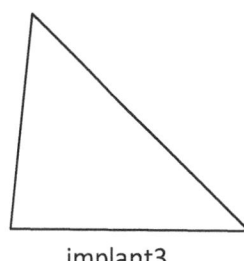

Implant 1   implant2   implant3

Source points in central plane for double plane triangular implant

Implant   *Quality*
a. implant 1   _____ i. slight deviation from rules, acceptable
b. implant 2   _____ ii. large deviation from rules, unacceptable
c. implant 3   _____ iii. as per the rules, ideal implant

31. In Paris system of interstitial implant, _____.
    a. larger spacing between wires are used for larger target thicknesses
    b. larger spacing between wires improves dose uniformity in the implant volume
    c. a spacing limit of 15–22 mm is made use of
    d. no minimum spacing is prescribed
    e. there is a sleeve of high-dose region surrounding each wire source
    f. the sleeve of high-dose region remains the same for all implants

32. A tissue target volume is to be treated as per the Paris system of dosimetry. Answer the following question from the given data:

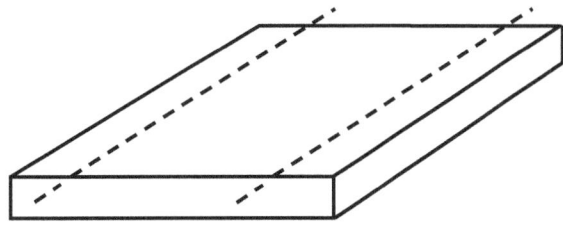

Target length L = 20 mm; target width = 15 mm; target thickness T = 5 mm;
Prescription dose to the target = 34 Gy; LAKS of the ^{192}Ir source = 0.85 µGy m²/hr m.
The implant in question is a _____.
    a. single-plane implant
    b. double-plane triangular implant
    c. double-plane square implant
    d. triple-plane triangular implant
    e. triple-plane square implant

33. For the figure implant shown above, the spacing between the wires (in mm) is _____.
    a. 16
    b. 15.3
    c. 10
    d. 8
    e. 4

34. For the figure implant shown above, the length of the ^{192}Ir wire (in mm) to be used is _____.
    a. 32.6
    b. 28
    c. 20
    d. 14.6
    e. 8.4

35. For the figure implant shown above, the lateral margin on either side (in mm) is _____.
    a. 6.0
    b. 4.9
    c. 3.7
    d. 2.0
    e. 1.4

36. For the figure implant shown below in the central plane, the treatment width (in cm) is given by _____.
    a. 20
    b. 17.4
    c. 10
    d. 9.5
    e. 5.5

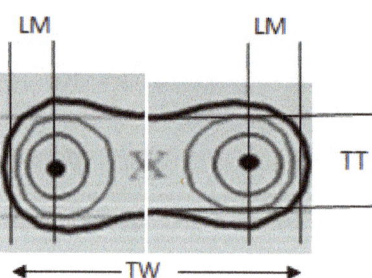

Treatment thickness (TT), treatment width (TW), and lateral margin (LM)
(Courtesy of the Nucletron Company.)

37. The treatment width covered by the prescription dose as shown in the central plane in the figure above is given (in mm) by _____.
    a. 10
    b. 13.4
    c. 17.4
    d. 15.0
    e. 20.0

38. A 2.8 cm length of ^{192}Ir 0.3 mm wire, with a linear air kerma strength of 1 U/cm, at 5 mm distance from the wire in the central plane gives a dose rate of 5.46 cGy/hr. _____ is the dose rate (in cGy/hr) at the same point for a wire of LAKS = 1 mCi/cm. (The unit conversion factor for the ^{192}Ir wire = 4.2 U/mCi.)
    a. 22.9
    b. 24.8
    c. 5.4
    d. 4.2
    e. 2.9

39. _____ is the mean basal dose rate (MBDR) in (cGy/hr) for two 2.8 cm length wires of linear activity of 1 mCi/cm 1 cm apart in the given implant (see the figure below). (The unit conversion factor for the ^{190}Ir wire = 4.2 U/mCi.)
    a. 56.9
    b. 45.8
    c. 34.6
    d. 22.6
    e. 11.3

Two-wire, single-plan implant

40. From the earlier problem, DR1 (W1) = DR2 (W2) = 22.9 cGy/hr and BDR = DR1 + DR2. _____ is the prescription dose rate (in cGy/hr) in the above problem.
    a. 46.2
    b. 42.8
    c. 38.9
    d. 26
    e. 22.6

41. If the desired prescription dose rate (PDR) is 60 cGy/hr, a wire of linear source strength = _____ U/cm wire would be required for the above implant. (For LAKS of 1 U/cm, the dose rate at 5 mm = 5.46 cGy/hr.)
    a. 10
    b. 6.47
    c. 5.46
    d. 4.3
    e. 2.2

42. If the desired prescription dose rate (PDR) is 60 cGy/hr, a wire of linear source strength = _____ mCi/cm would be required for the above implant. (For LAK of 1 U/cm, the dose rate at 5 mm = 5.46 cGy/hr. The unit conversion factor = 4.2 U/mCi).
    a. 22.6
    b. 5.4
    c. 4.2
    d. 1.5
    e. 1

43. An iridium wire implant consisting of 4 wires, (straight and parallel), each 50 mm long with separation of 15 mm, was performed on a patient (see the figure below). Wire LAKS = 0.45 U / mm. The basal dose rates (BDRs) at P1, P2, and P3 (in Gy/hr) was determined as 0.878, 0.926, and 0.878, respectively, for LAK of wires = 1 U/mm. The BDRs (in Gy/hr) of the implant at points P1, P2, and P3 are _____.
    a. 0.395, 0.417, 0.395
    b. 0.878, 0.926, 0.878
    c. 0. 439, 0.463, 0.434
    d. 0.247, 0.435, 0.943
    e. 0.146, 0.864, 0.943

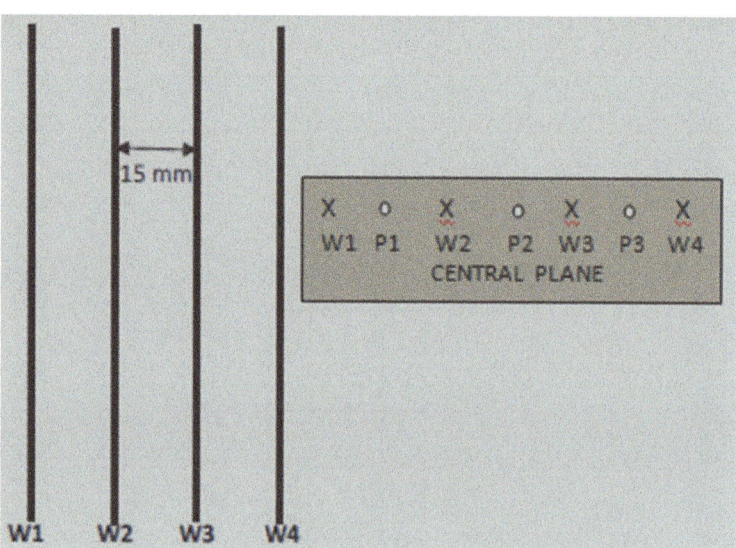

An interstitial Ir wire single plane implant in Paris system

44. For the figure above, the dose rates at P (P1, P2, etc.) is for a LAKS of 1 U/mm. If the LAKS of the source used here is 0.45 U/mm, _____ is the MBDR.
    a. 0. 417
    b. 0.402
    c. 0.395
    d. 0.201
    e. 0.1

45. For the figure above, _____ Gy/hr is the Reference Dose Rate (RDR).
    a. 0. 402
    b. 0.396
    c. 0.342
    d. 0.321
    e. 0.25

46. For the figure above, _____ is the treatment time (in hours) if a reference dose (RD) of 65 Gy has to be delivered to the implant.
    a. 200
    b. 190
    c. 150
    d. 100
    e. 85

47. A two-plane triangular implant was performed on a patient. The implant had a total of 5 wires, each 70 mm long and with a source separation 20 mm (see the figure below). The LAKS of the sources were assumed to be 1 U/mm. Using the tables for 1 U/mm LAKS wires of given length, the BDRs were calculated for P1, P2, and P3 as 0.877, 0.910, and 0.877 (Gy/h). _____ is the RDR of the implant.
    a. 0.755
    b. 0.877
    c. 0.888
    d. 1.05
    e. 1.85

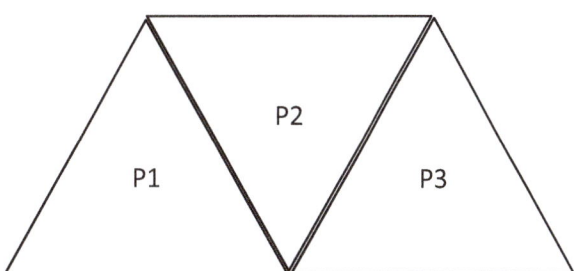

Source and BDR points in a ^{192}Ir wire double plane triangular implant in the Paris system

48. In problem number 45 above, if a dose of 25 Gy has to be delivered to the implant in 2.5 days, _____ LAKS (in U/mm) wires must be chosen for the sources used in the implant.
    a. 2
    b. 0.910
    c. 0.552
    d. 0.247
    e. 0.100

49. Match the various dosimetric parameters of the two-plane triangular implant to the numbers marked on the central plane implant configuration shown in the figure below.

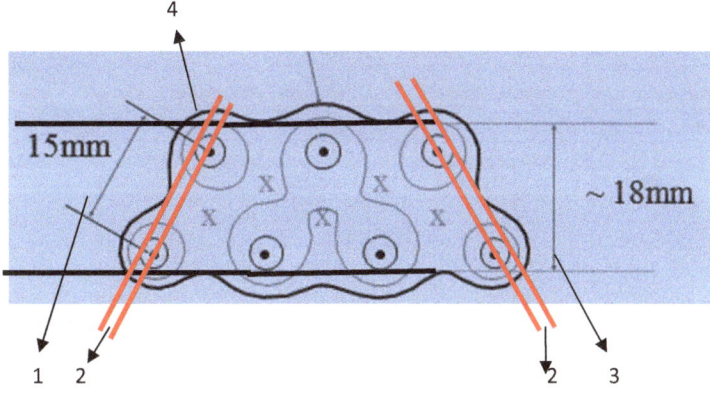

^{192}Ir wire double plane triangular implant
(Figure courtesy of Nucletron Pvt Ltd.)

Number	*Dosimetric Parameter*
a. 1	_____ i. lateral margin
b. 2	_____ ii. source separation
c. 3	_____ iii. reference isodose curve
d. 4	_____ iv. treatment thickness

50. Match the various dosimetric parameters of the single-plane implant to the numbers marked on the implant configuration shown in the figure below.

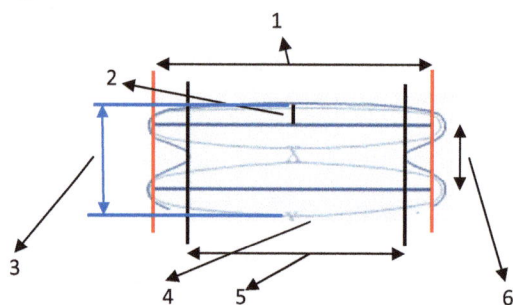

^{192}Ir wire source plane of a single plane implant
(Figure courtesy of Nucletron Pvt Ltd.)

Number		*Dosimetric Parameter*	
a.	1	_____ i.	lateral margin
b.	2	_____ ii.	source length
c.	3	_____ iii.	reference isodose curve
d.	4	_____ iv.	treatment width
e.	5	_____ v.	source separation
f.	6	_____ vi.	treatment length

51. The typical sources used in brachytherapy these days for permanent implants are _____.
  a. ^{222}Rn
  b. ^{192}Ir
  c. ^{60}Co
  d. ^{198}Au
  e. ^{125}I
  f. ^{192}Pd

52. The typical sources used in brachytherapy these days for permanent implants are _____.
  a. ^{137}Cs
  b. ^{125}I
  c. ^{32}P
  d. ^{90}Sr
  e. ^{226}Ra
  f. ^{192}Pd

53. The mean life for a ^{103}Pd seed source (half-life = 17.0 days) is given by _____.
  a. 60.20 days
  b. 34.78 days
  c. 27.00 days
  d. 24.48 days

54. The mean life for a ^{125}I source ($\lambda = 0.0115$ days^{-1}) is given by _____.
  a. 86.70 days
  b. 61.72 days
  c. 128.46 days
  d. 30.10 days

55. Permanent implants can be used _____.
    a. as monotherapy
    b. as a boost with external beam therapy
    c. as a boost with temporary implants
    d. in multiple fractions with repeated implants
    e. for any tumor size

56. Which of the following statements is true for permanent implants?
    a. Patients are discharged after implantation or within a couple of days.
    b. Seeds are not removed from the patient.
    c. Dose is delivered at low and decreasing dose rates.
    d. The bladder receives zero dose in prostate implants.
    e. Any radioactive source can be used.

57. The radiation level was measured as 1.5 µGy/h at 1 m from the ^{125}I permanent implant patient who was about to be discharged. What advice would be given to the patient so that people around him are protected from radiation?
    a. Pregnant women and persons younger than 18 must always stay at 1 m distance.
    b. The patient can freely interact with all without any restriction.
    c. Pregnant women and persons 18 or younger must stay at 1 m distance for about first two months.
    d. Only women must stay away from him for 3 months.

58. The initial dose rate of a ^{203}Pd implant is 29.5 cGy/hr. _____ cGy is the planned total dose for the target volume ($T_1/2 = 17$ days).
    a. 10,000
    b. 17,331
    c. 21,486
    d. 29,500
    e. 32,674

59. Unused ^{125}I seeds can be discarded after storing them for a period of _____.
    a. 10 half lives
    b. 20 half lives
    c. 100 half lives
    d. 20 months
    e. 10 years

60. The rapid falloff in dose with respect to distance is mainly due to _____.
    a. lower energies
    b. higher attenuation in tissues
    c. attenuation in source encapsulation
    d. beta rays
    e. the inverse square law

61. The number of seeds required for the permanent implant is determined _____.
    a. before target volume analysis
    b. after the preplan
    c. during the implant
    d. after the implant

62. The total treatment time of temporary implants depends on the _____.
    a. number of sources
    b. source strength
    c. pattern of distribution of sources
    d. experience of the radiation oncologist
    e. patient's health condition

63. A patient who had an ^{125}I permanent implants with total AKS of the implant = 86 U (about 30 mCi) comes to the hospital after a span of four years to have some new procedure done on him. Which of the following statements are NOT true?
    a. The patient cannot be examined again since he already has an implant in his body.
    b. The patient poses significant radiation risk to the physician.
    c. The physician must wear a lead apron while being with the patient.
    d. The earlier implant must be removed before examining the patient again.
    e. All the above are true.

64. A ^{125}I seed decays by about ____% in a single day.
    a. 10
    b. 5
    c. 3
    d. 1
    e. 0.1

65. A batch of ^{125}I seeds ordered by a hospital had a stated activity of 0.525 mCi per seed on arrival. The activity will be 0.486 mCi per seed after _____ days.
    a. 30
    b. 15
    c. 7
    d. 5
    e. 3

66. _____ dose (in %) would have been delivered by a ^{203}Pd implant after 51 days ($T_1/2 = 17$ days).
    a. 100
    b. 87.5
    c. 75.0
    d. 50.0
    e. 25.5

67. It takes approximately _____ to deliver more than 90% of the total dose in an ^{125}I implant ($T_{1/2} = 60$ days).
    a. 4 months
    b. 7 months
    c. 1 year and 3 months
    d. 1 year and 9 months
    e. 2 year and 4 months

68. A ^{203}Pd implant was planned to deliver 120 Gy. _____ dose (in cGy) remains to be delivered after 34 days ($T_{1/2} = 17$ days).
    a. 120
    b. 80
    c. 65
    d. 31
    e. 12.5

69. A ^{203}Pd implant was planned to deliver 124 Gy. _____ dose (in cGy) would have been delivered after 34 days ($T_{1/2} = 17$ days).
    a. 120
    b. 93
    c. 25
    e. 12.5

70. For an iodine implant of prostate gland for which 145 Gy was prescribed, a V90 of 92% means that _____.
    a. 92% of the gland received 90% of the prescription dose
    b. 90% of the gland received 92% of the prescription dose
    c. 92% of the gland received 145 Gy
    d. 10% of the gland received 90% of the prescription dose
    e. 10% of the gland received 145 Gy

71. In a prostate implant, D90 of 140 Gy means that _____.
    a. 90% of the prescription dose was received by the whole gland
    b. 90% of the gland received 140 Gy or more
    c. 90% of the gland received 90% of 140 Gy
    d. 10% of the gland received 140 Gy
    e. 10% of the gland received 90% of 140 Gy

72. Templates used in interstitial brachytherapy _____.
    a. control the direction of the needle insertion
    b. control the geometry of the implant
    c. hold needles in place during treatment
    d. improve treatment accuracy
    e. all of the above

73. A ^{192}Ir HDR interstitial treatment is delivered in 5 daily fractions. Which of the following statements are true?
    a. The AKS remains the same.
    b. The dwell time must remain the same for all fractions.
    c. The AKS × dwell time will remain constant.
    d. The dwell time must decrease as the AKS decreases

G. Intracavitary Therapy Dosimetry

Choose the right answer(s) (more than one may be correct):

1. Intracavitary therapy generally refers to treatment of cancers of _____.
    a. the cervix
    b. the vagina
    c. the uterine body
    d. lungs

2. The reference point for dose prescription in intracavitary therapy is _____.
    a. point A
    b. point B
    c. point on uterus
    d. point on vaginal wall

3. The dose prescribed for an intracavitary application was 7500 cGy. The point A dose rate is 50 cGy/h. The treatment time, in hours, is given by _____.
   a. 4
   b. 50
   c. 150
   d. none of the above

4. Conventional MAL treatment technique in intracavitary therapy normally involves _____.
   a. placing an intrauterine applicator and two ovoid applicators into the uterus and vagina, respectively
   b. standard loading of applicators after proper positioning
   c. prescription of dose to point A
   d. none of the above

5. With modern remote afterloading equipment, _____.
   a. the dose distribution can be customized
   b. the bladder and rectal doses can be reduced
   c. the point A dose has to be obtained only from the standard Manchester tables
   d. the treatment time can be reduced by increasing source activities

6. The organs at risk in gynecological treatments are _____.
   a. lungs
   b. bladder
   c. rectum
   d. none of the above

7. For an ideal intracavitary insertion using Fletcher Tandem and Ovoids, what is the dose rate at Point A given the following data: Tandem 15 – 10 – 10 mg-Ra-eq; Ovoids 15 mg each.
   a. 100
   b. 80
   c. 55
   d. 36
   e. 18

8. The ICRU report that deals with dose and volume specification in intracavitary brachytherapy is _____.
   a. IRCU 24
   b. ICRU 38
   c. ICRU 51
   d. ICRU 58
   e. ICRU 64

9. The total mg-Ra-eq of an intracavitary insertion is $(\text{mg-Ra-eq})_{\text{total}}$, or _____ in terms of total reference air kerma rate (TRAKR) in $\mu\text{Gy.m}^2$.
   a. $(\text{mg-Ra-eq})_{\text{total}} \times 8.25$
   b. $(\text{mg-Ra-eq})_{\text{total}} \times 8.25 \times 0.95$
   c. $(\text{mg-Ra-eq})_{\text{total}} \times 8.25 \times 0.95 / 0.722$
   d. $(\text{mg-Ra-eq})_{\text{total}} \times 8.25 \times 0.876$
   e. $(\text{mg-Ra-eq})_{\text{total}} \times 8.25 / 0.876$

10. An intracavitary therapy treatment of cervix involved 800 mg-Ra-eq hours of treatment. The TRAK for this treatment (in μGy.m^2) = _____.
    a. 5781.6
    b. 6000
    c. 6783.9
    d. 8000
    e. 9654.7

11. According to classical system of intracavitary brachytherapy (ICBT), the prescribed dosage is 7200 mg-hrs. This is equivalent to a TRAK (in μGy.m^2) of _____.
    a. 50,096.2
    b. 52,034.4
    c. 54,000.8
    d. 56,875.8
    e. 59,847

12. The important properties of TRAK are it is _____.
    a. a measure of treatment dose delivered at 10 to 20 cm distance from the intracavitary source configuration
    b. a modern SI units equivalent of the historical quantity mg-hr
    c. easy to calculate when source strengths are specified in AKS
    d. indicative of absorbed dose to the target volume
    e. a useful index of radiation protection of staff from the gynecological cancer patients

13. The ^{192}Ir HDR source AKS calibration of a well chamber is given by N_{aks} U/amp. (1 U = 1 μGy.m^2/hr) at 22 °C and 760 mm Hg. The saturation correction for the chamber is unity. The ^{192}Ir source at the calibration depth in the user hospital reads I amp at 24 °C and 756 mm Hg. The AKS of the source (in U) is given by _____.
    a. N_{aks} I
    b. N_{aks} / I
    c. N_{aks} I [(273.2 + 24)/ 295.2] $\times$ [760 / 756]
    d. N_{aks} I [(295.2 + 24)/ 297.2] $\times$ [756 / 760]
    e. N_{aks} I / [(273.2 + 24)/ 295.2] $\times$ [760 / 756]

14. The ^{192}Ir HDR source AKS calibration of a well chamber is given by 3.859×10^{11} U/amp. (1 U = 1 μGy.m^2/hr) at 22 °C and 760 mm Hg. The saturation correction for the chamber is unity. The ^{192}Ir source at the calibration depth in the user hospital reads 55.3777 nA at 24 °C and 758 mm Hg. The AKS of the source (in U) is given by _____.
    a. 20,648
    b. 21,562
    c. 22,347
    d. 23,276
    e. 24,223

15. The AKS of a ^{192}Ir HDR source was measured at the brachytherapy department using a well chamber as 17,961 U. _____ is the activity of the source in Ci (AKS to Ci conversion factor for the source = 4028 U/Ci).
    a. 4.459
    b. 4.634
    c. 4.729
    d. 5.212
    e. 5.364

16. The AKS of a ^{192}Ir HDR source was measured at the brachytherapy department using a well chamber as 40,796 U. The source certificate value corrected for decay to the measurement date is 10.229 Ci. Does the measured value agree with the certificate value? (The source manufacturer used a conversion factor of 4028 U/Ci.) PTV assumes additional significance in external beam therapy since the radiation fields are imposed from outside the body. In brachytherapy the sources are embedded in the target volume and so no additional margin is generally required to define a PTV.
    a. No, the two values are not the same.
    b. AKS cannot be converted to Ci, so we cannot compare.
    c. Source activity cannot be expressed in Ci.
    d. The measured value agrees with the certificate value.
    e. No, the difference is not within acceptable limits.

17. GTV is concerned with the identification of the actual tumor volume, so it is as important as it is in external beam therapy. According to ICRU 38, _____.
    a. the reference isodose surface for covering the target volume is 60 Gy for monotherapy
    b. for boost therapy the reference isodose is still the same
    c. there is no other method of specifying the intracavitary treatments
    d. the reference volume can be defined by the height, width, and thickness of the reference volume
    e. the reference volume must be theoretically determined

18. A treatment of cervix cancer involves a dose delivery of 40 Gy be external beam therapy followed by a boost of brachytherapy. The brachytherapy treatment volume, according to ICRU 38, would be enclosed by _____ Gy isodose at low dose rate.
    a. 60
    b. 40
    c. (60 + 40)
    d. (60 − 40)
    e. (60 + 40) / 2

19. To give a uniform dose to the surface of a cylindrical applicator in an HDR treatment, the dwell times at all dwell positions must be _____.
    a. the same
    b. lower at the ends
    c. higher at the ends
    d. higher at the upper end and lower at the lower end
    e. alternatively high and low

V. Brachytherapy ANSWERS

A. Basic Concepts

1. No   Ra sources were used in the form of needles (for interstitial therapy) and as tubes (for intracavitary or surface mold therapy).

2. Yes

3. No   They have been withdrawn from hospitals in most countries because of the radiation hazards involved in case of source leak.

4. Yes

5. Yes   With remote afterloading units, higher activities can be handled without any radiation protection problems for the staff. Afterloading allows more time for the radiation oncologists to place brachytherapy sources, thus improving the dosimetric accuracy.

6. Yes   For the higher-energy photon sources (e.g., ^{226}Ra, ^{60}Co, ^{137}Cs, etc.) the scatter buildup in the tissue medium approximately compensates for the medium attenuation, making the dose fall governed by the inverse square law. However, with low-energy photon sources, this is not holding true.

7. No   They are usually doubly encapsulated to contain their radioactivity and the beta emissions.

8. Yes

9. No ^{192}Ir sources are fabricated in the form of a ^{192}Ir alloy (30% ^{192}Ir and 70% Pt), for better strength. It is enclosed in a 0.1 mm thick Pt sheath.

10. Yes

11. No   Total length = active length + inactive length.

12. Yes

13. Yes

14. No   The rectal dose can be measured using rectal monitors.

15. No

16. No   PTV assumes additional significance in external beam therapy since the radiation fields are imposed from outside the body. In brachytherapy the sources are embedded in the target volume, so no additional margin is generally required to define a PTV.

17. No   GTV is concerned with the identification of the actual tumor volume, so it is as important as it is in external beam therapy.

18. Yes   Hence, MRI-based image-guided brachytherapy can lead to optimal tumor control and therefore dose escalation for larger tumors

19. Yes

20. No   3D dose optimization in brachytherapy gives better curative dose coverage of target volume and sparing of OAR compared to conventional x-ray-based planning, improving the therapeutic ratio.

21. Yes

22. Yes

23. No   Lead aprons are good for kV x-ray energies, but not for 0.6 MeV gammas. The attenuation would be negligible.

24. No   The source is usually calibrated during its replacement, and using the decay chart the daily AKS values must be calculated and checked on the treatment console display.

25. No   The only ones in clinical use must be tested.

26. No   It should, otherwise the tumor would be underdosed in the peripheral regions

27. No   The linear distribution of radioactivity must be uniform and identical for all wires in the Paris system.

28. No   It can be applied to high-activity sources, too (e.g., ^{192}Ir) but the prescription dose needs to be modified to account for the differences in the biological effects (or the dose rate effects).

29. Yes

30. No   The quantity expresses source output per unit source strength, so it depends on the source length, encapsulation, etc. A ^{192}Ir pellet and an Ir wire of a given length or ^{125}I seeds from different manufacturers will have different Λ values.

31. Yes   As in the case of prostate implants or breast or head and neck implants.

32. Yes   Assaying all seeds will take a long time.

33. No   They are highly anisotropic. The anisotropy is determined by the source energies and not the source size.

34. No   For sources with energies less than about 200 keV (e.g., ^{192}I and ^{203}Pd) the ratio is 1.01.

35. Yes.   For sources like ^{226}Ra, ^{137}Cs, and ^{192}Ir, the attenuation and scatter in tissue medium approximately cancel each other and only the inverse square law governs the fall of dose with distance for up to 5 cm.

36. Yes   Again, for both sources it is only the ISL that governs the dose fall with distance. This made it possible to use Paterson-Parker tables with ^{137}Cs sources by expressing their activities in terms of mg-Ra-eq.

37. Yes

38. Yes   They can be applied by making small corrections to the table and converting mg hrs / 1000 R Manchester tables into µGy hrs / Gy tables.

39. No   In the Manchester system, planar implants are single-source plane implants with dose specified in dosimetry planes at ±5 mm from the source plane.

40. No   This point has been defined as 2 cm lateral to the uterine canal and 2 cm from the mucous membrane of the lateral superior fornix of the vagina in the plane of the uterus.

41. Yes

42. No   It is a well-recognized complication in cervix treatment with tandems.

43. No   The source is at larger distance from the surface for the largest ovoids, so the inverse square law effect across 1 cm tissue is lesser, making the depth dose with respect to the surface less steep. Larger ovoids will deliver much more even and higher dose to mucosal surface, so the maximum size ovoids that can be accommodated in a patient must be used.

44. Yes

45. No   3D treatment planning analysis has shown maximum doses to bladder and rectum are significantly greater than the ICRU 38 reference point doses.

46. Yes

47. No   Never used in palliative cases.

48. a, b, c

49. a, b, d

50. a, b, c

51. b, c, d

52. a, b, c, d, e

53. b

54. d

55. a

56. d

57. a, b, c

58. a, b, d

59. b   Since the P-P table rules are empirical and based on radium emissions, they can be used only for sources of comparable source energies (e.g., ^{192}Ir or ^{137}Cs and not ^{125}I or ^{203}Pd, etc.). The Manchester system requires nonuniform distribution of source strengths to get uniform dose distribution, and this requires sources of varying strengths.

60. a, b, c

61. b $K_a(d) \propto (1/d^2)$; $180 \times (10 \text{ cm})^2 = 20 \times (d \text{ cm})^2$ gives d

62. a, b, c

63. a, b, c, e

64. a, b, c

B. Brachytherapy Source Characteristics

1. Yes

2. No   It does since EC turns a proton into a neutron in the nucleus thus changing the element.

3. No   The dosimetric characteristics of the source depend on its size, shape, and design.

4. No   It can be performed as the principal treatment mode or as one accompanying an external beam treatment.

5. No   The implantation is done only once, and if it is not exactly according to the preplan as revealed by the post-implantation imaging, the dosimetry is done as per the existing source configuration and a post-plan is developed and implemented.

6. Yes   The dose received will be a very small dose (NCRP Report 37).

7. No   Iodine is absorbed onto silver rods encapsulated in titanium.

8. Yes

9. Yes

10. Yes   The dose or dose rate at 10 cm or 20 cm from the implant in the pelvis or abdomen can be easily calculated from the TRAK. Also the dose at 1 m from the patient can be easily calculated neglecting the attenuation by the body tissues.

11. No   Brachytherapy sources are now specified in terms of their output (like reference air kerma, source strength, etc.) and activities are used only for regulatory and radiation protection purposes.

12. Yes   This quantity is analogous to mg h or mg-Ra-eq h and gives the total air kerma at the point of interest for the source in question. It can also be used for an implant by integrating the RAK of all the sources and their exposure times, i.e., $K_{ref} = \Sigma S_{k,j} t_j$.

    The American Brachytherapy Society (ABS) has proposed this quantity following the recommendation of ICRU to replace all old quantities like mg h or mg-Ra-eq h by air kerma-based quantities. This is numerically equal to TRAK recommended by ICRU.

13. Yes

14. No

15. d

16. c

17. b

18. a, b

19. a   For AKS, the output must be referred to 1 m distance. Doubling the distance from 50 cm will reduce the output by four times.

20. b

21. a

22. a. iv.   b. i.   c. ii.   d. iii.

23. b

24. a, b, c

25. a

26. a

27. b, c

28. a 20×8.5 R cm^2 / h. 3.26 R cm^2 / mCi h = $20 \times (8.5 / 3.6)$ mCi. It is easy to guess where the terms appear in the equation from the units of the quantities. For instance, the terms cannot appear as $(3.6 / 8.25)$ since mCi will then appear in the denominator, which is incorrect.

29. f

30. b   As the figures below show, a seed ribbon of 5 seeds with 0.8 cm spacing between the seed centers (= 4 cm equivalent length) and an ^{192}Ir wire of 4 cm length (= 5×0.8 cm) will have equivalent dose distribution.

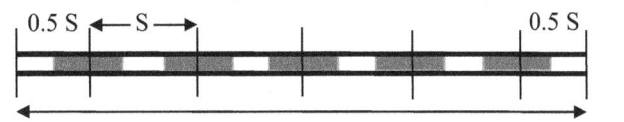

Equivalent Length (EL) = N x S = 5 x 0.8 = 4 cm (S = separation between seeds)

Active length (AL) of a line source (e.g. ^{192}Ir wire)

Equivalence of a line source and a source train of discrete sources

31. a. ii.   b. v.   c. i.   d. iii.   e. iv.

32. a. iii.   b. v.   c. iv.   d. ii.   e. i.

33. a. ii.   b. ii.   c. ii.   d. i.   e. i.

34. c, d

35. a, b, c, d

36. a, b, c

37. a. iii. and iv.   b. ii.   c. i. and ii.   d. v.   e. v.

38. b

39. b

40. a

41. a, b

42. d   From the definition of Γ, the output of ^{192}Ir (per mg-Ra-eq) will be the same as that of (8.25 / 4.69 = 1.75) mCi of ^{192}Ir. So, 0.75 mg-Ra-eq corresponds to 1.31 mCi of ^{192}Ir.

43. c   (8.25 / 3.6) gives equivalent mCi per mg-Ra-eq; $\times$ 15 mCi for 15 mg-Ra-eq source.

44. d

45. c   (3.6 / 8.25) gives equivalent mg-Ra-eq / mCi; $\times$ 38 mg-Ra-eq for 38 mCi source.

46. c

47. d   This is from the definition of mg and the reference air kerma rate.

48. c   Exposure rate (ER) at 1 m = $3.3 \times 10^3 \times (5^2 / 100^2)$ = 8.25 mR/h; 1 mg-Ra-eq at 1 m gives an ER of 0.825 mR/h (from the definition of Γ factor of Ra). Source in mg-Ra-eq = 8.25 / 0.825.

49. a $[30 \times (\Gamma_{x,Cs})] / 8.25 = (30 \times 3.26) / 8.25$

50. c   Use exposure to air kerma conversion factor from the equation $K_a = [X (W/e) / (1-g)]$.

51. c   Use exposure (in R) to air kerma (in cGy) conversion factor from the equation $Ka = \{X k_u (W/e) / (1-g)\}$, which gives a value of 0.876 cGy/R. k_u is the unit conversion factor from R to Coul./kg = 2.58×10^{-4} (C/kg)/R and (W/e) = 33.97 J/Coul. $g \approx 0$ for kV x-rays and 0.003 for Co sources, giving a value of 0.876 for the conversion factor, assuming g = 0.

52. a

53. a, b, c

54. b   The units of Γ_x and its value (3.26 R.cm^2 / mCi-hr) for ^{132}Cs must be known for this. Simple calculation: $3.26 \times 30 / 50^2$.

55. d $\Gamma_{x,Ir}$ = 4.69 R cm^2 / mCi-hr; Formula: $X(d) = A \Gamma_x / d^2$

56. b

57. a

58. d   One mg-Ra-eq corresponds to an exposure rate of 8.25 R cm^2/hr or 8.25×0.876 cGy cm^2/hr = 7.227 cGy.cm^2/h; 7.227 μGy.m^2/h = 7.227 U or 1 U = (1 / 7.227) mg-Ra-eq. We have earlier converted mg-Ra-eq into mCi, so one can also relate U to mCi as the next example shows.

59. d   One mg-Ra-eq of ^{138}Cs = (8.25/3.26) mCi; so 0.138 mg-Ra-eq (= 1 U) = (8.25 / 3.26) $\times$ 0.138 mCi.

60. d   Recollect the units of RAKR.

61. b

62. c

63. d   Two things should be noticed here. By definition, AKS = $\{dK_a(d) / dt\} \times d^2$ with units of μGy.m^2/h. For RAKR, d = 1m, and we get RAKR = $\{dK_a(at\ 1m) / dt\}$ μGy / h at 1m. Both RAKR and AKS are numerically equal but differ only in their measurement details.

The SI unit of the air kerma rate is Gy, but for the purposes of source specification, it is more convenient to use μGy for LDR sources.

64. a, e

65. b   Integrated Reference Air Kerma (IRAK) of a source configuration is given by:

$$IRAK = \sum_i S_{k,i} t_i \, \mu Gy \, m^2$$

where S_k and t_i refer to the air kerma strength and duration of exposure for the ith source. The output of a 1 mg-Ra-eq source = 7.22 μGy m^2/hr (recollect the definition of Γ factor). So if the source configuration is specified in terms of mg-Ra-eq-hr (instead of air kerma strength-hr), a multiplication factor of 7.22 must be made use of.

66. b

67. a

68. d

69. b

C. Dose Distributions

1.   a, b, c, d

2.   d

3.   a

4.   b, d

5.   a, b, c

6.   d   AKS specifies the source air kerma output at 1 m distance. The ISL factor refers the output to the specified distance. The medium perturbation corrects air kerma for medium attenuation and scattering effects. $(_m\mu_{en})_{w,a}$ converts air kerma to dose to water. This is the standard formalism.

7.   a

8.   b

9.   a, b

10. a

11. a

12. a, b, c

13. a, c

14. a, b, c

15. c

16. a, b, c

17. a, b, c, d

18. a, c

19. a, c

20. c

21. c, d

22. c

23. c

24. b

25. c

26. c

27. b

28. c

29. b

30. b

31. a, b, c

32. b

33. a, b, c, d

34. b

35. a, b

36. d

37. c

38. a, b, c, d

39. c

40. a

41. e

42. b, c, d

43. a, b

44. c

D. Dose Calculations

1. d   The geometry factor has not been given, but for distance >> source length, the geometry factor reduces to $1/r^2$ where r is given as 5 cm. The formula
$D(r) = S_k \Lambda G(r) g(r) \Phi_{an} = (0.5 \times 7.22) \times 1.12 \times (1/5^2) \times 0.996 \times 0.98$
gives the medium dose under point source approximation.

2. c   The dose rate $= A \Gamma_{x,Ra} \times M(5) \times f_{med} \times (1/r)^2$.

3. b

4. c, d

5. e

E. Source Localization

1. Yes

2. No   To adequately irradiate the periphery of the implant, the ends should be crossed as in Manchester system, or the source must extend beyond the length of the implant. The formula for knowing the total source length: Total Source length = Implant length / 0.7.

3. Yes

4. No

5. No

6. Yes

7. a, b, c, d

8. a, b

9. a, b, c

10. b

11. a

12. a, b, c

13. a, c

14. b

15. a

16. a

17. c

18. e

19. c

20. d

21. c

22. c

23. d

24. a

25. c

26. c

27. a

28. c

29. c, d, e

30. a. iv   b. i.   c. ii.   d. v.   e. iii.

31. b, c, d

F. Interstitial Dosimetry

1. a, b, c, d

2. c   To adequately irradiate the periphery of the implant, the ends should be crossed as in the Manchester system, or the source must extend beyond the length of the implant. The formula for knowing the Total source length is: Total Source length = Implant length / 0.7.

3. a, b, c

4. c

5. b

6. c

7. c, d

8. a, b, c

9. b, d

10. a, b, c,

11. c, d, e, f

12. b, d

13. a, b

14. d

15. c

16. a

17. c   From the definition of AKS and AKRC, AKS = $K_a(d) d^2 = A_{app} \Gamma_{Ka} / 1^2$ So by measuring the AKS of the source in the hospital, A_{app} can be derived and compared with the certificate value. The important point to remember here is that there are variations in the Γ_{Ka} values quoted.

18. a, b

19. a, b   The institution must purchase a well-type chamber with an AKS calibration factor and not apparent activity calibration. The well chamber can be calibrated for any type of source and, indeed, must be calibrated for all types of sources that are in use in brachytherapy.

20. c

21. b   Formula is "Implant length / 0.7" or about 30% longer than the target length.

22. a. iii.   b. i.   c. ii.

23. d   For single plane implant (≤12 m), (Implant thickness, T/ Source spacing, SS) = 0.5 or SS = 2T, for a two-wire implant.

24. a, b, c

25. c   For a two-plane triangular implant T ≈ 1.2 SS.

26. c   For a two-plane square implant T ≈ 1.5 SS.

27. b

28. b

29. a

30. a. ii.   b. iii.   c. i.

31. a, c, e Increasing spacing between the wires decreases dose in the central regions, increasing the dose nonuniformity. Hence, a limit of 15 mm has been suggested for the spacing of short wires (up to 50 mm) and a limit of 22 mm for long wires (>70 mm), in this system of dosimetry. Also, the sleeve of high-dose regions increases with increasing source separation and can cause tissue necrosis. Also, a minimum spacing of 8 mm is suggested in the Paris system from the point of view of ease of implantation (5 mm spacing may be used if it is possible for the clinician to do the implant).

32. a

33. c   (T/SS) = 0.5 or SS = T / 0.5 = 5 / 0.5 = 10 mm

34. b   Target volume length / 0.7 gives the source length. 0.7 may not be exact. The source is longer than the implant length by approximately 20% to 30% since the dose falls at the ends and will not give adequate dose at the periphery unless the ends are crossed, as in the Manchester system. Since the ends are not crossed, the length of the wire is increased beyond the target length. If one uses hairpin-type sources, one end is crossed, so only the other ends must be longer by 20% to 30%.

35. c   For single-plane implants, the safety margin = 0.37 × source spacing (10 mm).

36. b   Treatment width = 2 × (0.37 × 10.0) + 10.0 = 17.4 mm adding a safety.

37. c   The prescription dose must enclose the treatment width.

38. a

39. b   The basal dose rate is the dose rate at the midpoint of the pair of sources in the central plane. Each source (linear source strength = 1 mCi/cm) contributes 4.2 × 5.46 = 22.9 cGy/hr at the midpoint. So, the MBDR = 45.8 cGy/hr. Since there are only two line sources, there is only one midpoint, so the BDR is the MBDR.

40. c   The prescription dose rate (PDR) or the Reference dose rate (RDR) is defined as 0.85 × MBDR. So the prescription dose rate for this implant is 0.85 × 45.8 = 38.9 cGy/hr.

41. b   Here we have to work backwards. PDR / 0.85 (= 70.6 cGy/hr) gives the MBDR. So each wire contributes 35.3 cGy/hr at the midpoint. An ^{192}Ir wire of LAK strength (1 U/cm) gives at the midpoint (at 5 mm) 5.46 cG/h. So the required LAR strength for the implanted wire = 35.3 / 5.46 = 6.47 U/cm.

42. d   Here we have to express the LAK strength in terms of mCi/hr. The unit conversion factor is 4.2 U/mCi. So the linear activity of the source = 6.47 / 4.2 = 1.5 mCi/cm.

43. a   Here the midpoint or basal dose rates are given for wires of LAK strength 1 U/mm (10 U/cm). The implanted wires have LAR strengths of 4.5 U/cm, so the dose rates must be scaled accordingly.

44. b   MBDR = (0.395 + 0.417 + 0.395) / 3 = 0.402 Gy/hr.

45. c   RDR = 0.85 × MBDR Gy/hr.

46. b   Treatment time for the implant = RD / RDR = 65 / 0.342 = 190 hr.

47. a   MBDR = 0.888 Gy/hr; RDR = 0.85 × 0.888 = 0.755 Gy/hr.

48. c   To deliver 25 Gy in 2.5 days (60 hrs), RDR required = 25 / 60 = 0.417 Gy/hr.
          LAKS = RDR / DR / (U/cm) = 0.417 (Gy/h) / 0.755 (Gy/hr) / (U/cm) = 0.552 U/cm.

49.   a. ii.   b. i.   c. iv.   d. iii.

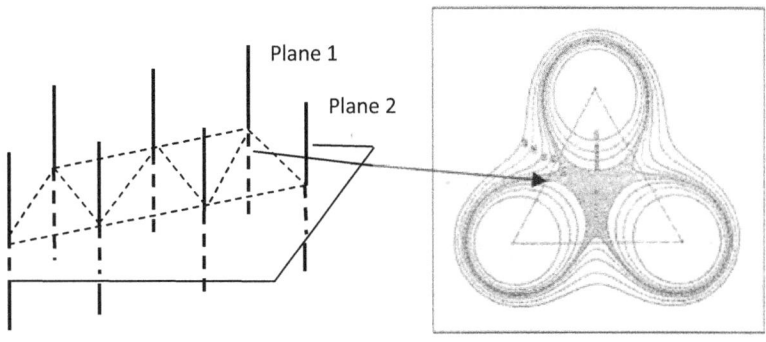

Double plane implant   Definition of basal dose (rate)

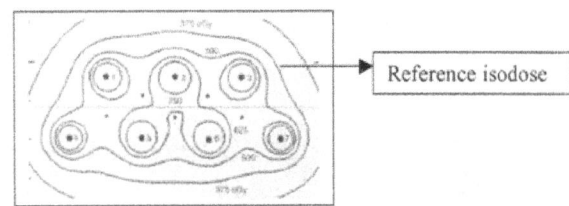

Concept of Reference isodose
Paris System of Dosimetry

50.   a. ii.   b. i.   c. iv.   d. iii.   e. vi.   f. v.

51.   e, f

52.   b, f

53.   d   Mean life = 1.44 × half life

54.   a   Half life = 0.693 / λ, from which mean life can be determined.

55.   a, b

56.   a, c

57.   c   The negligible dose criterion is 1 μGy/h. The half life of ^{125}I is about 59 days, so within 2 months the dose will be negligible and no restrictions would apply. Till that time, pregnant women and children must keep a distance from the patient.

58.   b $1.44 \times T_{av} \times (dD_0 / dt)$

59.   a, d   This is an AAPM TG-56 recommendation.

60.   e

61.   b

62.   a, b, c

63. e   The patient poses no radiation risk because the remnant AKS of the implant = $86 \times \exp(-0.693 \times 4 \times 365 / 60) \approx 4 \times 10^{-6}$ µGy m^2/h = 4×10^{-2} µGy/h at 1 cm which is about 0.4 mR/h, which will give a negligible exposure in the patient handling time. (By surveying the patient, one can easily see the radiation levels are low enough to examine the patient without much risk to the physician.)

64. d

65. c   0.486 / 0.525 = 0.922, or it would have decayed by approximately 7%. (Applies for $t \ll T_{1/2}$.)

66. b   51 days = 3 half lives when the activity would be 12.5%. i.e., 87.5% has decayed or 87.5% of the total dose has been delivered.

67. b   In 4 half lives (240 days) the activity would have dropped to 6. 25% or more than 90% of the decay would have occurred, i.e., more than 90% of the total dose would have been delivered in this time. So 7 months is the best answer.

68. d   34 days = two half lives. The activity would be 25% of initial activity, so 25% of 120 Gy = 30 Gy remains to be delivered.

69. b   After 34 days 25% of activity is left, i.e., 75% decayed or 75% of the dose = $0.75 \times 124 = 93$ Gy would have been delivered.

70. a

71. b

72. e

73. c

G. Intracavitary Therapy

1.   a, b, c

2.   a

3.   c

4.   a, b, c

5.   a, b

6.   b, c

7.   c

8.   b

9.   d

10.   a $\Gamma_{x,Ra}$ = 8.25 R m^2 / mg.hr or mg-Ra-eq.hr.
     800 mg-Ra-eq hr of treatment = $800 \times 8.25 \times 0.876 = 5781.6$ µGy m^2

11.   b

12.   a, b, c, e

13.   c

14.   b

15.   a

16.   d   Using the same conversion factor that the manufacturer used, the source activity comes out to be
             10.128 Ci, which agrees with the certificate value to within 1%. Any agreement to within 3% can be
             considered acceptable.

17.   a, d

18.   d

19.   c

VI. Radiation Protection

Introduction

The aim of radiation protection is to ensure that doses received by radiation workers (and the general public) remain as low as reasonably achievable and that under no circumstances exceed the relevant annual dose limits recommended by the regulatory authorities.

The dose received by a radiation worker or the public may be internal or external depending upon the type of exposure. Any radioactivity finding its way into the human body through inhalation, ingestion, or skin contamination gives rise to internal exposure. Any exposure from a source external to the body gives rise to external exposure. In a radiation oncology department, one has to guard only against external exposures. In the case of external beam therapy, there is a potential for external exposure, though the probability of exposure is very low if one follows the radiation safety rules. With brachytherapy sources, there is a greater likelihood of overexposure since one would be handing these sources. Here one should follow the basic principles of radiation protection to reduce individual exposures to well within acceptable levels. The order of annual dose received by a radiation worker in a radiation therapy department is about one tenth of the annual dose limit.

When a radiation worker is exposed to radiation from all directions—as happens in the case of staff working in radiation therapy departments—one is interested in estimating the "whole body" dose received by the workers. In special cases where only part of the body may have been exposed (for example, the skin or the eye), one is interested in doses to these specific tissues or organs as well.

It is difficult to measure with very high accuracy the doses as received by a radiation worker to the whole body or to the individual organs in the course of his or her work with radiation. However, when the doses received by radiation workers are much less than the annual dose limits, it is enough if these doses are estimated to an accuracy of about 30%. Only when the doses reach the annual limits does one need a much higher accuracy in the dose estimate.

Radiation workers are usually provided with personnel monitors to record the whole-body or partial-body doses they receive. These personnel monitors are calibrated in such a way that the monitor reading gives a conservative estimate of the doses received by the radiation worker, which is what is needed in most cases. If the annual doses reach the regulatory limits, extensive investigations must be carried out to derive the exact doses received by the worker.

The maximum permissible dose limits discussed here are according to ICRP 60 and NCRP 116 recommendations. Recent ICRP draft recommendations have revised the dose limits for the eyes from 150 mSv to an average dose limit of 20 mSv/y (the maximum not exceeding 50 mSv/y). This change makes a difference to interventional cardiologists and interventional radiologists whose eyes are at greater risk compared to other medical radiation workers, but it makes very little difference to radiation workers in radiation oncology. We will continue to adopt NCRP 116 recommendations until they are revised.

A. General Radiation Protection

Circle the right answer (Yes or No):

1. (Yes / No)   Radiation harm can be classified under stochastic and deterministic effects.

2. (Yes / No)   Radiation-induced cancer is a deterministic effect.

3. (Yes / No)   Skin reaction to radiation is a deterministic effect.

4. (Yes / No)   The aim of radiation protection is to prevent the occurrence of deterministic effects and to reduce the incidence of stochastic effect to levels as low as practicable.

5. (Yes / No)   No practice involving radiation should be adopted unless the benefits to society or the individuals outweigh the radiation risks.

6. (Yes / No)   The practice involving radiation should be optimized so that the doses received, the number of people exposed, and the likelihood of incurring exposures are as low as reasonably achievable, taking economic and social factors into account.

7. (Yes / No)   The exposure received by the public in a therapy department is classified as "Medical Exposure."

8. (Yes / No)   The exposure received by radiation workers in a therapy department is classified as "Occupational Exposure."

9. (Yes / No)   The exposure received by a patient in diagnosis or therapy is classified as "Medical Exposure."

10. (Yes / No)   There is a maximum permissible dose limit for medical exposures.

11. (Yes / No)   Maximum permissible dose limits exist for the staff working with radiation and for the general public who are not radiation workers.

12. (Yes / No)   Medical exposure received by a radiation worker (say, for some diagnostic purpose) must be taken into account to determine the annual occupational dose received by him or her.

13. (Yes / No)   The equivalent dose (EQD) received by an organ is expressed as "organ dose × radiation weightage factor."

14. (Yes / No)   The whole body dose received by a radiation worker is known as "Equivalent Dose."

15. (Yes / No)   The radiation risk to an organ depends not only on the equivalent dose to the organ, but also on the type of tissue involved.

16. (Yes / No)   For radiation protection purposes, dose equivalent, a point quantity, is the quantity of interest rather than physical dose.

17. (Yes / No)   Dose equivalent (H) and absorbed dose (D) are numerically the same.

18. (Yes / No)   Internal exposure occurs when radioactivity gets into the body through inhalation, ingestion, or through the skin.

19. (Yes / No)   Internal exposure is very unlikely in a radiation therapy department.

20. (Yes / No)   External exposure can be controlled by time, distance, and shielding.

21. (Yes / No)   Ordinary brick-walled rooms are enough for ^{60}Co or linac treatment rooms to reduce radiation levels outside the room to acceptable levels.

22. (Yes / No)   The effective dose received by a radiation worker can be monitored using a personnel monitor.

23. (Yes / No)   The (gamma) radiation levels in a room cannot be monitored.

24. (Yes / No)   Monitoring of workplaces is carried out to ensure that the working conditions are under control and to alert staff regarding any unanticipated exposure.

25. (Yes / No)   Survey monitors must be calibrated in terms of air kerma or exposure.

26. (Yes / No)   Individual monitoring is carried out to assess the equivalent dose to any part of the body or to the whole body.

27. (Yes / No)   Ion chamber-based survey monitors have better accuracy compared to G-M based systems.

28. (Yes / No)   Ion chamber-based survey monitors have better sensitivity compared to G-M based systems.

29. (Yes / No)   Gamma survey monitors can always detect beta radiation as well.

30. (Yes / No)   Neutron survey monitors are based on the detection of thermal neutrons by their (n, α) type interactions.

31. (Yes / No)   A film badge can measure the dose received from fast neutrons.

Choose the right answer(s) (more than one may be correct):

32. The unit for dose equivalent, H, is expressed in _____,
    a. sieverts
    b. grays
    c. rads
    d. roentgens

33. The energy absorption corresponding to 1 Sv of H is _____.
    a. 1 J/kg
    b. 1 erg/g
    c. 100 ergs/g
    d. none of the above

34. 1 Sv is equal to _____.
    a. 1 rem
    b. 100 rad
    c. 100 rem
    d. none of the above

35. Match radiation type to radiation weighting factor.

Radiation Type		*Weighting Factor*
a. photons	_____	i. 1
b. electrons	_____	ii. 5
c. protons (>2 MeV)	_____	iii. 10
d. heavy ions		

36. Match organs to tissue weighting factors.

Organs	Tissue Weighting Factors
a. gonads	____ i. 0.04
b. red bone marrow	____ ii. 0.12
c. thyroid	____ iii. 0.08
d. lung	

37. Match the quantity to the SI units.

Quantity	SI Units
a. activity	____ i. C/kg
b. exposure	____ ii. Bq
c. dose	____ iii. Gy
d. dose equivalent	____ iv. Sv
e. equivalent dose	

38. One TVL corresponds to a transmission of _____.
    a. 10%
    b. 50%
    c. 1%
    d. none of the above

39. One TVL is _____.
    a. 5 HVL
    b. 3.3 HVL
    c. 2 HVL
    d. none of the above

40. From an internal hazard point of view, the most hazardous radiation is _____.
    a. alpha rays
    b. beta rays
    c. gamma rays
    d. neutrons

41. A suitable shielding material for beta radiation is _____.
    a. lead
    b. depleted uranium
    c. steel
    d. aluminum

42. The exposure-rate at 10 m from a source is 20 mR/h. After two half lives, the exposure rate at 5 m (in mR/h) would be _____.
    a. 2
    b. 5
    c. 20
    d. 80

43. A radiation worker was wearing a pocket dosimeter at a place for 12 minutes. The reading on the pocket dosimeter showed 24 mR. The exposure rate (in mR/h) is _____.
    a. 2
    b. 24
    c. 100
    d. 120

44. The constant radiation level at a place is 40 mR/h where four persons are standing. The exposure received by each individual in one hour (in mR) is _____.
    a. 10
    b. 20
    c. 30
    d. 40

45. By increasing the distance between a person and a point source of radiation by a factor of three, the exposure rate will _____.
    a. decrease by a factor of 18
    b. increase by a actor of 3
    c. decrease by a factor of 16
    d. decrease by a factor of 9

46. The radiation level at 4 m from a source is 200 mR/h. After one half life, the exposure rate at 2 m from the source if a 2 HVL shielding is interposed is _____.
    a. 100 mR/h
    b. 200 mR/h
    c. 400 mR/h
    d. 10 mR/h

47. Medical investigation with radiation should be chosen only when necessary because _____.
    a. it is always expensive
    b. regulatory bodies prohibit the use of radiation
    c. of radiation risk to the individual
    d. none of the above

48. "Medical exposure" means the dose received by _____.
    a. patients undergoing diagnosis/therapy
    b. the relatives of patients
    c. the medical staff
    d. the general public in the vicinity of the hospital

B. Maximum Permissible Dose Equivalent (Dose and Dose Limits)

Choose the right answers:

1. The annual effective dose limit (AEDL) (i.e., the whole body dose) to a radiation worker, as per recent NCRP Report No. 116 recommendations (in mSv), is _____.
    a. 20
    b. 50
    c. 100
    d. 1000

2. The cumulative EDL (in mSv year) for a radiation worker is _____.
    a. 10 mSv × age (y)
    b. 50 mSv × age (y)
    c. 10 mSv × (age − 18) y
    d. none of the above

3. The annual equivalent dose limit (AEQDL) to lens of eye (in mSv) is _____.
   a. 20
   b. 50
   c. 150
   d. 1000

4. The annual equivalent dose limit to skin, hands, and feet (in mSv) is _____.
   a. 50
   b. 150
   c. 500
   d. 1000

5. The annual effective dose limit to the public in the case of continuous exposure is _____.
   a. 1/10th of the occupational worker limit
   b. 1/50th of the occupational worker limit
   c. 1/100th of the occupational worker limit
   d. the same as the occupational worker limit

6. The annual effective dose limit to the public (in mSv) in the case of infrequent exposure is _____.
   a. 1
   b. 5
   c. 50
   d. 0

7. The annual equivalent dose (AEQD) to the public to parts of the body is _____.
   a. 1/10th of the occupational worker limit
   b. 1/50th of the occupational worker limit
   c. 1/100th of the occupational worker limit
   d. the same as the occupational worker limits

8. The average AED received from natural background (in mSv) is _____.
   a. about 2–3
   b. about 0
   c. about 10
   d. none of the above

9. The average AED received by staff in a radiotherapy department (in mSv) is _____.
   a. about 2–4
   b. about 10
   c. about 20
   d. none of the above

10. Match the annual equivalent dose limits (in mSv/year) to the body parts for an occupational worker.

    Annual Dose Limits   *Body Parts*
    a. 500   _____   i.   eye lens
    b. 150   _____   ii. skin
    c. 6   _____   iii. hands and feet
                                   _____   iv. fetus

11. A medical radiation worker's occupational dose limit is 50 mSv. He receives 15 mSv while undergoing a medical examination. _____ is the maximum dose (in mSv) he can receive in his occupation during that year.
    a. 50
    b. (50 – 15)
    c. (50 – 15) – dose from natural background radiation
    d. (50 + 15)
    e. none of the above

12. There are maximum permissible dose limits for an occupational worker or the public, but not for a patient (medical exposure) because _____.
    a. a patient can tolerate any level of exposure
    b. a patient is sick and can be subjected to any dose
    c. a patient undergoes risk but also gets benefit which far outweighs his risk
    d. staff can accept only a certain amount of risk since the benefit goes to society

13. A radiation worker working in a high natural background region receive an annual dose of 10 mSv. He also receives a dose of 5 mSv in a medical examination. His occupational dose limit is 50 mSv, so he can receive a maximum dose of _____ mSv during the year.
    a. 50
    b. 50 – 10
    c. 50 – 5
    d. 50 – 10 – 5
    e. 50 + 10 – 5

C. Time, Distance, and Shielding (and ALARA Concepts)

The concepts of time, distance, and shielding play central roles in optimizing the dose received by exposed individuals. The dose received by a radiation worker is proportional to the time during which he is exposed, how far (or distant) he is from the source, and how much shielding is between the source of radiation and the radiation worker. All three parameters come into play while handling brachytherapy sources. Time and distance play an important role for the staff working inside fluoroscopy rooms performing procedures on a patient under fluoroscopic guidance. Shielding greatly controls the exposure of radiation workers in radiotherapy.

Circle the right answer (Yes or No):

1. (Yes / No)   Radiation dose received by staff working with radiation must be as low as reasonably achievable.

2. (Yes / No)   Adhering to the regulatory dose limits ensures that the doses received are at ALARA levels.

3. (Yes / No)   A radiation worker receives a dose of 20 mSv every year. Is his exposure within legal limits?

4. (Yes / No)   A medical radiation worker receives 20 mSv from the age of 18 until he retires at the age of 60. Does he exceed the NCRP recommended cumulative exposure for the radiation worker?

5. (Yes / No)   A medical radiation worker receives 20 mSv from the age of 18 until he retires at the age of 64. Does he exceed the ICRP recommended cumulative exposure for the radiation worker?

Choose the right answer(s) (more than one may be correct):

6. The concept of ALARA is based on the conservative assumption involved in the current philosophy of radiation protection that _____.
    a. radiation at any level involves an element of risk
    b. low doses do not cause any harm
    c. low doses are beneficial to the exposed individual
    d. it is easy to protect against low doses but not high doses
    e. low doses increases the efficiency of the radiation worker

7. The basic philosophy of radiation protection that stems from the ALARA concept is _____.
    a. justification
    b. optimization
    c. dose limitation
    d. training
    e. regulation

8. "As low as reasonably achievable" means taking _____ factors into account to decide on the ALARA.
    a. social
    b. technical
    c. economic
    d. practical
    e. all the above

9. Dose equals _____.
    a. dose rate
    b. dose rate / time
    c. dose rate × time
    d. dose rate × time^2
    e. none of the above

10. A radiation worker stands in a place where the radiation level in air kerma rate is 10 µGy/h. If he is in this field for 15 minutes, he receives a total exposure of _____ µGy.
    a. 1
    b. 2.5
    c. 5
    d. 10
    e. 40

11. A technologist can stay _____ in a 10 mSv/hr radiation field if he wishes to limit his dose to 1 mSv.
    a. 1 hr
    b. 30 min
    c. 10 min
    d. 6 min
    e. 10 sec

12. Increasing one's distance from a radioactive source _____.
    a. increases one's exposure
    b. increases one's exposure as the square of the distance from the source
    c. decreases one's exposure
    d. decreases one's exposure as the distance from the source (α 1/d)
    e. decreases one's exposure as the square of the distance from the source (α 1/d^2)

13. If the ambient dose equivalent H*(10) at 1 foot from the source is 5 mSv/h, then _____ would be the H*(10) value (in mSv/h) at 3 feet from the source.
    a. 5
    b. 5 / 3
    c. 5×3
    d. 5×3^2
    e. 5 / 9

14. The amount of shielding required in radiation protection depends on the _____.
    a. type of radiation
    b. output of the source (a radioactive source or a radiation-generating device)
    c. size of the radiation facility or source container
    d. acceptable dose level outside the shielding
    e. occupancy factors beyond the shielding
    f. nature of occupancy
    g. all of the above

15. In radiation oncology, the staff doses are optimized mainly by _____.
    a. the design of the equipment
    b. the design of the radiation installation (treatment room)
    c. the use of lead aprons
    d. increasing the staff's distance from the source
    e. all of the above

16. The unit for dose equivalent, H, is expressed in _____.
    a. sieverts
    b. grays
    c. rads
    d. roentgens

17. The energy absorption corresponding to 1 Sv of H is _____.
    a. 1 J/kg
    b. 1 erg/g
    c. 100 ergs/g
    d. none of the above

D. Brachytherapy Source Handling and Storage

Circle the right answer (Yes or No):

1. (Yes / No) Institutions require no special clearance from the state or national regulatory bodies to work with radioactive sources.

2. (Yes / No) All radiation workers of a radiotherapy department must have training and certification in radiation safety appropriate for their duties.

3. (Yes / No) Brachytherapy sources can be stored in any place in a radiation oncology department.

4. (Yes / No) A source movement register must be maintained to record sources leaving or returning to the storage room each day to avoid loss of sources.

5. (Yes / No)   Brachytherapy sources can be carried to the patient in any container available from the oncology department.

6. (Yes / No)   When the sources are removed from a patient, they must be tallied against the sources implanted in the patient (and recorded in the patient report and the log book).

7. (Yes / No)   In the case of temporary implants, the patient must be surveyed after the removal of the implant to ensure that no source is left behind in the patient.

8. (Yes / No)   The linen may be removed and sent to the laundry after the patient leaves.

9. (Yes / No)   Decayed brachytherapy sources must be returned to the supplier or the appropriate authorities as per the transport regulations.

10. (Yes / No)   The integrity of the source cannot be easily checked.

11. (Yes / No)   Short half-life (e.g., ^{125}I) sources can be disposed of when the survey meter readings near these sources are not above background levels.

12. (Yes / No)   A record of each source from procurement to final disposal must be kept by the radiation safety officer.

13. (Yes / No)   The source strength supplied in the source certificate must be used in the treatment planning system.

14. (Yes / No)   Brachytherapy sources can be handled by wearing gloves.

15. (Yes / No)   One should not stand near a patient undergoing intracavitary or interstitial treatment.

16. (Yes / No)   Nursing personnel must wear lead aprons while attending to a brachytherapy patient.

17. (Yes / No)   Survey monitors can be used for the calibration of brachytherapy sources.

18. (Yes / No)   An emergency storage container must be available in the brachytherapy patient room.

19. (Yes / No)   Radiation workers in a brachytherapy department need not wear a personnel monitoring badge.

20. (Yes / No)   Zone monitors or area monitors must be installed in the patient treatment room or in the vicinity.

21. (Yes / No)   The operating room must be carefully surveyed after the permanent implant patient leaves the room.

22. (Yes / No)   Brachytherapy sources must be prepared behind lead barriers (L-bench) for implants.

23. (Yes / No) ^{125}I sources are used only in permanent implants.

24. (Yes / No)   All brachytherapy sources emit gamma rays.

25. (Yes / No)   Sealed radioactive sources generally retain the source integrity and do not generally present any contamination problem.

26. (Yes / No)   After the temporary implants have been removed from the patient, the patient does not pose any radiation hazard.

27. (Yes / No)   A personal monitoring badge provides protection from radiation exposure.

28. (Yes / No)   A sealed source rather than an open (unsealed) source presents a potential contamination hazard.

29. (Yes / No)   The patient remains radioactive as long as the temporary implant is not removed from the patient.

30. (Yes / No)   Patients' visitors can be permitted to sit close to brachytherapy patients to make them comfortable.

31. (Yes / No)   Visitors must be provided with a dosimeter to monitor their dose.

32. (Yes / No)   All staff dealing with brachytherapy patients must wear a personal dosimeter badge.

33. (Yes / No)   If the brachytherapy room is provided with an area monitor, is it necessary to survey the patient before discharge?

34. (Yes / No)   The radiation survey needs to be done only after removing the source from the patient and not after the implantation procedure.

Choose the right answer(s) (more than one may be correct):

35. The radiation levels (in mrem/h) on a source transport container used for in-house source movement must not exceed _____.
    a. 200
    b. 10
    c. 10 at 1 m from surface
    d. 1 at 1 m from the source

36. Short half-life brachytherapy sources (half-life of a few months) can be disposed of after storing them for _____ half-lives.
    a. 10
    b. 1
    c. 20
    d. none of the above

37. External exposure can be controlled by _____.
    a. time
    b. shielding
    c. distance
    d. wearing a personnel monitor

38. The design value of annual dose equivalent in the controlled area of a linac room is 5 mSv/y. _____ will be the design value (in mSv) for monthly limit of annual dose equivalent.
    a. 60
    b. 50
    c. 2
    d. 0.4
    e. 0.01

39. Radiation protection problems that are likely to arise in brachytherapy are _____.
    a. external (whole body) exposure
    b. radioactive contamination
    c. loss of sources
    d. none of the above

40. A brachytherapy source would be considered leaky or contaminated if _____.
    a. the monitor reads more than the background level
    b. the activity detected is more than 185 Bq
    c. the source is more than 6 months old
    d. the source has not been tested before for leakage or contamination

41. The bed shield must be kept _____.
    a. as close to the patient bed as possible
    b. as far from the patient bed as possible
    c. at 1 m from the patient bed
    d. at any distance from the patient

42. An L-bench used for ^{137}Cs source handling has a thickness of 5 cm (TVT = 2.2 cm). The radiation transmitted by the L-bench is about _____.
    a. 0.5%
    b. 1%
    c. 5%
    d. 10%

43. The maximum dose equivalent rate outside the storage safe used in the source room must be around _____.
    a. 1 μSv/h
    b. 25 μSv/h
    c. 1 mSv/h
    d. 25 mSv/h

44. The air kerma strength of a brachytherapy source on the source preparation table is 180 μGy m^2/h. _____ is the dose equivalent rate near the hands (about 30 cm from source).
    a. 2 μSv/h
    b. 20 μSv/h
    c. 200 μSv/h
    d. 2 mSv/h

45. _____ brachytherapy sources would require you to limit the time spent with the patient because it is a gamma-emitting radiation source.
    a. ^{137}Cs
    b. ^{192}Ir
    c. ^{32}P
    d. ^{125}I
    e. ^{103}Pd

46. The amount of shielding required in radiation protection depends on the _____.
    a. type of radiation
    b. output of the source (a radioactive source or a radiation generating device)
    c. size of the radiation facility or source container
    d. acceptable dose level outside the shielding
    e. occupancy factors beyond the shielding
    f. nature of occupancy
    g. all of the above

47. In radiation oncology, the staff doses are optimized mainly by _____.
    a. the design of the equipment
    b. the design of the radiation installation (treatment room)
    c. the use of lead aprons
    d. increasing the staff's distances from the source
    e. all of the above

48. A nurse caring for brachytherapy patients may _____.
    a. place a bed shield near the patient
    b. wear a personal monitoring badge
    c. put a radioactivity caution sign on the room door
    d. sit near the patient all the time to offer any assistance that may be required
    e. let patients' relatives attend to the patients if they desire

49. If the a sealed source comes out of the catheter and falls on the floor, the brachytherapy nurse must _____.
    a. immediately close the room door and wait until the physician comes to attend to the patient
    b. pick up the source using the tongs and place it inside the emergency container available in the room
    c. immediately inform the RSO or the concerned physician regarding the incident
    d. ask the patient to go home
    e. inform the patient's relative

50. The nurse, in the case of visitors, _____.
    a. will restrict the visiting time as advised by the RSO
    b. may deny admission if visitors are less than 18 years of age
    c. may deny admission if visitors are pregnant
    d. will deny admission if visitors are very old (>70 years)

51. Sealed sources that are no longer in use _____.
    a. must be disposed of in special garbage cans
    b. must be sent to an authorized person or agency or source manufacturer for disposal
    c. can be kept in the department without disposing of
    d. must be disposed of only after making a record of the source and its disposal details
    e. can be stored carefully in the office of the radiation safety officer

52. Brachytherapy patients with radioactive implants _____.
    a. can be treated in general wards along with other patients
    b. must be having bed shields so that any adjacent bed brachytherapy patient does not get unjustified exposures
    c. should be surveyed to know the air kerma rates at 1 m from the patient so that nursing times and visitor times can be calculated
    d. can be discharged any time from the hospital

53. The actual activity content of a source compared to its apparent activity will be _____.
    a. nearly zero
    b. greater
    c. smaller
    d. equal
    e. smaller or greater depending on the encapsulation thickness

54. Patients treated with _____ sources shall not be discharged from the hospital until the sources have been removed.
    a. ^{60}Co
    b. ^{192}Ir
    c. ^{137}Cs
    d. ^{125}I
    e. ^{198}Au

E. Structural Shielding

Circle the right answer (Yes or No):

1. (Yes / No) Primary radiation is the radiation emitted by the source or the source head (teletherapy head).

2. (Yes / No) A primary barrier is the wall that receives primary radiation.

3. (Yes / No) The designated doses surrounding the treatment rooms or source handling rooms depend on the nature of occupancy of these areas.

4. (Yes / No) A controlled area around a radiation room is one that will be occupied only by the employees of the installation and not the visitors or public.

5. (Yes / No) While shielding the radiation treatment rooms, minimum exposure conditions are made use of.

6. (Yes / No) X-rays shielding with concrete is adequate for the shielding of electron beams and secondary neutrons.

7. (Yes / No) The rooms surrounding radiotherapy treatment rooms must be preferably of high occupancy and must be preferably for children.

8. (Yes / No) For megavoltage installations, the leakage barrier usually far exceeds that required for the scattered radiation.

9. (Yes / No) For ^{60}Co therapy installations, the leakage barrier usually far exceeds that required for the scattered radiation.

10. (Yes / No) In a simulator, the primary beam is significantly attenuated by the patient and the image intensifier.

11. (Yes / No) The simulator room door must have a lead lining of 1 to 2 mm.

12. (Yes / No) The design value for a linac controlled room is 10 mSv/y. This is usually approximated to 0.2 mSv/wk for shielding calculation purposes.

13. (Yes / No) The design value for a linac controlled room is 0.1 mSv/y. This dose cannot be further reduced for shielding thickness calculations.

14. (Yes / No) The design value for an uncontrolled area outside the linac wall is 0.5 mSv/y. This is usually approximated to 0.01 mSv/wk for shielding calculation purposes.

15. (Yes / No) The design value for an uncontrolled area is 1 μSv/y. This dose cannot be further reduced for shielding thickness calculations.

16. (Yes / No)   Scatter fraction, a(α, 20 x 20) is the fraction of the primary dose that scatters from the patient at a particular angle α for a field size 20 cm x 20 cm at the isocenter.

17. (Yes / No)   While computing the workload, physics workload can be safely neglected.

18. (Yes / No)   The TVL for leakage radiation and scatter radiation is the same for a linac.

Choose the right answer(s) (more than one may be correct):

19. If the thickness of the secondary barrier is about the same for leakage and scatter, _____.
    a.   use a thickness = larger thickness + 1 HVL
    b.   use a thickness = smaller thickness + 1 HVL
    c.   use the lesser thickness
    d.   use the larger thickness
    e.   use the sum of the two thicknesses

20. If the thickness of the secondary barrier for leakage and scatter differ by > 1 TVL, then _____.
    a.   use a thickness = smaller thickness + 1 TVL
    b.   use a thickness = larger thickness + 1 TVL
    c.   use the larger thickness
    d.   use the lesser thickness
    e.   use the sum of the two thicknesses

21. The typical leakage radiation from a linac head is assumed to be _____% of the primary beam at the isocenter.
    a.   0.01
    b.   0.1
    c.   1.0
    d.   5.0
    e.   10.0

22. Typical leakage for a linac is 0.1% of the primary beam. If W is the workload in mGy m^2/wk, then the leakage radiation from the linac head is given by _____.
    a.   W
    b.   0.01 W
    c. 10^{-3} W
    d.   W / 0.01
    e.   W / 10^{-3}

23. The scatter fraction for an angle of scatter α for a field size F cm x F cm, a(α, F x F), is given by _____.
    a.   a(α, 20x20)
    b.   a(α, 20x20) $\times$ (F $\times$ F)
    c.   a(α, 20x20) $\times$ [(F $\times$ F) / (20 $\times$ 20)]
    d.   a(α, 20x20) $\times$ [(20 $\times$ 20) / (F $\times$ F)]
    e.   (F $\times$ F) / (20 $\times$ 20)

24.   The following data are given for a linac treatment room (see the figure below).

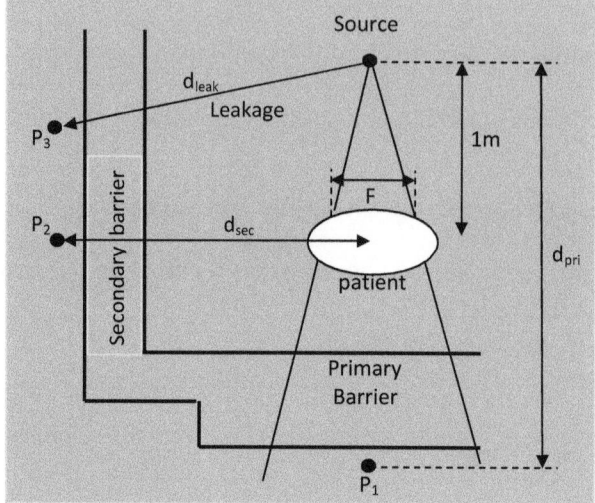

Shielding principles for a linac treatment room

Workload = W; secondary barrier use factor = 1; occupancy at P_2 = 1; d_{sec} = distance from isocenter to P_2 the point of interest. Scatter fraction for 90 deg. Scatter = a(90°, 20 x 20). The field size at isocenter = 30 cm x 30 cm. Permitted dose at = P. The dose reduction factor (RF) required for the secondary barrier is _____.

a. $W \times 1 \times [1/ P (d_{sec})^2]$
b. $\{[a(90°, 20 \times 20) \times \{30^2 / 20^2)] \times (W \times 1 \times 1)\} / [P (d_{sec})^2]$
c. $[a(90°, 20 \times 20) \times \{30^2 / 20^2)] \times (W \times 1 \times 1) \times [P (d_{sec})^2]$
d. $[P (d_{sec})^2] / \{a(90°, 20 \times 20) \times [30^2 / 20^2) \times W \times 1 \times 1]\}$
e. $[P (d_{sec})^2] / \{a(90°, 20 \times 20) \times W \times 1 \times 1\}$

25.   For the above problem if the linac leakage is L and the distance of point P_2 from linac target is d_{leak}, the dose reduction factor (RF) required at P_2 for the secondary barrier for the leakage radiation from the linac head is _____.

a. $W \times 1 \times 1 / \{P (d_{leak})^2\}$
b. $L W / \{P (d_{leak})^2\}$
c. $L W \times \{P (d_{leak})^2\}$
d. $\{P (d_{leak})^2\} / L W$
e. $L (d_{sec})^2 / P W$

26.   Secondary radiation is defined as _____.
a.   another name for the primary radiation
b.   radiation scattered by the scattering of the primary radiation beam by beam limiting devices, air volumes, walls, etc.
c.   the leakage radiation from the source head through the shielding
d.   the sum of the scatter and leakage radiation
e.   rays reflected by a field-defining mirror system

27.   The width of the primary barrier is equal to an area projected on the primary wall by _____.
a.   a 10 x 10 cm field
b.   a 30 cm x 30 cm field
c.   the maximum collimator field size
d.   the maximum collimator field size plus an extra margin of 30 cm on each width
e.   the maximum MLC field size

28. The parameters considered for radiation treatment room calculations are _____.
    a. primary radiation
    b. scatter radiation
    c. leakage radiation
    d. natural background radiation
    e. workload

29. Examples of controlled areas are _____.
    a. linac treatment rooms
    b. brachytherapy treatment rooms
    c. rooms housing treatment equipment operation consoles and accessories
    d. source handling areas
    e. rooms occupied by the radiation safety officer

30. The radiation output at 1 Gy/wk (or R/wk) from the source is known as _____.
    a. beam output
    b. output factor
    c. workload
    d. weekly load
    e. integral dose

31. The dose received by a person occupying any region around a linac treatment room depends on the _____.
    a. shielding provided by the appropriate wall
    b. use factor of the wall
    c. age of the person
    d. probability of occupation
    e. all of the above

32. The fraction of the beam-on time during which the primary beam is directed toward a particular barrier is known as the _____.
    a. use factor
    b. occupancy factor
    c. workload factor
    d. beam incident time
    e. all of the above

33. Match typical use factor to primary barriers as used in conventional shielding calculations.

Primary Barrier	Use Factor
a. floor	_____ i. zero
b. ceiling	_____ ii. 0.25
c. side wall 1	
d. side wall 2	
e. front and back walls	

34. The use factor for scatter radiation is _____.
    a. 0
    b. 0.125
    c. 0.25
    d. 0.50
    e. 1.0

35. The use factor for leakage radiation is _____.
    a. 0
    b. 0.125
    c. 0.25
    d. 1.0
    e. c or d

36. The use factor for walls of a brachytherapy treatment room is _____.
    a. 0
    b. 0.125
    c. 0.25
    d. 1.0
    e. depends on the source type

37. Match typical occupancy factors to occupied areas (full occupancy (FO) = 1, partial occupancy (PO) = 1/4) for a linac treatment room.

    Occupied Area   *Occupancy Factor*
    a. offices   _____ i.   0.25
    b. control room   _____ ii. 1
    c. corridors, employee lounges
    d. waiting rooms
    e. floor
    f. ceiling

38. In the figure below, match the numbers to the rays and barriers of the radiation installation.

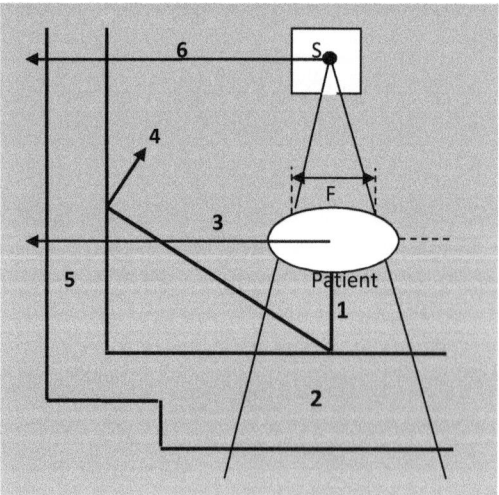

Shielding principles for a linac treatment room

Numbers in the Figure   *Rays or Barriers*
1   _____ i.   leakage radiation
2   _____ ii. multiply scattered radiation
3   _____ iii. primary barrier
4   _____ iv. scatter radiation
5   _____ v.   secondary barrier
6   _____ vi. primary radiation

39. A linac unit treats 30 patients per day, 5 days a week, and the dose per fraction is 200 cGy at 1 m from the focal spot. The workload (in mGy m^2/wk) is given by _____.
    a. 5×200
    b. $30 \times 200 / 5$
    c. $30 \times 200 \times 5$
    d. $30 \times 2000 \times 5$
    e. $30 \times 7 \times 200 \times 5$

40. For radiotherapy simulators, the workload is expressed by _____.
    a. Gy/wk
    b. R/wk
    c. mA-min/wk
    d. A-sec/wk
    e. Sv/wk

41. On a simulator unit, 10 patients are simulated per day in a 5-day week, with maximum patient simulation time of 1 minute and a maximum tube current of 5 mA. The workload of the simulator (expressed per week) is given by _____.
    a. 10
    b. 10×5
    c. $10 \times 5 \times 1 / 5$
    d. $10 \times 5 \times 5 \times 1$

42. _____ is the important factor that distinguishes the maze design of a high-energy (>10 MV) linac room compared to a low-energy (<10 MV) linac room.
    a. The size of the room
    b. Higher leakage of the higher-energy linac
    c. Primary wall thickness
    d. Neutron production which demands a special door that takes care of neuron shielding
    e. All of the above

43. The objectives of shielding a radiation treatment room are _____.
    a. to protect the expensive treatment delivery equipment from theft or misuse
    b. to reduce doses received by occupants of these areas to within their legal dose levels
    c. to reduce doses received by occupants of these areas to within their legal dose levels and to ALARA levels
    d. to protect the patients from excessive radiation
    e. all of the above

44. The primary shielding materials used for linac or HDR treatment rooms are _____.
    a. concrete / heavy concrete
    b. steel
    c. lead
    d. aluminum
    e. plywood

45. The maximum activity of the HDR ^{192}Ir source is A_{max} Ci. The maximum treatment time per patient is t_{max} minutes. The number of patients treated per week is N. The exposure rate constant (expressed at 1 m) = $\Gamma_{x,Ir}$ R m^2 / Ci h. The f factor (exposure to air kerma conversion factor) is f Gy/R. The workload W (in Gy/h @ 1 m) is given by _____.
    a. $(\Gamma_{x,Ir}\ f)\ A_{max}\ (t_{max}\ N)$
    b. $(\Gamma_{x,Ir}\ f)\ A_{max}\ (t_{max}\ N\ /\ 60)$
    c. $(\Gamma_{x,Ir}\ f)\ A_{max}\ (t_{max}\ N) \times 5$
    d. $(\Gamma_{x,Ir}\ f)\ A_{max}\ /\ (t_{max}\ N\ /\ 60)$
    e. $(t_{max}\ N\ /\ 60)\ /\ (\Gamma_{x,Ir}\ f)\ A_{max}$

46. Emergency situations that may arise in a radiation oncology department include _____.
    a. a lost source
    b. a source drawer getting stuck or not moving smoothly in a ^{60}Co head, or a brachytherapy source not returning to the container when retracted
    c. source contamination
    d. patient / staff overexposures by accident
    e. all the above

47. A typical linac room design and the required shielding data are shown in the following figure.

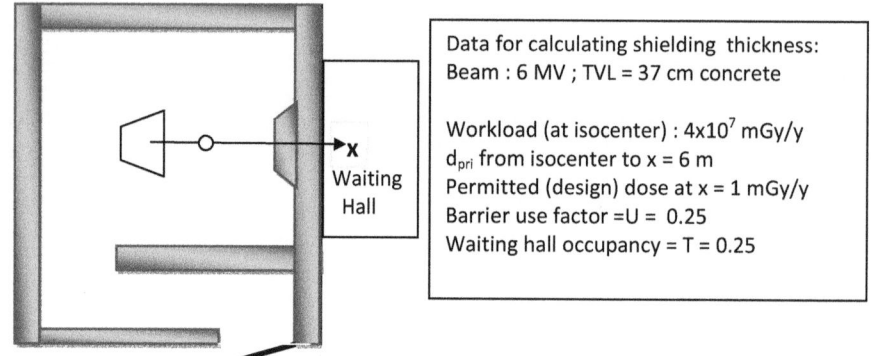

Shielding principles for a linac treatment room

Data for calculating shielding thickness:
Beam: 6 MV; TVL = 37 cm concrete; Workload (at isocenter): 4×10^7 mGy/y;
d_{pri} from isocenter to x = 6 m; Permitted (design) dose at x = 1 mGy/y;
Barrier use factor = U = 0.25; Waiting hall occupancy = T = 0.25.
The approximate reduction factor (RF) for the barrier shown in the figure is given by _____.
    a. $4 \times 10^7\ /\ 1$
    b. $4 \times 10^7\ /\ \times 0.25 \times 0.25 \times 1$
    c. $[4 \times 10^7\ /\ 6^2]\ /\ 1$
    d. $4 \times 10^7 \times 0.25 \times 0.25\ /\ 1$
    e. $[4 \times 10^7 \times 0.25 \times 0.25]\ /\ [1 \times 6^2]$

F. Radiation Monitoring

Circle the right answer (Yes or No):

1. (Yes / No) Neutron survey is important for x-ray beam qualities >10 MV.

2. (Yes / No) Shielding the ceiling of the radiation treatment room is not very important if there is no occupancy above the ceiling.

3. (Yes / No) If the radiation treatment room shielding plan has been correctly designed, it is not necessary to survey the installation following its construction since the design automatically takes care of radiation levels surrounding the installation.

4. (Yes / No) Personal monitors measure both the internal and external exposures.

5. (Yes / No) The radiation monitors used for individual monitoring purposes are known as personal dosimeters.

6. (Yes / No) The monitors used for workplace monitoring are known as work place meters.

7. (Yes / No) The personal dosimeters can also be directly calibrated in terms of effective dose.

8. (Yes / No) The survey meter must be calibrated free in air and the personal dosimeters must be calibrated on a phantom.

9. (Yes / No) Personal monitoring badges must be read by the user at the end of the day.

Choose the right answer(s) (more than one may be correct):

10. _____ personnel monitoring systems are used for measuring the x, γ, and β radiation.
    a. a TLD dosimeter
    b. an ionization chamber
    c. film badges
    d. radiochromic film

11. Personnel monitors help to _____.
    a. verify the effectiveness of radiation control practices in workplaces
    b. detect changes in workplace radiation levels
    c. provide information in case of accidental exposures
    d. reduce personal exposure

12. The film badge _____.
    a. is not affected by environmental factors
    b. requires complex calibration and calibration curves to determine dose in a mixed field (x, gamma, betas, etc.)
    c. is energy independent
    d. can be reused

13. The TLD badge _____.
    a. is not affected by environmental factors
    b. gives doses in a mixed field (x, gamma, betas, etc.)
    c. can be tissue-equivalent or non-tissue-equivalent
    d. can be reused

14. The direct reading personnel monitors _____.
    a. can reveal doses on the spot
    b. can track doses in the course of one's activity
    c. cannot read very low doses
    d. require some time to display the readings

15. The survey meters used for checking radiation levels are based on _____.
    a. gas-filled detectors
    b. scintillation-based detectors
    c. semiconductor-based detectors
    d. films

16. The gas-filled detector used in beta-gamma survey monitors is _____.
    a. an ion chamber
    b. a G-M detector
    c. proportional counters
    d. none of the above

17. The ^{60}Co treatment room or an HDR brachytherapy room must have a zone monitor to _____.
    a. indicate any potential exposure
    b. estimate the dose near the treatment room door
    c. monitor the functioning of the unit during treatment
    d. comfort the patient

18. The objective of personnel monitoring is to _____.
    a. assess the effective doses received by radiation workers
    b. demonstrate compliance with the state or the country regulations
    c. keep a record of annual doses received by the occupational workers for regulatory or legal purposes
    d. assure the radiation worker that his doses are indeed low and he has no cause for worry

19. A suitable radiation monitor for determining the dose received by a radiation worker immediately after exposure is _____.
    a. a personnel dosimeter
    b. a G-M survey meter
    c. an ionization chamber
    d. a pocket dosimeter

20. The most efficient device used for detection of light in TLD readers is _____.
    a. a scintillator
    b. a photomultiplier tube (PMT)
    c. a photocell
    d. an ion chamber

21. REM meters are used for the measurement of _____.
    a. neutron fluence
    b. neutron absorbed dose
    c. neutron dose equivalent
    d. photon dose equivalent

22. The efficiency of a β-γ survey meter is _____.
    a. less than a few percent for gamma
    b. less than a few percent for betas entering the sensitive volume
    c. more than a few percent for alphas entering the sensitive volume
    d. 0% for gammas

23. G-M counters are more sensitive compared to ionization chambers because _____.
    a. they operate under higher pressure
    b. they have much larger volume
    c. there is signal amplification due to ionization by collision
    d. they have much thinner walls

24. A personnel monitor _____.
    a. must be worn only by the person to whom it was issued
    b. can be worn during nonoccupational exposures (e.g., during his diagnostic tests)
    c. can be stored anywhere

25. The errors that have been noticed to give high radiation levels around correctly planned radiation installations are _____.
    a. improper density for the walls
    b. construction of the walls not conforming to the plan
    c. inadequate shielding for the ceiling
    d. improper positioning of the equipment
    e. all of the above

26. The survey of regions around the radiation treatment room must be conducted using _____.
    a. a 10 cm x 10 cm field size
    b. a 20 cm x 20 cm water phantom
    c. the maximum electron energy mode
    d. sensitive survey meter
    e. all of the above

27. The important place to survey for the adequacy of neutron shielding is _____.
    a. directly behind the primary barrier
    b. directly behind the secondary barrier
    c. near the door
    d. close to the linac
    e. in the corridors outside the treatment room

28. The optimal design for the door to give adequate neutron shielding is to provide _____.
    a. thick steel doors
    b. thick lead doors reinforced with steel frames
    c. a borated polyethylene door
    d. a borated polyethylene door lined outside with steel or lead
    e. any of the above

29. To conform to the recent recommendations in the field of radiation protection, the area monitors must be calibrated in terms of _____.
    a. exposure
    b. air kerma
    c. absorbed dose to water
    d. ambient dose equivalent
    e. none of the above quantities

30. Radiation monitoring is required for _____.
    a. assessing the safety of workplace conditions
    b. assessing staff doses
    c. monitoring sudden changes in radiation levels in workplaces
    d. maintaining dose records
    e. ensuring regulatory compliance
    f. all of the above

31. Gas-filled survey meters used to monitor x-ray radiation levels around radiotherapy treatment rooms operate in the _____.
    a. ionization region under normal pressure
    b. ionization region under high pressure
    c. proportional region
    d. limited proportional region
    e. GM region

32. Survey meters used to monitor neutron radiation levels around radiotherapy treatment rooms operate in the _____.
    a. ionization region under normal pressure
    b. ionization region under high pressure
    c. proportional region
    d. limited proportional region
    e. GM region

33. Neutron detectors used in neutron survey meters may contain _____.
    a. pure oxygen
    b. argon
    c. BF_3
    d. He^3
    e. a boron lining

34. Arrange the following survey meters in terms of sensitivity, highest (S1) to lowest (S4).

Survey Meters		Sensitivity
a. ionization based	_____	i. S1
b. high pressure ionization based	_____	ii. S2
c. scintillation based	_____	iii. S3
d. G-M based	_____	iv. S4

35. Match the dosimeters to the quantity they are calibrated to measure.

Dosimeters		Measurement Quantity
a. pocket dosimeter	_____	i. ambient dose equivalent H*
b. survey meter	_____	ii. personal dose equivalent Hp
c. TLD badge		
d. film badge		
e. area or zone monitors		

36. The dosimeter used for monitoring eye dose must be calibrated to read _____.
    a. H*(10)
    b. H*(3)
    c. H*(0.07)
    d. $H_p(10)$
    e. $H_p(3)$
    f. $H_p(0.07)$

37. The dosimeter used for monitoring skin doses must be calibrated to read _____.
    a. H*(10)
    b. H*(3)
    c. H*(0.07)
    d. $H_p(10)$
    e. $H_p(3)$
    f. $H_p(0.07)$

38. The dosimeters generally being used for personal monitoring purposes are _____.
    a. film badges
    b. TLD badges
    c. OSL badges
    d. chemical dosimeters
    e. all of the above

39. A film badge _____.
    a. provides permanent legal record of exposure
    b. monitors x, gamma, and beta radiation
    c. can discriminate between types of radiation
    d. is expensive
    e. is unaffected by environmental changes

40. The dose range covered by a film badge or a TLD badge is _____.
    a. 0.1 mSv to 10 Sv
    b. 1 μSv – 1 Sv
    c. 10 mSv – 100 mSv
    d. 100 mSv – 1000 Sv
    e. >1 Sv

41. Electronic personal dosimeters (EPDs) _____.
    a. are quite small and convenient to use
    b. show instantaneous or cumulative dose received at any time during usage
    c. give readings that are legal records of exposure received
    d. use semiconductors or miniature G-M detectors as radiation sensors
    e. are not easily available for purchase

42. A TLD badge _____.
    a. is more accurate and sensitive than film badges
    b. is reusable
    c. is very sensitive to temperature and humidity
    d. can be read multiple times

43. To conform to the recent recommendations in the field of radiation protection, personal monitors must be calibrated in terms of _____.
    a. exposure
    b. air kerma
    c. absorbed dose to water
    d. ambient dose equivalent
    e. personal dose equivalent

VI. Radiation Protection ANSWERS

A. General Radiation Protection

1. Yes

2. No   It is a probabilistic or stochastic effect.

3. Yes

4. Yes

5. Yes

6. Yes

7. No   It is public exposure.

8. Yes

9. Yes

10. No

11. Yes

12. No

13. Yes   For the same organ dose, the biological harm depends on the pattern of dose distribution in the organ. The latter depends on the linear energy transfer (LET) of radiation.

14. No   The whole body dose is known as the "Effective Dose."

15. Yes   Because all tissues do not have the same radiosensitivity.

16. Yes   The protection instruments must be calibrated in terms of dose equivalent and not exposure or air kerma. Standards have been established for this quantity by many standards laboratories.

17. No   H = D × Quality factor (QF) and QF = 1 only for x, gamma, and electrons.

18. Yes

19. Yes

20. Yes

21. No   These beams are very penetrating and have enormous output. The walls of the treatment room, therefore, must be constructed of concrete with sufficient thickness to reduce radiation levels outside the room to acceptable levels.

22. Yes   The personnel monitor calibrated in Personal dose equivalent, H_P, gives a conservative estimate of the effective dose received by the user. This suffices in the majority of cases where the dose received is much less than the annual effective dose limit.

23. No   The ambient radiation levels can be monitored using a zone monitor or survey meter.

24. Yes

25. No   The survey meter must be calibrated in terms of Ambient Dose Equivalent (ADE).

26. Yes   But special monitors (e.g., finger dosimeters to assess finger dose) must be used to assess the equivalent dose received by, say, hands, eye, fingers, etc.

27. Yes

28. No

29. No   They must have a beta window to register betas.

30. Yes   The neutron counter is surrounded by a thermalizing material (e.g., any hydrogenous material). The fast neutrons are slowed down and thermalized before reaching the thermal neutron counter.

31. No   A film badge can measure only thermal neutrons. However, special fast neutron badges are available for measuring the fast neutron dose.

32. a

33. a

34. c

35. a. i.   b. i.   c. ii.   d. iii.

36. a. iii.   b. and d. ii.   c i.   Source of tissue weighting factors: ICRP 103 (2007).

37. a. ii.   b. i.   c. iii.   d. and e. iv.

38. a.

39. b.

40. a

41. d

42. c

43. d

44. d

45. d

46. a   After one half-life, exposure reduces to 100 mR/h at 4 m. At 2 m the exposure rate becomes 400, and with 2 HVL it reduces to 100 mR/h.

47. c

48. a

B. Dose and Dose Limits (See NCRP 116)

1. b   NCRP 103 recommends a five-year average of 20 mSv/y and a limit of 50 mSv in any single year. Some countries recommend a lower limit of 30 mSv/y (e.g., India). While each country must abide by the regulations, practically speaking it makes very little difference due to the practice of ALARA. A very minute fraction of radiation workers receive more than 20 mSv, and most radiation workers in radiation oncology receive just a few mSv in a single year.

2. a

3. c   Here again ICRP 103 recommendations are more restrictive, applying the whole body or effective dose limits for the eyes as well.

4. c

5. b   As per NCRP 116 recommendations, if the public exposures are infrequent, a dose exceeding 1 mSv/y may be allowed (but not exceeding 5 mSv in a year). However, the exposures must be justifiable and cannot occur often in a lifetime.

6. b

7. a   Hands, feet, skin, and eyes are likely to receive exposures in certain situations (e.g., interventional cardiologists in cath labs, staff handling brachytherapy sources, etc). For hands, skin, and feet, the occupational annual equivalent dose limit is 500 mSv, and the limit for the public is one tenth of the occupational dose limits.

8. a   The average effective dose received from natural background radiation in the United States is about 1 mSv/y, excluding radon exposure. The dose can be 10 times higher in high natural background regions.

9. a

10. a. ii., iii.   b. i.   c. iv. The fetus dose limit is actually 0.5 mSv/month.

11. a.   Medical exposure, occupational exposure, and natural background exposure belong to different "bank accounts" and cannot be in any way connected with one another since the justifications for these exposures are different.

12. c, d.   The staff work for the benefit of the society and so their risks should be of the order of risk involved in any "safe" industry (not high-risk industries).

13. a

C. Time, Distance, Shielding (and ALARA Concepts)

1. Yes   ALARA is not only a safety principle, but it is a regulatory requirement for all radiation safety programs.

2. No.   The average annual regulatory dose limit for the medical radiation worker is 20 mSv, but medical radiation workers receive only a few mSv in a year, so unoptimized doses are totally unacceptable.

3. Yes   But totally unacceptable since the ALARA qualified dose limit is much less than the legal limit for almost all radiation workers.

4. Yes

5. No

6. a

7. a, b, c

8. e

9. c

10. b

11. d

12. c, e

13. e

14. g

15. a, b

16. a

17. a

D. Brachytherapy Source Handling and Storage

1. No   Institutions do need licenses to work with radioactive sources.

2. Yes

3. No   Brachytherapy sources must be stored in a special storage room with controlled access.

4. Yes

5. No   Brachytherapy sources must be carried to the patient room in a transport container that has adequate shielding and its lid securely closed so that in case of any incident, the source would not fall off.

6. Yes

7. Yes

8. No   The linen must be surveyed before removal to ensure no source has fallen on the linen.

9. Yes

10. No

11. Yes

12. Yes

13. No   The air kerma strength of the source as measured in the department using a chamber with traceability to an accredited calibration laboratory must be used in the TPS.

14. No   They must always be handled with tongs.

15. Yes

16. No

17. No

18. Yes   If a source gets dislodged, it must be immediately picked up by the nurse using tongs or long forceps and placed inside the storage container.

19. No   They must.

20. Yes

21. Yes

22. Yes

23. Yes

24. No   There are many pure beta emitters, too.

25. Yes   Thorough periodic leak tests are necessary since any undetected contamination would lead to very messy cleanup procedures later on.

26. Yes

27. No

28. No

29. No   The gamma energies of brachytherapy sources are much less than the nuclear binding energies and, hence, cannot induce radioactivity.

30. No   The visitor must stay about 6 feet away from the patient and should not go close to the patient.

31. No   Visitors are not monitored for the dose received. Visiting time restrictions ensure they are adequately protected.

32. Yes   In addition the staff dealing with implant sources, they must wear a wrist badge to monitor dose to the hands.

33. Yes   There have been accidents due to malfunctioning of area monitors, or the area monitor alarm being mistaken for false alarm, leading to a discharge of a patient with the dislodged source.

34. No

35. b

36. c

37. a, b, c

38. d   One can conveniently divide the annual design limits by 10 or 50 to get monthly or weekly dose equivalent limits. It will not make too much of a difference in the cost of construction.

39. a, b, c

40. b

41. a

42. a

43. b

44. d   Air kerma of 1 Gy is approximately 1 Sv for radiation protection purposes.

45. a, b

46. g

47. a, b

48. a, b, c

49. a, b

50. a, b, c

51. b, d

52. b, c

53. b

54. a, b, c

E. Structural Shielding

1.   Yes

2.   No   Only the area of the wall that receives primary radiation is termed a primary barrier. All the other portions of walls see only scatter and leakage, making them secondary barriers.

3.   Yes

4.   No

5.   No

6.   Yes

7. No   They must be preferably of low occupancy and not areas for children.

8. Yes Because the leakage radiation is more penetrating than the scattered radiation.

9. No   Scatter is more significant for ^{60}Co even if leakage radiation is more penetrating, so barrier thickness for scatter radiation exceeds the thickness for leakage radiation.

10. Yes

11. Yes

12. Yes There are 52 weeks in a year, but using 50 simplifies calculation and gives a conservatively higher estimate of the weekly dose.

13. No

14. Yes The same argument applies as for controlled areas.

15. No   A reduction in the design value is always theoretically possible. An optimal value is chosen depending upon the cost and benefit.

16. Yes

17. No   The total workload (clinical + physics) must be considered or we may underestimate shielding thicknesses.

18. No   Leakage energy is the primary beam energy and, hence, is more penetrating or has larger TVL.

19. a

20. c

21. b

22. c

23. c

24. b   Usually, transmission factors are determined in shielding calculations. It is much easier to calculate the dose reduction factor (RF), which is the reverse of the transmission factor. If the dose at the point of interest, with and without shielding is H (representing dose equivalent) and P (the design value of the dose equivalent), the RF = H/P. For example, if RF = 10^4, 4 TVL will be required for the shielding.

    W is defined at 1 m distance from the target. For other distances of the patient (scatterer) W / $(d_{sca})^2$ gives the incident dose. At any angle of scatter, a(θ) is the scatter coefficient for a 20 x 20 incident field. For any other field size F, the scatter fraction will be: a $\times$ ($F^2/20^2$). So scatter radiation at 1 m from the scatterer is given by [a $\times$ ($F^2/20^2$) WUT]. At the point of interest P_2, the dose is [a $\times$ ($F^2/20^2$) WUT] / P $(d_{sec})^2$.

25. b   Leakage is assumed to be isotropic, so the use factor for the walls = 1, and T is given as 1.

26. b

27. d

28. a, b, c, e

29. a, b, c, d

30. c

31. a, b, d

32. a

33. a. ii.   b. ii.   c. ii.   d. ii.   e. i.
    Use factor for a primary barrier depends on the fraction of the workload the beam is directed toward that barrier. Use factors generally depend on the situation. In isocentric units, if any barrier is not preferentially used, it is conventional to apportion the use factor equally for all barriers, or U = 0.25 for each of the barriers. But if a particular barrier is used more than the others (e.g., TBI or tangential breast irradiation) different values of use factors would be assumed. Some prefer to use U = 1 for the floor and 0.25 for other barriers as more realistic. Since the linac can rotate only in one plane, the front and back walls cannot receive primary radiation, so U = 0.

34. e   For 100% of the workload, all walls receive scatter radiation, so U = 1 for scatter radiation.

35. d   For 100% of the workload, all walls receive leakage radiation, so U = 1 for leakage radiation.

36. d   For 100% of the workload, all walls receive scatter radiation, so U = 1 for scatter radiation.

37. a. ii.   b. ii.   c. i.   d. i.   e. ii.   f. i.

38. 1. vi.   2. iii.   3. iv.   4. ii.   5. v.   6. i.

39. c

40. c

41. d

42. d

43. c

44. a, b, c

45. b

46. e

47. e

F. Radiation Monitoring

1.   Yes   It is particularly important for 15 MV x-rays most frequently used in hospitals.

2.   No   In the case of inadequate shielding of the ceiling, radiation scattered by the air and the stray radiation reaching below (and called the skyshine radiation) will give rise to unnecessary dose to the public. So shielding of the ceiling is equally important.

3.   No   How will you ensure the shielding plan is correct? You can do that only by radiation survey.

4. No   They can measure only external exposures.

5. Yes

6. No   These monitors are commonly known as area monitors or survey meters.

7. No

8. Yes   The survey meter is used to measure radiation levels in air, and the personal dosimeter is worn on the body. Hence, these monitors are calibrated in the same way.

9. No

10. a, c

11. a, c

12. a, b   A film badge is very much affected by environmental factors, like any film. The film is inherently energy dependent, but in a badge a filter system is incorporated to flatten the energy response at lower photon energies.

13. a, b, c, d

14. a, b

15. a, b, c

16. a, b

17. a

18. a, b, c, d

19. d

20. b

21. c

22. a

23. c

24. a

25. e

26. d

27. c   Concrete shielding takes care of neutron shielding in all regions except the door region, and it is here the door needs to be specially designed.

28. d

29. d

30. f

31. a, b

32. c

33. c, d, e

34. a. iv.   b. iii.   c. i.   d. ii.

35. a. ii.   b. i.   c. ii.   d. ii.   e. i.

36. e   Often $H_P(0.07)$ calibration of the monitor is used for a conservative estimate of eye lens dose

37. f

38. a, b, c

39. a, b, c But it is inexpensive, economical, and sensitive to temperature, humidity, shock, etc., though not easily damaged if handled properly.

40. a

41. a, b, d

42. a, b   TLD badges are not sensitive to temperature or humidity since deeper traps are used for reading. The badge can be read only once since the signal is lost during reading.

43. e

VII. Quality Assurance

A. Treatment Delivery Equipment (External Beam Therapy and Brachytherapy)

Circle the right answer (Yes or No):

1. (Yes / No)   Each model of radiation therapy equipment, simulator, or remote afterloading brachytherapy equipment must be type approved before granting approval for patient use.

2. (Yes / No)   Any type of approved equipment need not be individually acceptance tested at hospital sites before putting them to use on patients.

3. (Yes / No)   Acceptance tests are carried out to ensure that the equipment conforms to some national or international performance standards set for the equipment.

4. (Yes / No)   It is the responsibility of the manufacturer to demonstrate compliance of its equipment to international standards (e.g., IEC, IAEA, AAPM, etc.).

5. (Yes / No)   Radiation therapy (or brachytherapy) equipment that complies with any national or international standard would be safe for patient treatment.

6. (Yes / No)   Before carrying out acceptance tests, a radiation protection survey of the installation room must be carried out.

7. (Yes / No)   It is not necessary to carry out periodic QA on radiation therapy equipment that has been type approved and acceptance tested for patient use.

8. (Yes / No)   The beam data of any machine can be obtained from any hospital having the same model of the equipment.

9. (Yes / No)   All the QA tests on radiation therapy equipment must be carried out with the same frequency.

10. (Yes / No)   Beam quality of an accelerator beam must be established only during commissioning measurements. It need not be monitored periodically.

11. (Yes / No)   The beam output and beam uniformity must be checked with respect to gantry angle.

12. (Yes / No)   The wedge transmission factors given by the manufacturer need no verification.

13. (Yes / No)   The collimator axis does not shift with respect to collimator rotation.

14. (Yes / No)   The dwell position accuracy of a remote afterloading ^{192}Ir HDR unit can be checked by taking an auto radiograph.

15. (Yes / No)   The air kerma strength of a brachytherapy source can be measured using a calibrated well-type ionization chamber.

16. (Yes / No)   In a remote afterloading brachytherapy unit, the source travel time is independent of the source catheter length.

17. (Yes / No)   The accuracy of the timer of a remote afterloading ^{192}Ir HDR unit is difficult to verify.

Choose the right answer(s) (more than one may be correct):

18. Acceptance tests are carried out on radiation therapy or brachytherapy equipment to ensure that the equipment conforms to _____.
    a. state regulations
    b. standards set by any professional body in the country's protocol (e.g., AAPM)
    c. manufacturer's specifications
    d. none of the above

19. After acceptance testing and following commissioning, _____ measurements must be carried out.
    a. beam data
    b. beam calibration
    c. radiation protection installation survey
    d. none of the above

20. Functional performance of radiation therapy equipment can _____.
    a. not vary with time
    b. change due to malfunctions of system electronics or components
    c. change due to wear and tear on the equipment
    d. change due to environmental conditions

21. The main features of radiation therapy equipment that need be tested are _____.
    a. mechanical characteristics
    b. electrical characteristics (including emergency beam off switches and door interlocks)
    c. radiological characteristics
    d. none of the above

22. _____ in order to ensure radiation safety of patients, staff, and the public.
    a. Warning signs have to be displayed in radiation areas
    b. Radiation on/off lights must be in place at the treatment room door
    c. Emergency beam off switches must be tested daily
    d. Door interlocks must be tested daily

23. Field uniformity can be checked by measuring _____.
    a. output
    b. field size dependence of output
    c. flatness
    d. symmetry

24. The isocenter of a clinical linac can be checked by _____.
    a. optical means
    b. mechanical means
    c. radiation means
    d. none of the above

25. The important QA tests associated with the light field are _____.
    a. field size accuracy
    b. beam output
    c. light and radiation field coincidence
    d. none of the above

26. The gantry of the accelerator is adjusted to be vertical with the beam central axis normal to the couch top. The tests to be carried out on the treatment couch are invariance of the beam central axis (within specified tolerance) with respect to the couch's _____.
    a. vertical motion
    b. rotational motion
    c. table top lateral displacement
    d. collimator rotation.

27. The target to patient skin distance is measured using a/an _____.
    a. meter scale
    b. SSD rod
    c. ODI indicator
    d. none of the above

28. The beam output calibration for clinical photon beams must be carried out _____.
    a. in a water phantom for a 10 cm x 10 cm field
    b. using a Farmer-type ionization chamber
    c. at depth of dose maximum
    d. monthly

29. The QA tests on external beam therapy equipment must be performed _____.
    a. periodically
    b. only during acceptance testing
    c. daily
    d. after any major repair to the equipment

30. The isocenter of a linac is indicated by _____.
    a. ODI
    b. the SSD rod
    c. lasers installed in the room
    d. the collimator

31. The absence of a machine QA program in a hospital could lead to _____.
    a. overdosing or underdosing of patients
    b. fatal accidents
    c. better quality of treatment
    d. none of the above

32. The tolerance for the localizing laser or the ODI indicator (at normal clinical distance) for conventional treatment is _____.
    a. 1 mm
    b. 2 mm
    c. 5 mm
    d. none of the above

33. The constancy of linac output for photon and electron beams (of at least one energy) must be checked
    a. daily
    b. weekly
    c. monthly
    d. only during commissioning

34. The constancy of linac output, checked daily for photon and electron beams (of at least one energy), when compared to the benchmark values established during commissioning, must be within _____.
    a. ±1%
    b. ±3%
    c. ±5%
    d. none of the above

35. The monitor used in a linac to control treatment duration must be checked for _____.
    a. any end effect
    b. linearity
    c. physical damage
    d. none of the above

36. The symmetry and flatness of the clinical photon beam can be checked using _____.
    a. film
    b. a beam profiler
    c. a calorimeter
    d. none of the above

37. Using graph paper, one can easily check the _____ of a linac.
    a. orthogonality of the jaws
    b. parallelism of the jaws
    c. field size
    d. symmetry of the jaws

38. QA tests on beam therapy equipment must be carried out for _____.
    a. any one position of the gantry
    b. different gantry angles
    c. vertical gantry position only

39. The x-ray beam flatness must be within about _____.
    a. ±1%
    b. ±3%
    c. ±5%
    d. none of the above

40. The electron beam flatness must be within about _____.
    a. ±1%
    b. ±3%
    c. ±5%
    d. none of the above

41.   The x-ray and electron beam symmetry must be within _____.
      a.   ±1%
      b.   ±3%
      c.   ±5%
      d.   none of the above

42.   The x-ray beam flatness is specified for _____.
      a.   10 cm depth in water
      b.   the central 80% of the field width
      c.   the x, y, and diagonal axes
      d.   none of the above

43.   The x-ray beam symmetry is specified for _____.
      a.   10 cm depth in water
      b.   points equidistant from the central axis
      c.   the x, y, and diagonal axes
      d.   none of the above

44.   The electron beam flatness is specified for _____.
      a.   10 cm depth in water
      b.   the central 80% of the field width
      c.   the x, y, and diagonal axes
      d.   none of the above

45.   The electron beam symmetry is specified for _____.
      a.   10 cm depth in water
      b.   points equidistant from the central axis
      c.   the x, y, and diagonal axes
      d.   none of the above

46.   Disadvantages of custom blocking instead of MLC blocking are _____.
      a.   the time and effort required in fabricating custom blocks
      b.   the handling and placement of heavy blocks
      c.   the chemical hazards involved in the fabrication of Cerrobend™ blocks
      d.   its being more conformal than MLC shapes

47.   A quality assurance program helps _____.
      a.   to detect dosimetry errors or errors in targeting the tumor and the critical targets during treatment delivery
      b.   to avoid accidents or reduce the likelihood of their occurrence in radiotherapy, which can be very serious
      c.   in regulatory compliance
      d.   in reducing the occurrence of cancer
      e.   in all the above

48.   In external beam radiotherapy, _____ is considered misadministration.
      a.   having no written directive
      b.   getting the wrong patient
      c.   getting the wrong site
      d.   weekly dose exceeding 30%
      e.   total dose exceeding 20%
      f.   all of the above
      g.   none of the above

49. For stereotactic radiosurgery/radiotherapy, _____ are considered misadministration.
    a. having no written directive
    b. getting the wrong patient
    c. getting the wrong site
    d. weekly dose exceeding 30%
    e. total dose exceeding 10%
    f. all of the above
    g. none of the above

50. The accuracy of IMRT delivery depends on _____,
    a. mechanical accuracy of the linac
    b. beam stability
    c. MLC system accuracy
    d. beam output accuracy
    e. all of the above

51. The important features that must be considered in linac QA relate to _____.
    a. dosimetry
    b. reliability of mechanical movements
    c. radiation safety
    d. electrical features of the equipment
    e. all of the above

52. QA tests are usually categorized into daily, monthly, quarterly, semiannual, and annual by various national protocols The frequency of testing depends on the _____.
    a. time required for testing
    b. probability of change
    c. impact of the change on treatment
    d. stability of the parameter
    e. all of the above

53. All the linac QA tests must be performed _____.
    a. daily
    b. weekly
    c. monthly
    d. annually
    e. with different frequencies depending on the protocol recommendations parameters
    f. only once

54. _____ is the definition of misadministration in fractionated (>3fx) external beam treatment delivery.
    a. The wrong patient
    b. The wrong site
    c. A weekly dose (overdose) >30%
    d. A total dose (overdose) >20%
    e. Any of the above

55. _____ is the definition of misadministration in SRT/SRS (<3fx) external beam treatment delivery.
    a. Having no written directive
    b. The wrong patient
    c. The wrong site
    d. A total dose error >10%
    e. Any of the above

56. TG-40 recommends QA tests on _____ be performed on a linac on a daily basis.
    a. x-ray and electron output constancy
    b. localization lasers
    c. x-ray and electron energy constancy
    d. safety door interlocks
    e. collimator field size indication
    f. all of the above

57. _____ is the tolerance for x-ray and electron output constancy in daily (D) and monthly (M) testing, according to TG-142.
    a. 5% (D) and 5% (M)
    b. 5% (D) and 3% (M)
    c. 3% (D) and 3% (M)
    d. 3% (D) and 2% (M)
    e. 2% (D) and 1% (M)

58. The mechanical tests (tolerances) recommended for testing on a daily basis, according to TG-142, for non-IMRT machines are _____
    a. laser localization (2 mm)
    b. ODI (2 mm)
    c. door interlock (2%)
    d. collimator size indication (2 mm)
    e. light/radiation field coincidence (2% or 1%, whichever is greater)
    f. all of the above

59. The tolerance for daily laser localization tests, according to TG-142, for IMRT and stereotactic machines are _____.
    a. ±2 mm for all machines
    b. ±2 mm for IMRT machines and ±1.5 mm for stereotactic machines
    c. ±1.5 mm for all machines
    d. ±1.5 mm for IMRT and ±1 mm for stereotactic machines
    e. ±1 mm for all machines

60. According to AAPM TG-142 recommendations, x-ray output constancy checks should be performed _____.
    a. daily
    b. weekly
    c. every two weeks
    d. monthly
    e. annually

61. According to TG-142, the functionality of _____ safety features need to be ensured on a daily basis for non-IMRT and IMRT machines.
    a. door interlock
    b. door closing safety
    c. audiovisual monitor
    d. area monitor
    e. all of the above

62. According to TG-142, the tolerance for light/radiation field coincidence is _____mm or _____% on a side, whichever is larger.
    a. 3 or 3
    b. 2 or 2
    c. 2 or 1
    d. 1 or 2
    e. 1 or 1

63. As per TG-142 recommendations, collimator rotation at isocenter tolerance is _____.
    a. ±1 degree
    b. ±2 mm
    c. 2%
    d. ±1 mm
    e. ±0.5 mm

64. The tolerance for x-ray beam flatness constancy, as per TG-142, is _____.
    a. 3%
    b. 2%
    c. 1.5%
    d. 1%
    e. 0.5%

65. The tolerance for electron beam flatness constancy, as per TG-142, is _____.
    a. 5%
    b. 3%
    c. 2%
    d. 1%
    e. 0.5%

66. The tolerance for an x-ray beam symmetry test, as per TG-142, is _____.
    a. 4%
    b. 2%
    c. 1%
    d. 0.5%
    e. 0.1%

67. The tolerance for MU linearity for x-ray and electron beam treatment modes, as per TG-142, for ≥5 MU is _____.
    a. ±5%
    b. ±3%
    c. ±2%
    d. ±1%
    e. ±0.5%

68. The tolerance for x-ray and electron beam output constancy with respect to gantry angle, as per TG-142, is _____.
    a. ±5%
    b. ±3%
    c. ±2%
    d. ±1%
    e. ±0.5%

69. According to TG-142, the tolerance for verifying wedge factors for all energies is _____.
    a. ±5%
    b. ±3%
    c. ±2%
    d. ±1%
    e. ±0.5%

70. The *three* sources of leaf leakages of MLC leaves are _____.
    a. leakage through the leaves
    b. leakage between the leaves
    c. leakage between abutting leaves
    d. leakage through primary collimators
    e. head leakage

71. How is dynamic MLC performance tested?
    a. Since the leaves are moving fast, this performance cannot be tested.
    b. This can be tested only by calculations.
    c. By measuring the dose, the dynamic MLC performance can be evaluated.
    d. Sweeping gap dynamic MLC plan moving at certain speed is used to test the performance.
    e. This test is usually not performed

72. What is a picket fence test?
    a. There is no such test defined in linac QA.
    b. It tests the accuracy of dynamic MLC performance.
    c. It tests image quality of on-board imaging systems.
    d. It tests the static MLC performance.
    e. None of the above are true.

73. _____ tests are recommended by AAPM TG-142 report for MLC annual QA.
    a. MLC transmission
    b. Leaf position repeatability
    c. Moving window IMRT
    d. MLC spoke shot
    e. Leaf interdigitization
    f. All of the above

74. Safety interlock tests must be carried out _____.
    a. daily
    b. weekly
    c. monthly
    d. semiannually
    e. annually

75. Patient dose from CBCT depends on _____.
    a. kV/mAs
    b. field size and beam quality
    c. the number of images in a rotation
    d. patient size and the irradiated body part
    e. all of the above

76. Which of the following tolerances are correctly stated, according to TG-142 recommendations?
    a. Imaging-treatment isocenter coincidence (SRS only) must be ≤1 mm.
    b. Positioning/repositioning (SRS only) must be ≤1 mm.
    c. Imaging-treatment isocenter coincidence (SBRT only) must be ≤2 mm.
    d. Positioning/repositioning (SBRT only) must be ≤2 mm.

77. The main features of OBI workstations are _____.
    a. acquisition of kV images (radio and fluoro mode)
    b. 2D/ 2D automatic and manual matching of reference and acquired images
    c. 2D/ 2D automatic and manual matching of reference and acquired kV/MV images with implanted markers
    d. qualitative verification of RPM gating with fluoroscopy
    e. CBCT acquisition and matching of 3D CBCT images to reference 3D images
    f. all of the above

78. The accuracy of IMRT depends on _____.
    a. accuracy of treatment planning
    b. accuracy of treatment delivery
    c. plan verification by actual measurements
    d. choice of the patient for IMRT
    e. all of the above

79. IMRT QA to ensure accurate delivery as per plan is important because _____.
    a. treatment planning and treatment delivery are more complex
    b. there are inherent errors in treatment planning
    c. there are inherent errors in treatment delivery
    d. there are limitations to TPS and delivery systems
    e. all of the above

80. The inherent errors in TPS and treatment delivery systems are _____.
    a. beam modeling
    b. patient modeling
    c. dose calculation methods
    d. MLC calibration
    e. machine output
    f. all of the above

81. QA procedures are essential in brachytherapy to ensure _____.
    a. patient safety
    b. staff safety
    c. accuracy and quality treatment
    d. none of the above

82. A source inventory containing _____ information must be maintained by an institution.
    a. source identity
    b. source configuration
    c. source strength (at a certain time)
    d. none of the above

83. To control safety and security of sources, _____.
    a. the source storage facility must be under the control of an RSO
    b. a source movement register must be maintained
    c. an area monitor must be available in the storage area
    d. a survey meter must be available

84. Brachytherapy sources must be _____.
    a. handled using long-handled forceps
    b. measured in terms of AKS
    c. manipulated behind an L-bench
    d. transported to the administration room using long-handled forceps.

85. The exposure rate on the surface of source transport containers (used for transporting sources from the source preparation room to the administration room) should not be more than about _____.
    a. 1 mR/h
    b. 10 mR/h
    c. 1 R/h
    d. none of the above

86. In the case of ^{192}Ir HDR sources, when inputting source strength into the TPS, one must make use of _____.
    a. the measured value of AKS
    b. the certificate value of AKS
    c. the average of the measured and certificate values
    d. the certificate value if it agrees with the measured value (to better than 3%)

87. The brachytherapy source strength must be specified only in terms of AKS, and stating the strength in other older units must be avoided since _____.
    a. the use of different units has led to incidents in brachytherapy
    b. the use of older units has unnecessarily prolonged the switchover to new units
    c. it leads to inputting wrong values into the TPS
    d. none of the above

88. A dosimeter of adequate sensitivity, accuracy, and simplicity for the calibration of brachytherapy sources in terms of AKS is _____.
    a. the Farmer chamber
    b. film
    c. a well-type ionization chamber
    d. a parallel plate chamber

89. Discrete brachytherapy sources must be leak-tested _____.
    a. only once
    b. once a month
    c. semiannually
    d. annually

90. Brachytherapy sources must be calibrated _____.
    a. by the manufacturer only
    b. by an accredited dosimetry calibration laboratory only
    c. before being used for clinical dosimetry
    d. by a user utilizing a chamber with calibration traceable to an accredited dosimetry calibration laboratory

91. A brachytherapy source is considered leaky if the wipe test indicates an activity exceeding _____.
    a. 5 nCi
    b. 185 Bq
    c. 1 μCi
    d. 5 μCi

92. The integrity of sources used in remote afterloading equipment can be checked by _____.
    a. periodically taking out the sources and testing them for leakage
    b. observing any change in the strength of the sources (after correcting for decay)
    c. wiping the inside of the catheters with tissue paper dipped in alcohol and checking for any activity
    d. none of the above

93. The exact position of the source in the applicator can be checked by _____.
    a. visual inspection
    b. measuring the AKS strength of the source
    c. taking an auto radiograph
    d. none of the above

94. To ensure safe operation of remote afterloading equipment, _____ must be checked daily.
    a. treatment tube/catheter integrity
    b. indicator lights
    c. AKS of the source
    d. none of the above

95. After removing the source from a brachytherapy patient, _____.
    a. the patient must be surveyed using a survey meter
    b. the source container must be surveyed after source replacement
    c. the room may be surveyed
    d. none of the above procedures need to be followed

96. Patients with permanent implants can be released only when the dose rate at 1 m from the patient is less than _____.
    a. 2.5 mrem/h
    b. 5 mrem/h
    c. 50 μSv/h
    d. none of the above

97. When a brachytherapy source is found to be leaky, the user _____.
    a. must isolate the source
    b. can continue to use it until a replacement source becomes available
    c. must report the matter to the regulatory authority
    d. can repair the source and reuse it

98. In the case of brachytherapy sources, _____.
    a. all the sources must be leak tested
    b. short half-life sources need not be tested for leakage
    c. sources not in use need not be tested for leakage

99. In a brachytherapy treatment room, _____ must be available all the time.
    a. a radiation survey meter
    b. a temporary storage container
    c. a cutter and long-handled tongs

100. The AAPM task group reports that deal with brachytherapy are _____.
    a. TG-21
    b. TG-41
    c. TG-43
    d. TG-56
    e. TG-59
    f. all the above

101. The main objectives of HDR brachytherapy QA are ensuring _____.
    a. patient safety
    b. staff safety
    c. public safety
    d. treatment accuracy
    e. all of the above

102. The main safety aspects that must be considered in the practice of HDR brachytherapy are _____.
    a. installation safety
    b. equipment safety
    c. emergency response
    d. source safety
    e. all of the above

103. The physical aspects of dose delivery accuracy in HDR treatment depends on the accuracy of _____.
    a. source calibration
    b. dose data
    c. other relevant corrections like transit time, applicator attenuation, etc.
    d. the stage of the disease
    e. TPS calculation
    f. all of the above factors

104. The clinical aspects of dose delivery accuracy in HDR treatment depend mainly on the accuracy of _____.
    a. imaging
    b. implants with respect to the PTV
    c. source decay
    d. implant data transfer to TPS
    e. dwell time optimization
    f. all of the above

105. A simple rule to observe to ensure QA in HDR brachytherapy is to _____.
    a. check TPS QA thoroughly
    b. perform source strength calibration every week
    c. do the implant accurately
    d. check whatever can be checked
    e. check all the above factors

106. In the case of power failure or an electrical problem, how can the source be withdrawn from the patient?
    a. One has to wait for the resumption of power.
    b. There must be an emergency power option in the treatment room.
    c. There must be provision for manually withdrawing the source into the afterloader in case of an emergency.
    d. The applicator wire must be cut and the source transferred to a container by a medical physicist.
    e. The service engineer must be immediately called to the site.

107. How should one handle an emergency situation involving the failure of all source retraction mechanisms?
    a. No method has so far been devised to handle such an emergency.
    b. Such an emergency is impossible to arise due to the automatic retraction and manual retraction mechanisms available.
    c. The patient and the HDR unit must be moved to an isolated place.
    d. Written and tested emergency procedure must exist.
    e. A forceps, cutting player, survey meter, emergency container, etc. must be available for cutting the wire from the HDR unit and transferring the source into the emergency container.
    f. The service engineer must be immediately called for assistance.

108. A physicist or physician engaged in an emergency operation is likely to get a dose of about _____.
    a. 1 mSv
    b. 10 mSv
    c. 50 mSv
    d. 500 mSv
    e. 10 Sv

109. What happens if one programs an HDR unit and initiates a trial run without connecting the applicator, or if there is a kink or obstruction in the transfer tube, or if the indexer has not been locked?
    a. The source comes to the end of the transfer tube and stops.
    b. The source would fall out and must be loaded back into the afterloader.
    c. The source comes to the point of obstruction or to the end of the tube and returns to the afterloader.
    d. The system generates an error code indicating the problem, but the source would not be loaded into the channel.
    e. The system self corrects for the mistakes.

110. Is there a way to know that the source actually went to the treatment position or that it was actually retrieved after the treatment and did not remain in the patient?
    a. There is no way to know this.
    b. The system has a way of detecting these mistakes and giving the indication.
    c. It can be known only from patient follow-up.
    d. Only during routine QA testing can one know this through measurements.
    e. The service engineer can tell the user about this.

111. The source position accuracy must be verified to be ±_____.
    a. 0.1 mm
    b. 0.5 mm
    c. 1 mm
    d. 3 mm
    e. 5 mm

112. Is there a need and is there a way to check for the source integrity?
    a. There is no need, and there is no way to check.
    b. There is a need, but there is no way to check.
    c. The source integrity is guaranteed, and it need not be checked.
    d. The manufacturer has already checked the integrity and, hence, there is no need to check.
    e. It can be checked by checking the cable and applicator for contamination.

113. The HDR unit leakage under maximum loading conditions must conform to _____.
    a. <100 µGy/h at 5 cm from the surface
    b. <10 µGy/h at 1 m from the source
    c. >50 µGy/h at 5 cm from the surface
    d. >5 µGy/h at 1 m from the source
    e. <1 µGy/h at the door of the treatment room

114. Which of the following statements regarding the HDR source is true?
    a. Source calibration must agree with the manufacturer certificate value to within ±5%.
    b. The source needs no calibration, and the certificate value of the source strength must be entered into the TPS system.
    c. The source may be conveniently calibrated using a well type chamber.
    d. Source strength must be measured every day and must be entered into the TPS system before the start of treatment.
    e. Source strength must be calculated by the user and entered into the TPS system.

115. The brachytherapy source strength must be specified in terms of _____.
    a. air kerma strength (AKS) or exposure strength (ES)
    b. activity
    c. both AKS and activity
    d. size
    e. source energy

116. During the QA testing of the HDR unit, the "interrupt" and "emergency" buttons _____.
    a. need not be tested
    b. can be operated only when there is any problem with the treatment
    c. must always be tested to ensure that the source will be retracted when pressed
    d. cannot be tested since they are meant for operating in service mode only
    e. can be tested, but only one of them can be tested

117. When the treatment is "ON," if someone opens the door to enter the room, _____.
    a. the person will receive high radiation dose
    b. the door will not open since it is locked
    c. the source will be automatically retracted
    d. an alarm will go off alerting the person trying to enter the room
    e. the patient will switch off the unit

118. During the HDR treatment, the technologist will be able to observe the patient all the time. If the technologist sees the patient managing to pull the catheter out of the applicator, what measure can be taken to minimize any radiation hazard?
    a. Nothing much can be done.
    b. The patient can be instructed to put the source catheter into the applicator.
    c. The unit can be reprogramed to reduce the dose.
    d. Pressing the "emergency" button on the console will retract the source into the safe.
    e. The technologist must enter the room and put the catheter back into the applicator.

119. The monitors and emergency switches must be checked _____ in an HDR room.
    a. daily
    b. weekly
    c. monthly
    d. quarterly
    e. semiannually

B. Imaging Systems for Treatment Planning

Circle the right answer (Yes or No):

1. (Yes / No)   It is not very important to test the QA of a simulator since the risk of high radiation exposure is very low with this equipment.

2. (Yes / No)   The tolerances of simulator parameters need not be as stringent as that of radiation therapy equipment.

3. (Yes / No)   The choice of fields for optimal treatment can cause image distortion in the simulator film.

4. (Yes / No)   A simulator possesses all the mechanical and geometrical features of an external beam treatment unit.

5. (Yes / No) Delineators are used for defining the treatment field size in a simulator.

6. (Yes / No) Beam-limiting diaphragms define the treatment field size in a simulator.

7. (Yes / No) The important parts of a simulator are a) the x-ray source, b) patient support assembly or patient treatment couch, and c) the imaging system.

8. (Yes / No) A CT scan provides the patient's anatomical information existing in the scanned slice.

9. (Yes / No) A CT scanner can provide a scan slice in any plane of the patient.

10. (Yes / No) A CT scan can provide functional information.

11. (Yes / No) The two-dimensional image produced by a simulator showing the treatment area can be digitized and displayed or stored.

Choose the right answer(s) (more than one may be correct):

12. The collimator, the gantry, and the couch top of a simulator (or linac) rotate about their respective axes. The recommended tolerance for locating the isocenter in these rotations is a sphere with a dimension of _____.
   a. 2 mm radius
   b. 2 mm diameter
   c. 5 mm radius
   d. none of the above

13. The tolerances for the gantry, couch, and collimator motions of a simulator, compared to MV treatment machines, _____.
   a. must be at least as stringent
   b. can be much less stringent
   c. have no limits and can take any values

14. The constancy of CT numbers for tissue, air, and bone-like materials _____.
   a. must be checked at regular intervals
   b. need no checking
   c. must be enough to check once

15. Scan and couch positional accuracy for CT planning must be _____.
   a. <1 mm
   b. about 2 mm
   c. about 5 mm
   d. about 1 cm

16. The kVp set on a simulator _____.
   a. can be measured using a kVp meter
   b. must agree with the measured value to better than 5%
   c. can have any accuracy

17. To measure the kVp that is set on a simulator, the kVp meter was placed on the treatment couch and 60 kVp was set on the machine. mA = 100. For mAs set to 20, 60, and 80, measured kVp were found to be 61.5, 61.6, and 61.7, respectively. The kVp displayed by the machine is _____.
   a. within tolerance limit
   b. beyond tolerance limit
   c. depends on mAs

18. The reproducibility of simulator output must be better than _____.
    a. 1%
    b. 3%
    c. 5%
    d. 10%

19. The linearity of simulator output must be better than _____.
    a. 1%
    b. 3%
    c. 5%
    d. 10%

20. _____ x-ray beam parameters must be tested for a therapy simulator.
    a. kVp
    b. HVL
    c. Focal spot
    d. Dose (for typical mAs used in simulations)
    e. All of the above

21. The daily tests recommended for a treatment simulator are _____.
    a. functionality of safety switches and the door interlock
    b. lasers
    c. ODI
    d. collimator rotation isocenter
    e. cross hair centering

22. The tolerance for laser alignment is _____ mm.
    a. 5
    b. 3
    c. 2
    d. 1
    e. 0.5

23. The important monthly tests for a treatment simulator are _____.
    a. field size indication
    b. cross hair centering
    c. light/radiation field coincidence
    d. focal spot axis indicator
    e. all of the above

24. Following the preparation of a patient on the CT table for a simulation, which of the following steps are performed for CT simulation?
    a. A set of fiducial marks are placed identifying a reference plane.
    b. Thin CT slices are taken in the regions of interest, and a virtual patient is created on the computer system.
    c. The fiducial marks coincide with the treatment setup and identify the isocenter.
    d. Treatment fields are marked on the patient.
    e. All of the above are true.

25. The AAPM task group reports dealing with the QA of the CT simulator include _____.
    a. TG-25
    b. TG-51
    c. TG-62
    d. TG-66
    e. TG-101

26. The AAPM task group reports dealing with the QA of CT simulators include _____.
    a. TG-43
    b. TG-51
    c. TG-62
    d. TG-66
    e. TG-142

27. The AAPM task group report on the QA of CT simulators mainly deals with _____.
    a. localization
    b. imaging performance
    c. patient safety
    d. treatment dose calculations
    e. treatment delivery

28. _____ relates to the imaging performance of a CT sim.
    a. Spatial resolution
    b. Contrast resolution
    c. Field uniformity
    d. Laser positioning
    e. Image noise
    f. All of the above

29. According to TG-66, the tolerance for the CT number of water (in HU) is ±_____.
    a. 2
    b. 5
    c. 7
    d. 10
    e. 15

30. _____ distinguishes a diagnostic CT scanner from a CT simulator.
    a. A flat tabletop
    b. A larger bore opening with a larger field-of-view
    c. External patient positioning/marking lasers
    d. Virtual simulation software
    e. All the above

31. What is virtual simulation?
    a. Actually, no simulation is done by the simulator.
    b. It is treatment simulation done on the actual patient.
    c. It is treatment simulation on a virtual patient using virtual patient data.
    d. It is treatment simulation on a virtual patient generated by the actual patient data.
    e. None of the above are true.

32. For ensuring geometric accuracy, _____ must be tested in a CT simulator.
    a. alignment of gantry lasers with the imaging plane center
    b. orientation of gantry lasers with the imaging plane
    c. spacing of lateral wall lasers with respect to the gantry lasers
    d. orientation of both wall and ceiling lasers with respect to the imaging plane
    e. all of the above

33. The daily QA tests for CT Sim as per AAPM TG-66 recommendations include _____.
    a. laser alignment with respect to the center of imaging plane
    b. the CT number for water
    c. in-plane spatial integrity
    d. image noise
    e. all of the above

34. The monthly QA tests for CT Sim as per AAPM TG-66 recommendations include _____.
    a. laser alignment with respect to the imaging plane
    b. table orientation and motion
    c. image uniformity (at the most commonly used kVp)
    d. CT number accuracy (4–5 materials)
    e. x-ray tube leakage
    f. all of the above

35. The annual QA tests for CT Sim as per AAPM TG-66 recommendations include _____.
    a. gantry tilt
    b. table indexing and positioning
    c. scan localization (scout image accuracy)
    d. radiation profile or dose
    e. all of the above

36. The annual QA tests for CT Sim as per AAPM TG-66 recommendations include _____.
    a. electron density/CT # calibration
    b. spatial resolution
    c. contrast resolution
    d. virtual simulation software testing
    e. all of the above

37. According to TG-66, annual QA tests of _____ must be performed on the CT scanner.
    a. table indexing and position accuracy
    b. gantry tilt and position accuracy
    c. scan localization
    d. dose calculation accuracy
    e. image registration accuracy

38. An "alignment of gantry lasers with imaging plane center" test must be carried out _____.
    a. daily
    b. weekly
    c. monthly
    d. annual
    e. only during commissioning

39. According to TG-66, the tolerance for an "Alignment of gantry lasers with imaging plane center" test is ±_____ mm.
    a. 5
    b. 3
    c. 2
    d. 1.5
    e. 0.5

40. According to TG-66, monthly QA testing on the CT scanner must include _____.
    a. orthogonality of the CT scanner tabletop with the imaging plane
    b. tabletop vertical and horizontal motion according to digital indicators
    c. gantry tilt accuracy
    d. table indexing and position accuracy
    e. all of the above

41. According to TG-66, the tolerance for the "Orthogonality of the CT scanner tabletop with the imaging plane" test is ±_____ mm.
    a. 5
    b. 3
    c. 2
    d. 1.5
    e. 0.5

42. According to TG-66, the tolerance for the "Tabletop vertical and horizontal motion according to digital indicators" test is ±_____ mm.
    a. 5
    b. 3
    c. 2
    d. 1
    e. 0.5

43. According to TG-66, the tolerance for the "Table indexing and position" test and the "Scan localization" test are ±_____ mm and ±_____ mm over the scan range, respectively.
    a. 5 and 5
    b. 3 and 3
    c. 2 and 1
    d. 1 and 1
    e. 0.5 and 1

44. According to TG-66, the tolerance for the "Gantry tilt accuracy" test and the "Gantry tilt position accuracy" test are _____ and _____, respectively.
    a. ±5° and ±5°
    b. ±3° and ±3°
    c. ±2.5° and ±2.5°
    d. ±2° and ±2°
    e. ±1° and ±1° or 1mm

C. Treatment Planning System (TPS)

Choose the right answer(s) (more than one may be correct):

1. A 2D treatment planning system (TPS) incorporates _____.
    a. no patient-related information
    b. a computation algorithm for dose calculation and display in 2D
    c. treatment plan optimization
    d. sophisticated inhomogeneity corrections

2. A 3D TPS incorporates _____.
    a. patient information in 3D
    b. a computation algorithm for dose computation and display in 2D or 3D
    c. more accurate inhomogeneity corrections compared to earlier systems
    d. treatment plan optimization

3. While comparing measured and TPS computed data, _____.
    a. the margin for agreement is the same for all cases
    b. larger margins must be set for dose in the buildup region for oblique incidence or for complex fields
    c. in high dose gradient regions, margins should be set in terms of mm.

4. To obtain DVH, _____.
    a. 3D dose computations are necessary
    b. 3D definition of patient anatomy is not necessary
    c. 3D information on PTV and OAR is required
    d. 3D dose computation is necessary for all points in the patient.

5. TPS computed dose can be checked by _____.
    a. beam calibration
    b. manual dose calculation for specific cases
    c. *in vivo* dosimetry
    d. actual measurement for specific cases

6. The checksum utility of a TPS _____.
    a. ensures that the operator has not changed the program code or beam data
    b. ensures that the system data have not been corrupted
    c. must be run at each start of the TPS or at least weekly
    d. is not supplied by any vendor

7. QA tests on a TPS must be carried out _____.
    a. during acceptance testing
    b. every year
    c. at regular intervals
    d. with the same frequency of testing for all parameters.

8. Beam's-eye view (BEV) is _____.
    a. the view as sighted from the source position
    b. the collimator field size
    c. the CT cut along the beam's central axis
    d. none of the above

9.   The main AAPM documents available for TPS QA guidance are _____.
      a.   TG-53
      b.   TG-55
      c.   TG-23
      d.   TG 43
      e.   all of the above

10.   TPS system QA involves _____.
      a.   acceptance testing
      b.   commissioning
      c.   routine QA
      d.   algorithm modifications for accurate dose estimation
      e.   all of the above

11.   TPS acceptance testing _____.
      a.   is done following clinical use for at least one week
      b.   is required to confirm that the unit procured meets the purchase specifications quoted
      c.   must conform to the tests developed by the vendor and tests described in AAPM TPS protocols
      d.   must be repeated at every 6 months in the first year of its use
      e.   is true for all the above

12.   Acceptance test refers to _____.
      a.   tests performed by the vendor
      b.   routine QA of the device
      c.   tests performed to confirm that the device performs according to its specifications
      d.   tests to be performed following any repairs carried out on the device
      e.   none of the above definitions

13.   Acceptance testing of TPS systems involve three important components, namely _____.
      a.   computer hardware
      b.   computer software and functions
      c.   benchmark tests
      d.   input data
      e.   output data

14.   TPS commissioning mainly involves _____.
      a.   performing all beam data measurements that need to be input into the TPS system
      b.   testing TPS dose distribution calculations against actual measurements
      c.   planning of actual patients on the TPS
      d.   checking the accuracy and linearity of I/O devices
      e.   accuracy and limitations of algorithm verification for the range of clinical use
      f.   all of the above

15.   Some of the Monte Carlo codes used in computations are _____.
      a.   EGS4
      b.   MCNP
      c.   ICRP
      d.   NRPB
      e.   GEANT

16. The different types of calculation algorithms used in TPS systems include _____.
    a. semiempirical
    b. model based
    c. Monte Carlo based
    d. hybrid
    e. organ based
    f. all of the above

17. In TPS, algorithms are used for _____.
    a. MU calculations
    b. isodose distributions
    c. DVH generation
    d. DRR generation
    e. IMRT optimization
    f. all of the above

18. TPS QA is important for which of the following reasons?
    a. Unlike delivery errors, which are likely to cancel one another out to some extent, TPS errors act as systematic errors with respect to patient treatment and will affect all the patients in a facility.
    b. Computer programs cannot be considered as perfect.
    c. TPS users are not error free.
    d. TPSs are subject to usual electronic and mechanical failure, like any other computer system.
    e. The lack of proper TPS QA procedures has led to some serious accidents.
    f. All of the above are true.

19. Which of the following statements regarding TPS QA are true?
    a. The first AAPM report to discuss TPS QA was TG-40 published in 1994.
    b. TG-40 deals with 2D TPS systems.
    c. The TG-53 report of AAPM deals with the QA requirements of 3D TPS systems.
    d. The TG-62 report deals with 4D treatment planning for gated radiotherapy.
    e. All of the above are true.

20. Can the accuracy of the relation between the CT number and the relative electron density (RED) in the TPS system be checked?
    a. No, it must be assumed to be correct.
    b. It can be checked by imaging a patient.
    c. It can be checked by Monte Carlo computations.
    d. It can be checked with a phantom containing inserts of known REDs.
    e. It can't be checked by any of the methods suggested above.

21. _____ can affect the calculation of monitor units for a given treatment plan.
    a. Data corruption
    b. Software bugs
    c. Hardware failure
    d. The commercial TPS being used
    e. All of the above

22. The brachytherapy dosimetry in brachytherapy TPS is based on the formalisms of _____.
    a. TG-21
    b. TG-43
    c. TG-51
    d. TG 62
    e. all the above

23. The TG-43 formalism is based on _____ assumptions.
    a. acrylic medium
    b. superposition of single source dose distributions
    c. no inter-source attenuation effects
    d. full scatter conditions at dose calculation points
    e. all of the above

24. The dose rate tables of brachytherapy sources adopted in the TG-43 formalism are based on _____.
    a. comparisons of all published datasets
    b. Monte Carlo calculations
    c. experimental measurements
    d. analytical methods of dose evaluation
    e. all of the above

25. In brachytherapy TPS QA, the TPS calculations must be verified by _____.
    a. Monte Carlo computations
    b. actual measurements in a water phantom
    c. comparing with published dose rate tables
    d. user-developed algorithms for brachytherapy dose calculation
    e. all the above methods

26. The only source-specific parameter that must be entered by the user in the BT TPS system for a clinical brachytherapy source is its _____.
    a. dose rate constant
    b. decay constant
    c   half life
    d. air kerma or exposure strength
    e. size

27. Reliable brachytherapy source data can be obtained from _____.
    a. TG-43 updates published in the journal *Medical Physics*
    b. AAPM report 229
    c. the ESTRO website
    d. the Radiological Physics Center at M.D. Anderson Hospital in Houston
    e. the source's manufacturer

28. Which of the following statements are true regarding the QA of BT TPS system?
    a. Both the hardware and the software of the TPS must be QA tested.
    b. The main hardware tests include checking the accuracy and linearity of input digitizers, output plotters, and printers.
    c. The main software tests relate to checking the accuracy of dose distributions for selected treatment conditions against published data or manual calculations.
    d. Logarithms used in the system must be individually tested by hand calculations.
    e. All of the above are true.

D. Record and Verify Systems

Choose the right answer(s) (more than one may be correct):

1. Which of the following statements regarding the RVS are true?
   a. RVS acts as an interface between the treatment planning system (TPS) and the treatment delivery system (TDS).
   b. RVSs were initially developed to reduce the risk of treatment errors.
   c. Early RVSs were used to check the consistency between prescription and actual setup.
   d. Modern RVSs almost fully control the treatment machine, allowing partial or fully automated setup with minimum operator intervention.
   e. Modern RVSs have been thoroughly tested and cannot cause any treatment errors.
   f. All of the above are true.

2. RVS systems _____.
   a. need no quality control since the RVS itself ensures error-free treatment.
   b. are medical devices that are subject to national and regional regulations
   c. must also be subjected to acceptance testing procedures like TPSs, linacs, etc.
   d. must comply with IEC recommendations and internationally accepted QA tests
   e. all the above are true

3. The errors that have been recognized as more likely to happen with the RVSs are _____.
   a. incorrect machine settings
   b. incorrect patient treatment parameter settings
   c. MUs calculated for one type of treatment but a different kind of treatment is delivered
   d. the treatment of the wrong patient, the wrong site, or the wrong number of fractions
   e. all of the above

4. The QA tests that must be performed on RVS can be categorized as _____.
   a. the RVS as a whole
   b. the TPS
   c. the TDS
   e. the patient
   f. all of the above

5. General tests specific to RVS testing include _____.
   a. checking that the system is able to identify each patient uniquely
   b. ensuring that treatment delivery is not possible without approval of either the treatment prescription in the system or its modifications by an authorized radiation oncologist
   c. end-to-end tests from TPS to TDS (treatment delivery system) for data compatibility and accuracy
   d. QA testing of the linac
   e. all of the above

E. QA in Advanced Treatment Techniques

Choose the right answer(s) (more than one may be correct):

1. _____ factors relating to IMRT planning will introduce uncertainties in IMRT treatment.
   a. MLC leaf modeling
   b. Beam modeling
   c. Grid size (used in calculations)
   d. Small field model (for dose calculations)
   e. All of the above

2. _____ factors relating to treatment delivery systems will introduce uncertainties in IMRT treatment.
   a. MLC design
   b. MLC leaf movement
   c. MLC leaf number
   d. Beam stability
   e. All of the above

3. Common detectors used for IMRT QA measurements include _____.
   a. ion chambers
   b. film
   c. 2D arrays of ion chambers, diodes, MOSFETs, or EPID
   d. chemical dosimeters
   e. all of the above

4. Common methods of IMRT QA measurements include the _____.
   a. beam by beam method
   b. composite beam by beam method
   c. true composite method
   d. virtual beam method
   e. all of the above

5. What are the advantages and disadvantages of ion chambers or diode-based 2D detector arrays compared to the use of film in IMRT QA?
   a. No pre-analysis (like film development) is involved and, hence, they are a much easier and quicker method for IMRT verifications.
   b. They can be used for absolute dose distribution verification.
   c. They may exhibit angular dependence, which may need correction—a disadvantage.
   d. They have limited resolution depending on the detector size and detector density compared to film—a disadvantage.
   e. All of the above are true.

6. The important characteristics of detectors used in IMRT QA are _____.
   a. close to a "point detector" for good spatial resolution
   b. angular independence or known angular dependence
   c. energy independence or known energy dependence
   d. usable only in "dose rate" mode
   e. linearity
   f. all of the above

7. Match the characteristics to the detector.

Characteristics	*Detector*
a. poorer resolution than diodes	_____ i. diamond detector
b. over-responds to low-energy photons	_____ ii. micro chamber
c. poorer resolution than diodes, expensive	_____ iii. p-Si diodes

8. The advantages of radiographic films in IMRT QA measurements are _____.
   a. easy availability
   b. can be cut to any size
   c. give 2D dose distribution in single exposure
   d. have excellent spatial resolution
   e. are less expensive than other 2D detector systems
   f. all of the above

9. The IGRT imaging techniques used in clinical radiotherapy are _____.
   a. kV orthogonal imaging
   b. kV cone-beam CT
   c. MV orthogonal imaging
   d. MV cone-beam CT
   e. Gamma Knife therapy

10. The main QA tests that need to be tested on an IGRT system are _____.
    a. image quality
    b. spatial accuracy (scaling)
    c. congruence of imaging and treatment isocenters
    d. accuracy of registration/couch movements
    e. imaging dose
    f. all of the above

11. In an IGRT technique, the gantry-mounted kV imaging system in a linac is used for _____.
    a. diagnostic imaging
    b. patient repositioning
    c. organ movement detection
    d. adaptive therapy
    e. all of the above

12. The safety tests to be performed on the gantry-mounted imaging systems are _____.
    a. radiation output
    b. fluoroscopic dead-man switch
    c. interrupt button
    d. door interlock
    e. all of the above

13. Commissioning of a RapidArc system requires testing the _____.
    a. accuracy of DMLC position
    b. ability to vary dose rate and gantry speed
    c. ability to accurately vary MLC leaf speed
    d. system from end to end
    e. CBCT imaging dose

14. Risk of collision increases when _____.
    a. the treatment isocenter is lateral to patient midplane
    b. the treatment isocenter is anterior
    c. a large immobilization device is used
    d. the couch is rotated
    e. treatment time is long

15. In RapidArc treatment delivery, the average prescription dose loss to target structure from couch attenuation is approximately _____.
    a. 1%
    b. 3%
    c. 5%
    d. 10%
    e. 20%

16. For VMAT patient-specific QA, the _____ QA tools can be used.
    a. PTW OCTAVIUS 4D
    b. Sun Nuclear ArcCheck
    c. Scandos Delta^4
    d. Sun Nuclear MapCheck
    e. all of the above

17. _____ are the basic requirements for SBRT treatment delivery.
    a. Immobilization
    b. Image guidance
    c. Motion management
    d. Gated beam-on devices
    e. All of the above

18. SBRT involves _____.
    a. very high dose per fraction
    b. few fractions
    c. more accurate patent positioning and treatment delivery procedures
    d. no motion management
    e. a larger treatment time
    f. all of the above

19. Which of the following statements regarding stereotactic body radiotherapy (SBRT) are true?
    a. SBRT is stereotactic treatment of extracranial targets delivered in less than or equal to about five fractions.
    b. SBRT involves a high target dose.
    c. SBRT exhibits steep dose gradient beyond the periphery.
    d. SBRT involves dose per fraction ≈2 Gy.
    e. SBRT has a reduced CTV-PTV margin compared to conventional therapy.
    f. All of the above are true.

20. Factors that may introduce uncertainties in IMRT planning include _____.
    a. the MLC beam and leaf end modeling used in the TPS
    b. the jaws, MLC, and penumbra modeling used
    c. the output and off-axis profiles for small fields
    d. collimator backscatter
    e. all of the above

21. In an SBRT treatment of the lung, 105% of the covered volume received the prescription dose. If the PTV volume was 40.2 cc, then _____ cc of the normal tissues bordering on the PTV received the prescription dose.
    a. 0.1
    b. 0.5
    c. 1.0
    d. 1.5
    e. 10

22. _____ are factors that may introduce uncertainties in IMRT treatment delivery.
    a. MLC leaf position errors
    b. MLC leaf acceleration or deceleration
    c. Beam stability
    d. Beam energy
    e. All of the above

23. _____ parameters must be tested *daily* in the routine Gamma Knife (model 4C or earlier) QA procedures.
    a. Door interlock
    b. Emergency off
    c. AV communications
    d. Radiation monitor
    e. Source output
    f. All of the above

24. _____ parameters must be tested *weekly* in the routine Gamma Knife QA procedures for the conventional Gamma Knife system (model 4C or earlier).
    a. Couch release handle
    b. Helmet microswitches
    c. Helmet trunions
    d. Automatic positioning system
    e. All of the above

25. _____ parameters must be tested *monthly* in the routine Gamma Knife QA procedures for the conventional Gamma Knife system (model 4C or earlier).
    a. Radiation output
    b. Computer output vs. measured output
    c. Emergency rod release
    d. Relative helmet factors (RHF)
    e. Timer constancy, linearity, and accuracy
    f. All of the above

26. The important tests to be carried out for the Perfexion Gamma Knife treatment delivery system include _____.
    a. coincidence of the mechanical isocenter of the patient positioning system (PPS) with the radiation focal point (RFP)
    b. agreement of measured beam profiles with Leksell GammaPlan for all the collimators
    c. beam calibration for the largest collimator
    d. checking relative output factors for smaller collimators
    e. all of the above

27. Which of the following statements regarding CyberKnife are true?
    a. CyberKnife delivers frameless radiosurgery.
    b. Patient position offsets are detected by image guidance, and the delivery is suitably modified.
    c. Robotic arm-mounted imaging systems are used for image guidance.
    d. Patient positioning is monitored using bony landmarks or implanted fiducial markers.
    e. All of the above are true.

28. The daily tests recommended for a CyberKnife system include _____.
    a. a system status check
    b. a linac output constancy check
    c. a safety interlock check
    d. an output calibration
    e. all of the above

29. The CyberKnife linac output must be measured _____.
    a. using a semiconductor diode
    b. under reproducible geometric conditions
    c. using temperature/pressure correction to get the output
    d. for several beam orientations
    e. complying with all the above

30. A CyberKnife system _____
    a. may involve 100–300 beams for a typical treatment
    b. uses highly conformal dose distribution
    c. is strongly image guided
    d. uses beam-by-beam tracking based on bony anatomy or fiducials
    e. all of the above are true

31. The functional (i.e., functioning or not) checks that need to be carried out on a daily basis for CyberKnife, according to AAPM TG-135 report, include the _____.
    a. safety interlocks
    b. CCTV cameras and monitors
    c. audio monitor
    d. collimator assembly collision detector
    e. radiation output
    f. all of the above

32. The nonfunctional daily QA checks (tolerances) for CyberKnife, according to AAPM TG-135 recommendations, include _____.
    a. accelerator output constancy (<2%)
    b. visual check of beam laser and a standard floor mark (<1 mm)
    c. auto QA (AQA test) <1 mm from baseline
    d. beam energy (2%)
    e. beam symmetry (<2%)
    f. all of the above

33. The monthly QA checks for CyberKnife, according to AAPM TG-135 recommendations, include _____.
    a. beam parameters
    b. robot pointing
    c. end-to-end test
    d. absolute beam calibration
    e. all of the above

34. Match the following tests to the frequency of testing (as per AAPM TG-135 recommendations).

    Tests   _Frequency_
    a. laser/radiation coincidence   _____   i.   quarterly
    b. spot checking of beam data   _____   ii. annual
    c. TPS beam data and calculation checks
    d. imaging system alignment

35. Image quality can be defined by _____ parameters.
    a. contrast
    b. spatial resolution
    c. noise or contrast-to-noise ratio
    d. uniformity
    e. all of the above

36. The daily QA checks for planar kV, MV (EPID), and cone-beam CT (kV and MV) imaging systems, according to AAPM TG-142 recommendations, include _____.
    a. collision interlocks
    b. position/reposition
    c. imaging and treatment coordinate coincidence (single gantry angle)
    d. spatial resolution
    e. all of the above

37. The monthly QA checks for planar kV and MV (EPID) imaging systems, according to AAPM TG-142 recommendations, include _____.
    a. scaling
    b. spatial resolution
    c. contrast
    d. uniformity and noise
    e. all of the above

38. The monthly QA checks for cone-beam CT (kV and MV) imaging systems, according to AAPM TG-142 recommendations, include _____.
    a. geometric distortion
    b. spatial resolution
    c. contrast
    d. HU constancy
    e. uniformity and noise
    f. all of the above

39. Which of the following factors regarding the imaging dose are true?
    a. Water kerma and water absorbed dose are NOT approximately the same for KV and MV imaging.
    b. The doses are measured in the same manner for all imaging modalities (i.e., kV, MV, and CT imaging).
    c. The dose delivered by different modalities cannot be summed up without any modifications.
    d. Compared to MV imaging, kV imaging delivers more skin dose.
    e. All of the above are true.

40. Which of the following statements regarding the x-ray unit of the kV imaging system are true?
    a. The performance of the imaging system depends on the stability of the x-ray unit (generator).
    b. The output can be measured in terms of air kerma or exposure in air (per unit mAs), under reproducible geometric conditions.
    c. The beam quality must be checked annually since the image quality depends on the beam quality.
    d. The imaging dose must be measured annually.
    e. All of the above are true.

41. Patient-specific pretreatment QA is important in IMRT and other advanced radiotherapy techniques because
_____.
    a. the treatments are complex and highly individualized, giving no clue that the treatment is being delivered as per the plan
    b. it verifies the treatment planning system's ability to calculate the dose accurately
    c. it verifies the delivery system's ability to deliver the dose accurately (as per the plan)
    d. it uniquely validates that the treatment is perfectly carried out as per plan
    e. it ensures all the above

42. The different patient-specific QA methods in use in radiotherapy are the _____.
    a. planar per-beam method
    b. planar cumulative method
    c. semi 3D method
    d. full 3D method
    e. 4D method
    f. all of the above

43. The detectors generally used in patient-specific QA include _____.
    a. film
    b. 2D arrays of ion chambers or diodes
    c. EPIDs
    d. polymer gels
    e. all of the above

44. The verification methods used in advanced treatment techniques like IMRT for patient-specific QA include the
_____.
    a. beam by beam method (BbB method)
    b. composite BbB method (CBbB method)
    c. true composite method
    d. differential method
    e. all of the above

45. The QA metric used to check the agreement between the TPS computed and measured dose distributions in pretreatment verification is the _____.
    a. absolute dose index
    b. relative dose index
    c. gamma method
    d. gamma index
    e. none of the above

46. The acceptance criteria for patient-specific QA verification agreement according to TG-119 recommendations is _____.
    a. 1%/1mm
    b. 2%/2mm
    c. 3%/3mm
    d. 4%/4mm
    e. 3%/4mm

47. In a gamma analysis of a patient QA verification, 5346 points were checked and 5132 points passed both for a passing rate of _____.
    a. 100%
    b. about 96%
    c. 90%
    d. 85%
    e. <50%

F. Display Monitor QA

Choose the right answer(s) (more than one may be correct):

1. The brightness of a display monitor is referred to as _____.
   a. luminance
   b. monitor brightness
   c. persistence
   d. radiance
   e. none of the above

2. An advantage of LCD monitors over CRT monitors is that LCD monitors _____.
   a. are less expensive
   b. are more durable.
   c. have lower measurable black levels
   d. can display more number of colors
   e. have less screen glare

3. Current ACR standards specify that grayscale monitors used for interpreting radiographs provide a minimum of _____ cd/m^2.
   a. 120
   b. 171
   c. 256
   d. 329
   e. 512

4. Localized blurring and stitching artifacts that occur with flat-panel receptors are caused by problems with the _____.
   a. plate or plate processor
   b. collimator
   c. wire-mesh screen
   d. individual detector elements
   e. none of the above

5. The display monitor should be checked _____ using the SMPTE test pattern.
   a. daily
   b. weekly
   c. monthly
   d. annually
   e. once in three years

6. An LCD monitor's resolution is determined by _____.
   a. bandwidth
   b. monitor size
   b. line spacing
   c. pixel density
   d. imaging accuracy

7. A display monitor _____.
   a. is not affected by noise
   b. is affected by noise
   c. can have image interpretation affected by unacceptable levels of noise
   d. can have image noise assessed visually using the TG18AFC test pattern
   e. needs no QA testing

8. LCD monitors have become the monitor of choice for radiology and oncology departments because they _____.
   a. consume less energy
   b. generate less heat
   c. have superior luminance
   d. have greater resolution
   e. have all the above characteristics

9. What is the need for QA testing of monitor displays?
   a. Subtle shades of gray must be properly represented on monitors.
   b. The display on different monitors must be uniform.
   c. Display characteristics must not change with time.
   d. Displays must conform to common standards.
   e. All of the above are true.

10. _____ deals with the testing and QA of display systems.
    a. AAPM TG 18
    b. IPEM Report 91
    c. DIN 6868-57
    d. ICRU 43
    e. IAEA Report Series 430

11. _____ are required for carrying out display monitor QA.
    a. Luminance meters
    b. Illuminance meters
    c. Test images (e.g., TG-18, SMPTE patterns)
    d. Dosimeters
    e. All of the above

12. Which steps are necessary for optimal viewing of images on display monitors?
    a. No special steps are necessary.
    b. The display screen must be spotlessly clean.
    c. There should not be external light falling on and reflecting from the surface.
    d. The ambient light must be within acceptable levels.
    e. The monitor must be kept away from magnetic fields (e.g., MRI rooms) for CRT screens.

13. What is the difference between luminance and illuminance?
    a. Both mean the same.
    b. Luminance gives the light emitted by the display monitor.
    c. Luminance cannot be measured, while illuminance can be measured.
    d. Illuminance gives the quantity of visible light striking the display monitor.
    e. All of the above are true.

14. _____ is the maximum luminance (in cd/m^2) recommended by AAPM for primary display monitors.
    a. 1000
    b. 500
    c. Not less than 170
    d. Not less than 100
    e. None of the above

15. _____ is the maximum luminance (in cd/m^2) recommended by AAPM for secondary display monitors.
    a. 1000
    b. 500
    c. Not less than 170
    d. Not less than 100
    e. None of the above

16. The contrast ratio (CR) for the display monitor is defined as _____.
    a. $L_{max} - L_{min}$
    b. $L_{max} + L_{min}$
    c. L_{max} / L_{min}
    d. L_{min} / L_{max}
    e. $(L_{max} - L_{min}) / (L_{max} + L_{min})$

17. _____ is the contrast ratio (CR) recommended by AAPM for primary display monitors.
    a. >1000
    b. ≥500
    c. ≥250
    d. >150
    e. None of the above

18. _____ is the brightness uniformity across the monitor recommended by AAPM for display monitors.
    a. 60%
    b. <50%
    c. <30%
    d. <15%
    e. <10%

G. Measurement Equipment

Circle the right answer (Yes or No):

1. (Yes / No)   Before connecting the chamber to an electrometer for carrying out measurements, it is important to check their compatibility.

2. (Yes / No)   The ionization chamber used for beam calibration is usually a sealed chamber.

3.　(Yes / No)　Humidity can affect chamber measurement.

4.　(Yes / No)　The electrometer used with the ionization chamber needs no separate calibration.

Choose the right answer(s) (more than once may be correct):

5.　The chamber used for beam calibration in case of high energy (>1 MV) photon beams _____.
　　a.　is a Farmer-type chamber
　　b.　is a plane parallel type chamber
　　c.　can be of any volume
　　d.　has a volume of about 0.6 cm^3.

6.　The ionization chamber can exhibit significant leakage _____.
　　a.　during rainy days
　　b.　if the chamber input connectors catch dust
　　c.　if chamber insulation deteriorates due to dust or moisture
　　d.　if it is not stored in cupboards

7.　The leakage exhibited by an ionization chamber measuring system could be due to _____.
　　a.　only the chamber
　　b.　only the electrometer
　　c.　only the connecting cable
　　d.　all three

8.　While making ionization measurements, it is important to check for _____.
　　a.　leakage
　　b.　saturation
　　c.　polarity effects
　　d.　warm-up time required

VII. Quality Assurance ANSWERS

A. Treatment Delivery Equipment (External Beam Therapy and Brachytherapy)

1.   Yes   Type approval is the process of a regulatory body approving any radiation-generating equipment, after testing, certifying that the equipment is safe to install for clinical use. In many countries only type-approved equipment can be installed for patient use.

2.   No   The safety of every piece of equipment has to be ensured before being put to clinical use.

3.   Yes

4.   Yes

5.   Yes

6.   Yes   It is always good practice to ensure radiation safety of any installation before switching on the unit for any measurement.

7.   No   Periodic QA is essential to ensure that the equipment continues to conform to the performance standards.

8.   No   The beam data are unique for each machine and so must be generated for each unit before it is put to clinical use.

9.   No   The frequency of each test depends on its relative stability and impact on patient treatment.

10.   No   The beam quality and output must be monitored on a daily basis since any drift in beam stability can affect these parameters and accelerator beams are not as stable as, say, a ^{60}Co beam.

11.   Yes

12.   No   The user must make use of experimentally measured wedge transmission factors.

13.   No   The collimator axis may shift with respect to collimator rotation but the shift must be within the specified limits (See AAPM TG-40 report).

14.   Yes

15.   Yes

16.   No

17.   No

18.   a, b, c

19.   a, b, c

20.   b, c

21. a, b, c

22. a, b, c, d

23. c, d

24. a, b, c

25. a, c

26. a, b, d

27. c

28. a, b

29. a, d

30. a, b, c

31. a, b

32. b   The laser localization accuracies required for advanced treatments like the IMRT or SRS/SBRT are much more stringent compared to non-IMRT treatment, as can be seen from AAPM's TG-142 report.

33. a

34. b

35. a, b

36. a, b

37. a, b, c, d

38. b

39. b   The flatness and symmetry criteria are given in various protocols. See AAPM TG-40 report, or Hendee and Ibbott (1996).

40. b   The tolerance for the daily testing of output constancy for both the photon and electron beams is a little more relaxed (3%) compared to monthly testing (2%). See AAPM protocols TG-60 and TG-142.

41. b

42. a, b, c

43. a, b, c

44. b, c

45. b, c

46. a, b, c

47. a, b, c

48. f

49. b, c, e   Here stereotactic radiotherapy is defined as a treatment course of <3 fractions. For >3 fractions, it must be considered as external beam radiotherapy. (See the presentation of Jack Yang on Linac QA, Barnabath Health, NJ, USA.)

50. e

51. a, b, c

52. b, c, d

53. e

54. e

55. e

56. a, b, d

57. d

58. a, b

59. d   For monthly testing, TG-142 recommends tighter tolerances, namely ±1 mm and <±1 mm, respectively.

60. a

61. e

62. c   In the case of asymmetric fields, the tolerance is 1 mm or 1%.

63. b

64. d

65. d   The tolerance and interpretation are the same (i.e., 1% with respect to the benchmark profile), while TG-40 allows a more liberal tolerance of 3%.

66. c

67. c

68. d   This is an important parameter. We measure output of the beam in a standard geometry, but that output obviously should not change when the gantry is rotated, as we are required to do while treating a patient.

69. c   This test must be carried out monthly. The tolerance applies to all wedge concepts. The check also verifies the wedge centering. The wedge angle needs to be checked only on an annual basis.

70. a, b, c

71. d

72. d

73. a, b, c, d

74. a   The testing of interlocks is a functional test and needs to be performed before the start of treatment every day.

75. e

76. a, b   The tolerances are the same for SBRT as well.

77. f

78. a, b, c

79. e

80. f

81. a, b, c

82. a, b, c

83. a, b, c, d

84. a, b, c

85. b

86. a

87. a, b, c

88. c

89. c

90. c, d

91. a, b

92. c

93. c

94. a, b

95. a, b, c

96. d

97. a, c

98. b, c

99. a, b, c

100. b, c, d, e

101. e

102. e

103. a, b, c, e

104. a, b, d, e

105. d   Always check whatever can be checked. If the catheter length is 950 mm—check. If the TPS computed dose is 1050 cGy—check....

106. c

107. d, e   Such a situation may be improbable, but not impossible to occur.

108. c

109. d

110. b

111. c

112. e

113. a, b

114. a, c

115. a

116. c

117. c

118. d

119. d

B. Imaging Systems for Treatment Planning

1.   No

2.   No   The tolerances for simulator QA parameters must be at least as stringent as that of the accelerator since the treatment accuracy depends on simulation data.

3.   Yes

4.   Yes

5.   Yes

6.   No

7.   Yes

8. Yes

9. No

10. No

11. Yes

12. b

13. a

14. a

15. b

16. a, b

17. a

18. c

19. c

20. e

21. a, b, c   The other two tests need to be done less frequently.

22. c   The tolerance for the ODI is also the same. The accuracy of the treatment simulator must be better than or at least as good as the treatment delivery equipment.

23. e

24. a, b, d

25. d

26. d

27. a, b, c

28. a, b, c, e

29. b

30. e

31. d

32. e

33. e

34. a, b, c, d

35. e

36. a, b, c

37. a, b, c

38. a   This must be done daily before patient simulation, but all the other laser alignment tests need to be performed only once a month or after any laser adjustment.

39. c   The tolerance for all the other monthly laser tests is also the same.

40. a, b   Tests c and d must be performed only annually.

41. c   The orthogonality must exist over the length and width of the tabletop.

42. d   The table top motions must be accurate and reproducible to this tolerance.

43. d

44. e   A gantry tilt position accuracy test verifies that the gantry accurately returns to normal position after tilting.

C. Treatment Planning System (TPS)

1. b

2. a, b, c, d

3. b, c

4. a, c

5. b, c, d

6. b, c

7. a, c

8. a

9. a, b, c

10. a, b, c

11. b, c

12. c

13. a, b, c

14. a, b, d, e

     Planning of clinical cases starts only after acceptance testing and commissioning of the TPS system.

15. a, b, e

16. a, b, c, d

17. f

18. f

19. a, b, c

20. d

21. a, b, c

22. b

23. b, c, d   TG-43 assumes water as the source medium and not acrylic.

24. a, b, c

25. c

26. d

27. a, b, c, d

28. a, b, c

D. Record and Verify Systems

1. a, b, c, d

2. e

3. e   The possible origin of these errors is listed with examples in the IAEA publication mentioned above.

4. a, b, c

5. a, b, c

E. QA in Advanced Treatment Techniques

1. e   The above factors are not independent but interrelated.

2. a, b, c

3. a, b, c

4. a, b, c

5. a, c, d   Films can give only relative dose and need proper calibration for dose interpretation.

6. a, b, c, e

7. a. ii.   b. iii.   c. i.
   All high-Z detectors will over-respond at low-energy photons unless it is accounted for (compensated for) in its design.

8. f

9. a, b, c, d

10. f   Special QA tools are required to perform these tests.

11. b, c, d

12. b, c, d These have to be checked daily. There are no tolerances since they are just functional tests.

13. a, b, c, d

14. a, b, c, d

15. b

16. a, b, c MapCheck requires a dedicated phantom for VMAT QA testing.

17. e

18. a, b, c, e
         Motion management is a very important component

19. a, b, c, e
         The use of multiple nonoverlapping beams causes the rapid dose falloff outside the PTV.

20. e   In addition to the above factors, dose calibration and grid size used in TPS also contribute to the
         planning uncertainty.

21. a

22. a, b, c

23. a, b, c, d

24. e

25. a, b, c, e

26. e

27. a, b, d The CyberKnife uses room-mounted orthogonal kV imaging systems for image guidance.

28. a, b, c The linac output constancy must be checked daily.

29. b, c   The output must always be measured using a Farmer-type chamber that has been vented to the
         atmosphere.

30. e

31. a, b, c, d

32. a, b, c

33. a, b, c

34. a. (i)   b. (ii)   c. (ii)   d. (i)

35. e

36. a, b, c

37. e

38. f

39. a, d

40. e

41. a, b, c

42. a, b, c, d

43. e

44. a, b, c

45. d

46. c

47. b

F. Display Monitor QA

1. a

2. e

3. b

4. d

5. a

6. c   Closeness of adjacent pixels

7. b, c, d   The display affects our ability to detect small- or low-contrast objects within a displayed image. This can lead to missing detail and misdiagnosis.

8. e

9. e

10. a, b, c

11. a, b, c

12. b, c, d, e

13. b, d

14. c

15. d

16. c

17. c

18. c

G. Measurement Equipment

1. Yes

2. No   It is unsealed and, hence, the measurements must be corrected for temperature and pressure.

3. Yes   In places of high humidity the chambers must be kept in desiccators.

4. No   It does need calibration but much less frequently compared to an ionization chamber.

5. a, d

6. a, b, c

7. d

8. a, b, c, d